Footprints in the Dust

FOOTPRINTS IN THE DUST

The Epic Voyages of Apollo, 1969–1975

Edited by Colin Burgess | Foreword by Richard F. Gordon

UNIVERSITY OF NEBRASKA PRESS • LINCOLN AND LONDON

Manufactured in the
United States of America
♾
Library of Congress
Cataloging-in-Publication Data
Footprints in the dust : the epic voyages of
Apollo, 1969–1975 / edited by Colin Burgess;
foreword by Richard F. Gordon.
p. cm.—(Outward odyssey: a people's history of
spaceflight)
Includes bibliographical references and index.
ISBN 978-0-8032-2665-4 (cloth : alk. paper)
1. Project Apollo (U.S.)—History. 2. Space
flight to the moon—History—20th century.
I. Burgess, Colin, 1947–
TL789.8.U6A53334 2010
629.45'4—dc22
2009047697

Set in Adobe Garamond.
Designed by R. W. Boeche and Kim Essman.

This book is dedicated to the men and women who have given their time and talents to help keep the United States at the forefront of science and technology through the Astronaut Scholarship Foundation (ASF). The Foundation, a 501(c)(3) nonprofit organization, annually funds a number of scholarships awarded to outstanding college students pursuing degrees in science and technology. These students exhibit motivation, imagination, and exceptional performance both inside and outside the classroom.

In recognition of such exemplary work the authors have donated their talents and individual proceeds from this book to the Astronaut Scholarship Foundation through the University of Nebraska Press.

On a far more personal note this book is also dedicated to the memory of Philip Baker from Hampshire, England, who lost a protracted battle with illness before this book was published. Philip was the author of the *Apollo 14* chapter, "Science, and a Little Golf." His cheery optimism and passion for spaceflight will be sorely missed by everyone associated with this book and within the space history community.

In Memoriam
Charles "Pete" Conrad Jr.
James B. Irwin
Alan B. Shepard Jr.
May their footprints in the dust last a million years.

Look downward on that globe, whose hither side
With light from hence, though but reflected, shines:
That place is Earth, the seat of man; that light
His day, which else, as th' other hemisphere,
Night would invade; but there the neighboring Moon
(So call that opposite fair star) her aid
Timely interposes, and, her monthly round
Still ending, still renewing, though mid-heaven,
With borrowed light her countenance triform
Hence fills and empties, to enlighten the Earth.

John Milton, *Paradise Lost*

Contents

Illustrations

Foreword

Certain difficult and controversial questions always seem to be asked whenever my fellow astronauts and I are quizzed about Apollo and subsequent space programs. One that perpetually chases us around is why we are spending all this money on space exploration when it could surely be put to far better use right here on Earth.

I always respond to this by saying that not a dime has been spent in space; it's spent right here on Earth. And not just on job creation: the incredible technological and lifesaving advances we now enjoy, as well as our current standard of living, have been principally driven and accelerated by our space program and those of other nations. The mere fact that you can pick up a telephone today and call anywhere in the world and the response sounds like it's coming from next door—that's space technology.

In an era when technology is fast allowing a merging and fusion of information science, telecommunications and computers, there are boundless opportunities for all of us in high-tech service industries, from electronic and fiber optic compression techniques to designer materials for specific applications, pollution control, and environmental monitoring. Think of global cellular phones, push-button banking, computers, and especially the Internet and e-mails; all have an unseen nervous system—a galaxy of satellites in both low Earth orbit and at geosynchronous altitudes. Much of what we take for granted these days has derived from space technology and exploration. The only other event that accelerates technology is war, and you know which one I would choose. I think you would too.

Of course when we think of space, we almost invariably tend to think of human space exploration; this natural quest of the soul, where we seek to go beyond our reach. It provides the underpinning for motivation, inquiry, inspiration, imagination, and our unquenchable search for knowledge and answers.

Back in 1859 an exciting new book was published. Written by the celebrated English novelist Charles Dickens, *A Tale of Two Cities* begins with the unforgettable first line, "It was the best of times; it was the worst of times." Those words were still very apropos over a hundred years later, in 1969, as I was in training to travel to the moon late that year with Pete Conrad and Al Bean on *Apollo 12*—the second lunar landing expedition.

In many respects it was indeed the worst of times, and maybe we'll also see it as the best of times, if we think of the way our world was back in the 1960s. In that unsettled era we had the very divisive nature of the Vietnam War, the Cuban missile crisis, the Bay of Pigs fiasco, campus riots, and bloody civil rights marches. We also shared in the grief and horrors of political assassination as one by one several great men were brutally gunned down—President John F. Kennedy, his brother Bobby Kennedy, Martin Luther King Jr., and Malcolm X. Yes, indeed, we had the worst of times. Conversely, looking farther afield, they were some of the best of times as well.

I am of the firm belief that in those otherwise turbulent times, social advances and our space program also gave us some of the best of times. But it didn't start out that way. Back in 1957 the Soviet Union had achieved the first giant step in the challenge of space flight, with the launch of its first Sputnik satellite. The following month it followed up this feat by sending a dog named Laika into orbit aboard *Sputnik 2*. Less than four years later, as the United States struggled to compete against an ever more sophisticated array of satellites circling our planet, Soviet scientists triumphed yet again, this time by launching cosmonaut Yuri Gagarin, the world's first human space traveler, into a single Earth orbit in 1961. It was definitely a time of uncertainty for our nation as we endeavored to play catch-up. Much to our increasing chagrin the Russians put the first woman into space, followed on the next flight by the first three-man crew, and then on the subsequent Voskhod flight a human eased out through an inflatable hatch to become the first person to walk in space—a person who later became a very good friend, Alexei Leonov. He may have been the first to walk in space, but it was later revealed that his return to the confines of *Voskhod 2* was fraught with extreme danger, and he came perilously close to losing his life. The United States was rapidly closing in, however, and would soon take the lead in the race to the moon, a lead we would not lose.

A colleague of mine from those days was quite correct when he said in some interview that we were at war back then. It was called the Cold War, an era when a very definite adversarial relationship existed between the Soviet Union and the United States of America. I had experienced that "cold war" from a personal perspective before my participation in the space program. I was a young navy carrier pilot assigned to the Sixth Fleet in the Mediterranean in 1954 and 1956 on six-month deployments and once again with the Seventh Fleet in Westpac [Western Pacific] on a nine-month cruise in 1962–63. Politically, economically, socially, it was a war. The Russians were desperately trying to get to the moon ahead of the United States. Through achievement and a good dose of propaganda they wanted to demonstrate to the world what they claimed was their superior technology and capabilities. And of course, so did we.

Following the successful conclusion of Project Mercury, the National Aeronautics and Space Administration (NASA) embarked on the Gemini program, which was created for a very specific reason. In October 1963 I became one of fourteen new astronauts selected by NASA, and we were immediately dedicated to the success of Gemini.

On 25 May 1961, just twenty days after Alan Shepard flew a suborbital flight, and with a total of only fifteen minutes of human spaceflight under our collective belts, President Kennedy said we should go to the moon. Just think about that greatest of all mission statements for a moment: how could it possibly be done? We hadn't even sent one of our astronauts into orbit. We still had to learn how to do that. As history shows, the Mercury program proved that people could survive and do useful work in space, while Gemini was created to learn the things we'd need to achieve in order to accomplish an Apollo lunar landing. Planning for the moon then centered on a ten-day mission—three days out, three days back, and an arbitrary four days in near-lunar vicinity. It meant learning how to survive long-duration flights of at least fourteen days, how to maneuver our spacecraft, how to rendezvous and dock with another orbiting spacecraft. We had to know how to do those things.

It began well enough: *Gemini 4*, *5*, and *7* were long-duration flights, while *Gemini 6* and *8* onward were all missions lasting three or four days that gave us the necessary experience in how to rendezvous, dock, and maneuver our spacecraft. In that very brief period of time, of ten manned flights in 1965–66, we accomplished all of the individual elements that we needed to go to the moon.

There were many highlights during the ten Gemini missions, carried out at an astonishing average of a manned flight every two months. Ed White, for example, aboard *Gemini 4*, conducted a great and attention-grabbing spacewalk. Unfortunately, Ed did too good a job and made extravehicular activity (EVA) look easy. It wasn't, and the experiences of those who followed him were entirely different. Gene Cernan on *Gemini 9*, Mike Collins on *Gemini 10*, me on *Gemini 11*—we simply hadn't learned how to utilize or negotiate that environment very well. Mine was a particularly frustrating experience, not all that successful, and I wish now that I somehow had the opportunity to do it all again. The tasks I had were very difficult and challenging ones to carry out, which I equate to trying to tie your shoelace with one hand. But these experiences taught us well; we learned to create handholds and restraint systems that would allow us to position and secure ourselves to conduct useful work. They were lessons that now allow our astronauts to carry out complex repairs and construction work on the exterior of the International Space Station (ISS).

We thought we were ready to go in January 1967, and then Gus Grissom, Ed White, and Roger Chaffee lost their lives in a spacecraft under routine testing conditions on the pad at Cape Kennedy. This appalling tragedy took our breath away and caused an immediate halt to Project Apollo. But knowing those three individuals and what they wanted us to do, we had to dig deep to find out what had happened, fix the problems, and then continue our plans to land men on the moon before the end of the decade.

There were two Apollo spacecraft variants at that time, known as Block 1 and Block 2. Gus, Ed, and Roger had been assigned to fly a very early version of the spacecraft in the program, one that was designed not to go to the moon but to launch what we called a "shakedown" test flight in Earth's orbit. But we had become, as a team, very complacent about the environment in which we were operating. One hundred percent oxygen, with flammable material within the spacecraft. Over-pressurized. Above sea level pressure during this particular test. All were factors that would combine to create an environment ripe for disaster. All that was needed was a source of ignition. The hatch design was a very poor one. Ed White was a physically strong individual, but it would have taken three or four Ed Whites to get that hatch open. On that day we lost three friends and colleagues, and the nation lost three heroes.

Following the resultant redesign of the Block 2 spacecraft, the 100 percent oxygen environment was eliminated, along with flammable materials. The brutally heavy hatch was redesigned, together with the wiring system within the spacecraft. It was still a very difficult time for everyone associated with the program. We had been all the way through Mercury and Gemini, and all of a sudden this terrible thing had happened. In a pragmatic sense the loss of our friends allowed us to go to the moon and accomplish the goal set out by President Kennedy in 1961 to which we had been committed ever since. It is a very difficult thing to say, but without their loss, there may have been more disasters—perhaps even on the way to the moon. It was certainly a very trying time, not only for the families but for the rest of us in the program.

It took us more than a year and a half, until October 1968, to solve all the problems we had encountered with *Apollo 1*. Finally, *Apollo 7*—we filled up those other numbers with launched unmanned Saturn vehicles.

Apollo 8 was a very daring event and one that I think had the highest risk element of any flight we have flown, including the first lunar landing. We sent three astronauts to the moon aboard a Command and Service Module without a Lunar Module, orbited our nearest neighbor ten times, and returned the crew safely to the Earth at the tail end of 1968. Particularly for us in the space program, we had endured an agonizing period following the launch pad tragedy, and now we had this tremendous, even audacious, accomplishment. *Apollo 9* then carried the first Lunar Module into Earth's orbit, and *Apollo 10* became a magnificent dress rehearsal for going to the moon and landing.

In July 1969 *Apollo 11* accomplished the goal set just eight years earlier by President Kennedy, and I had the great distinction and privilege of flying the next mission, *Apollo 12*. In November 1969 Pete Conrad and Al Bean carried out the second lunar landing in the Lunar Module *Intrepid* and explored the Ocean of Storms, while I flew as Command Module pilot aboard the good ship *Yankee Clipper*. Not only did we (an all-navy crew) achieve the goal that President Kennedy had set for our nation, but we did it twice before the decade of the 1960s was over. The motivation that President Kennedy gave us as a people was exactly what we needed to accomplish those goals.

The Apollo program then proceeded with the flight of *Apollo 13*, a troubled flight with which we are all familiar, successfully bringing home three astronauts who came perilously close to losing their lives. Their safe return was doubtless one of NASA's greatest accomplishments. *Apollo 14*, with Alan Shepard in command, became the next flight to take two crew members to the moon, taking over the flight plan that *Apollo 13* had originally been assigned. The last three Apollo flights—*15*, *16*, and *17*—had the added benefit of the Lunar Roving Vehicle, designed to be driven across the lunar surface. The culmination of that program was *Apollo 17* in December 1972, when the crew traveled in excess of twenty-one miles around the area of Taurus-Littrow aboard the Lunar Rover, in the process gathering several hundred pounds of lunar surface material.

The Apollo program returned 2,196 rock samples, 842 pounds in all, for distribution to the scientific community. Varying in age between 3.1 and 4.7 billion years old, these samples greatly enhanced our knowledge of the moon and its characteristics. Twelve men spent close to 160 hours exploring the surface of the moon, collectively traveling some sixty miles by foot and aboard the Lunar Rovers.

As for me, I came off *Apollo 12* and was assigned the role of backup commander for *Apollo 15*. At that time the lunar flights were manifested to extend through to *Apollo 20*. By extrapolation of the normal rotation process in use then, I was therefore in line to command *Apollo 18* and return to the moon, this time to set foot on the lunar surface as mission commander. Much to my extreme disappointment and that of my fellow crew members it was not to be; severe congressional budget cuts descended on the Apollo program, and it was announced that *Apollo 17* would be the final manned lunar landing mission. *Apollo 18* never flew—so tantalizingly close yet abandoned in an overt haste to save a few dollars. It was a devastating blow at the time for the nine men who might have flown the three canceled missions and in retrospect a massive waste of resources and potential scientific results. If you can think of a way I can get those last sixty miles under my belt, please let me know—I'd still be glad to go.

Three Skylab flights ensued that utilized the remaining Apollo hardware. Then interestingly enough, in 1975 we flew the Apollo-Soyuz Test Project (ASTP). Our former Cold War adversaries came on board with the United States, and we flew a joint mission—a long way from the titanic and hard-fought space race of the previous decade. Today we are seeing that multinational cooperation continue with the International Space Station, with some sixteen nations participating in the program, and we have now made a major transition from the exploration of space to the exploitation of space.

Apollo was a true epoch of the Space Age, a golden era of scientific endeavor, advancement, and incredible discovery, and I was proud to have been a participant, along with some of the finest people one could ever have the privilege to know. I share a great treasure with my astronaut colleagues, for we have seen the splendor of our home planet from the vastness of space—a blue and white ball totally surrounded by the blackness and infinity of space. Seeing it, you are instantly struck with the beauty and wonder of it all. To paraphrase T. S. Eliot, man was made to explore, and when his exploration days are over he will return from whence he came and know the place for the very first time.

I have also been one of just twenty-four people who have traveled from the Earth to the moon, and this book celebrates not only those fantastic voyages of discovery but the evolution of Apollo into an Earth-orbiting program of scientific discovery and finally a vehicle that welcomes aboard a former adversary in a flight of international détente. This is the story of possibly the greatest adventure of all time. It is in our collective nature to expand our horizons, to seek new knowledge and explore. The frontier of space is without doubt the toughest challenge we have ever faced and one that is never breached without danger. Yet the human spirit is indomitable, and as we now make plans to return to the moon and travel beyond, a whole new generation of dreams waits to be realized.

Footprints in the Dust is a monumental account of an epic program and one in which the authors have truly captured the crackling excitement of those amazing times. Reading through these pages, I feel like less of a participant and more of a wholly interested witness to the events that characterize the drive and can-do spirit prevailing in that part of the twentieth century. The book has reacquainted me with the lively personalities who inhabited that historic program and made Project Apollo not only possible but a reality.

Personally, I am delighted that our dream of space flight prevails and remains alive within us and that we will continue to explore other worlds in the majestic spirit of Apollo.

Capt. Richard F. Gordon Jr., USN *(ret.)*
Pilot, Gemini 11
Command Module Pilot, Apollo 12

Acknowledgments

Considering one irrefutable fact—that there have been literally dozens of books written about the Apollo program—it took a mighty effort to come up with new angles and fresh material in approaching this well-known and certainly well-chronicled human endeavor. The genesis for this book must therefore be acknowledged as a by-product of the creative genius of screen actor and producer Tom Hanks and his extraordinary HBO miniseries *From the Earth to the Moon*. He envisioned the exciting concept of telling the Apollo story through the creative talents of a number of different directors, each given a mandate to produce an episode based on a different facet of an Apollo mission.

Originally, this was to have been a single-author book, but when circumstances forced a change and a bold new approach was needed, the inspiring lesson of Tom Hanks was recalled. Eventually, this concept brought together a cadre of spaceflight enthusiasts with the skills, knowledge, and talents necessary to produce the book.

Equal thanks must also go to the University of Nebraska Press (UNP), which set forth on this innovative series some years back with a steadfast enthusiasm and an in-house team ready to lend support. Their willingness to create a magnificent series dealing with the social or human history of space exploration still prevails through its administrative, acquisitions, editorial, marketing, publicity, and production staff, and though a collective thank-you is offered, we must sincerely acknowledge the tremendous and untiring work of UNP acquisitions editor Rob Taylor. In this book and others in the Outward Odyssey series he has not only given the authors great confidence and reassurance in their own abilities but has always been there for guidance and to answer a steady stream of questions emanating from the far corners of the globe. The authors are sincerely grateful for the gentle but incisive hand of copyeditor Elizabeth Gratch, who, through her diligence and professionalism (and love of the subject), gave our work that little extra thrust and polish it really needed.

Special thanks are also extended to the *collectSPACE* Web site (www.collectspace.com) so wonderfully administered by one of this book's authors, the tireless Robert Pearlman. Several of the authors for this volume and other books in the series came about through this brilliant forum, daily populated by some of the finest, most knowledgeable, and friendly spaceflight enthusiasts imaginable. May it last forever.

In researching Apollo-related photo illustrations and associated captions for this or any other publication, there is no finer online source than Kipp Teague's amazing Web site *Project Apollo Archives* (www.apolloarchive.com), which is in turn a companion to Eric Jones's equally comprehensive *Apollo Lunar Surface Journal* (www.hq.nasa.gov/alsj).

As usual, there are a host of people to thank for their kind assistance in researching this book. Help may sometimes come in small packages or large, but in all cases the authors are extremely grateful to those people who selflessly helped in a number of ways. This is especially true of many spaceflight historians who contributed their thoughts, words, and expertise to this project, sometimes unknowingly. They are Michael Cassutt, Francis and Erin French, Rex Hall, Brian Harvey, Bart Hendrickx, Ed Hengeveld, Larry McGlynn, Bruce Moody, James Oberg, David J. Shayler, Ingemar Skoog, Charles Vick, Bert Vis, David R. Woods, and Anatoly Zak.

We also extend our thanks to those who were actual participants in the Apollo story. Not just the astronauts but the ground support crews, the designers, the engineers, the controllers, and many, many others. In fact all of those who not only made Project Apollo the most outstanding scientific and engineering feat in human history but who brought all the astronauts back home safely. They graciously consented to provide us with their time and memories and filled in many of the missing parts of this story by once again delving into the past, and we are extraordinarily grateful to them. They are John Aaron, Arnold Aldrich, Alan Bean, Jerry Bostick, Vance Brand, Ed Buckbee, Gene Cernan, Charlie Duke, Farouk El-Baz, Dick and Linda Gordon, Jim Hannigan, Christopher Kraft, Gene Kranz, Bob Legler, Sy Liebergot, Glynn Lunney, Harrison H. Schmitt, David Scott, and Al Worden.

Last, though certainly far from least, we acknowledge our families and friends, old and new. We cherish them for their love, patience, encouragement, support, and interest throughout the entire inception, research, associated travel, writing, and production of this work. It has meant a lot to us, and each of you has played an undeniably large part in giving pride, soul, and expression to this book. Our thanks to David Barry; Phil Baxter; Kaky Berry; Pat Burgess; Nancy Conrad; Dick Conway; Matt Erickson; Sandy, Joe, and Jennifer Estep; Deborah Evans Price; Mike Evans; Jesse Florea; Mike Fresina; Tad and Jodi Geschickter; Jimmy and Una Graham; Walt Green; Artie, Cindy, and Bryce Greer; Pierre, Carol, and Katy Hamel; Les Hanks; Brian Hedrick and the rest of the past and present staff at *Stand Firm*; Jeanie, Adam, Jesse, Richard, Doug, and Larry Houston; Charlene Hubbard; Gene and Emma Hubbard; Doug Hurley; Allison Johnson; Junior Johnson and the Breakfast Gang; Fred, Judy, and Joe Knight; Jerry Lankford; Christopher Larkins (Penwal Industries); the entire Maplewood Baptist Church family; National Corvette Museum; Jeff Owens; Beryl, Damien, Lisa, and Tony Phelan; Tom and Jean Reavis; James, Lib, and Jamie Reynolds; Bruce Roney; Briana Smith; Time Life / Getty Images; Alessia, Brian, and Margaret Vaughan; Steve Waid; and Deb Williams. Philip Baker also wished to express his appreciation, "as always," to his wife, Helen, and his parents for their ongoing support.

Rice University Address by President John F. Kennedy

At 10:00 a.m. on 12 September 1962, a typically sweltering late summer morning in Houston, President John F. Kennedy delivered an eloquent speech before an openly perspiring but enthusiastic audience of some thirty-five thousand people gathered together on the football field at the William Marsh Rice University. Previously known as the Rice Institute, the university had acted as an intermediary in the transfer of a large tract of flat, windswept cow pasture land, carried out between the Humble Oil and Refining Company and the United States's civilian space agency, the National Aeronautics and Space Administration, or NASA. This, the first of two land transfers, had enabled the construction of the $60 million Manned Spacecraft Center, later to be renamed the Johnson Space Center, or JSC, which would have strong, continuing ties to the university in regard to research and instruction.

On that day the president was at Rice to speak about the nation's space effort. In his stirring oratory he cited scientific progress as irrefutable evidence that the exploration of space was not only inevitable but that the United States should lead such space efforts in order to preserve a position of leadership on Earth.

Although he would be cruelly assassinated just fourteen months later, on 22 November 1963, President Kennedy's dream of winning the space race was fulfilled with outstanding competence on 20 July 1969, when a fragile landing craft named *Eagle* successfully set down on the powdery surface of the moon with two men named Neil and Buzz.

The president's inspiring and wonderfully articulate speech at Rice University served to galvanize the country in a way that few such deliveries have in human history. It is therefore fitting that in the same way his im-

1. President John F. Kennedy addresses a large audience assembled at Rice University on 12 September 1962. Courtesy NASA.

passioned words launched the United States on an engineering and scientific feat of unprecedented proportions, so too are his words used as a metaphoric launchpad for the story of this great endeavor.

In prefacing his speech, President Kennedy acknowledged the presence of Dr. Kenneth Pitzer, president of Rice University; U.S. vice president Lyndon B. Johnson; Price Daniel, governor of Texas; Representative Albert Thomas of Texas; Senator Alexander Wiley of Wisconsin; Representative George P. Miller of California; James E. Webb, NASA administrator; and David E. Bell, director of the Bureau of the Budget.

Address at Rice University on the Nation's Space Effort
President John F. Kennedy
Houston, Texas, 12 September 1962

President Pitzer, Mr. Vice President, Governor, Congressman Thomas, Senator Wiley, and Congressman Miller, Mr. Webb, Mr. Bell, scientists, distinguished guests, and ladies and gentlemen:

I appreciate your president having made me an honorary visiting professor, and I will assure you that my first lecture will be very brief. I am delighted to be here and I'm particularly delighted to be here on this occasion.

We meet at a college noted for knowledge, in a city noted for progress, in a State noted for strength, and we stand in need of all three, for we meet in an hour of change and challenge, in a decade of hope and fear, in an age of both knowledge and ignorance. The greater our knowledge increases, the greater our ignorance unfolds.

Despite the striking fact that most of the scientists that the world has ever known are alive and working today; despite the fact that this Nation's own scientific manpower is doubling every twelve years in a rate of growth more than three times that of our population as a whole; despite that, the vast stretches of the unknown and the unanswered and the unfinished still far outstrip our collective comprehension.

No man can fully grasp how far and how fast we have come, but condense, if you will, the 50,000 years of man's recorded history in a time span of but a half-century. Stated in these terms, we know very little about the first forty years, except at the end of them advanced man had learned to use the skins of animals to cover them. Then about ten years ago, under this standard, man emerged from his caves to construct other kinds of shelter. Only five years ago man learned to write and use a cart with wheels. Christianity began less than two years ago. The printing press came this year, and then less than two months ago, during this whole fifty-year span of human history, the steam engine provided a new source of power.

Newton explored the meaning of gravity. Last month electric lights and telephones and automobiles and airplanes became available. Only last week did we develop penicillin and television and nuclear power, and now if America's new spacecraft succeeds in reaching Venus, we will have literally reached the stars before midnight tonight.

This is a breathtaking pace, and such a pace cannot help but create new ills as it dispels old; new ignorance, new problems, new dangers. Surely the opening vistas of space promise high costs and hardships, as well as high reward.

So it is not surprising that some would have us stay where we are a little longer to rest, to wait. But this city of Houston, this State of Texas, this country of the United States was not built by those who waited and rested and wished to look behind them. This country was conquered by those who moved forward—and so will space.

William Bradford, speaking in 1630 of the founding of the Plymouth Bay Colony, said that all great and honorable actions are accompanied with great difficulties, and both must be enterprised and overcome with answerable courage.

If this capsule history of our progress teaches us anything, it is that man, in his quest for knowledge and progress, is determined and cannot be deterred. The exploration of space will go ahead, whether we join in it or not, and it is one of the great adventures of all time, and no nation which expects to be the leader of other nations can expect to stay behind in the race for space.

Those who came before us made certain that this country rode the first waves of the industrial revolutions, the first waves of modern invention, and the first wave of nuclear power, and this generation does not intend to founder in the backwash of the coming age of space. We mean to be a part of it; we mean to lead it. For the eyes of the world now look into space, to the moon and to the planets beyond, and we have vowed that we shall not see it governed by a hostile flag of conquest, but by a banner of freedom and peace. We have vowed that we shall not see space filled with weapons of mass destruction, but with instruments of knowledge and understanding.

Yet the vows of this Nation can only be fulfilled if we in this Nation are first, and, therefore, we intend to be first. In short, our leadership in science and in industry, our hopes for peace and security, our obligations to ourselves as well as others, all require us to make this effort, to solve these mysteries, to solve them for the good of all men, and to become the world's leading space-faring nation.

We set sail on this new sea because there is new knowledge to be gained, and new rights to be won, and they must be won and used for the progress of all people. For space science, like nuclear science and all technology, has no conscience of its own. Whether it will become a force for good or ill depends on man, and only if the United States occupies a position of preeminence can we help decide whether this new ocean will be a sea of peace or a new terrifying theater of war. I do not say that we should or will go unprotected against the hostile misuse of space any more than we go unprotected against the hostile use of land or sea, but I do say that space can be explored and mastered without feeding the fires of war, without repeating the mistakes that man has made in extending his writ around this globe of ours.

There is no strife, no prejudice, no national conflict in outer space as yet. Its hazards are hostile to us all. Its conquest deserves the best of all mankind, and its opportunity for peaceful cooperation may never come again. But why, some say, the moon? Why choose this as our goal? And they may well ask why climb the highest mountain? Why, thirty-five years ago, fly the Atlantic? Why does Rice play Texas?

We choose to go to the moon. We choose to go to the moon in this decade and do the other things, not because they are easy, but because they are hard; because that goal will serve to organize and measure the best of our energies and skills, because that challenge is one that we are willing to accept, one we are unwilling to postpone, and one which we intend to win, and the others, too.

It is for these reasons that I regard the decision last year to shift our efforts in space from low to high gear as among the most important decisions that will be made during my incumbency in the office of the Presidency.

In the last twenty-four hours we have seen facilities now being created for the greatest and most complex exploration in man's history. We have felt the ground shake and the air shattered by the testing of a Saturn C-1 booster rocket, many times as powerful as the Atlas which launched John Glenn, generating power equivalent to 10,000 automobiles with their accelerators on the floor. We have seen the site where five F-1 rocket engines, each one as powerful as all eight engines of the Saturn combined, will be clustered together to make the advanced Saturn missile, assembled in a new building to be built at Cape Canaveral as tall as a forty-eight-story structure, as wide as a city block, and as long as two lengths of this field.

Within these last nineteen months at least forty-five satellites have circled the Earth. Some forty of them were "made in the United States of America," and they were far more sophisticated and supplied far more knowledge to the people of the world than those of the Soviet Union.

The Mariner spacecraft now on its way to Venus is the most intricate instrument in the history of space science. The accuracy of that shot is comparable to firing a missile from Cape Canaveral and dropping it in this stadium between the forty-yard lines.

Transit satellites are helping our ships at sea to steer a safer course. TIROS satellites have given us unprecedented warnings of hurricanes and storms, and will do the same for forest fires and icebergs.

We have had our failures, but so have others, even if they do not admit them. And they may be less public.

To be sure, we are behind, and will be behind for some time in manned flight. But we do not intend to stay behind, and in this decade, we shall make up and move ahead.

The growth of our science and education will be enriched by new knowledge of our universe and environment, by new techniques of learning and mapping and observation, by new tools and computers for industry, medicine, the home as well as the school. Technical institutions, such as Rice, will reap the harvest of these gains.

And finally, the space effort itself, while still in its infancy, has already created a great number of new companies, and tens of thousands of new jobs. Space and related industries are generating new demands in investment and skilled personnel, and this city and this state, and this region, will share greatly in this growth. What was once the furthest outpost on the old frontier of the West will be the furthest outpost on the new frontier of science and space. Houston, your city of Houston, with its Manned Spacecraft Center, will become the heart of a large scientific and engineering community. During the next five years the National Aeronautics and Space Administration expects to double the number of scientists and engineers in this area, to increase its outlays for salaries and expenses to $60 million a year; to invest some $200 million in plant and laboratory facilities; and to direct or contract for new space efforts over $1 billion from this center in this city.

To be sure, all this costs us all a good deal of money. This year's space budget is three times what it was in January 1961, and it is greater than the space budget of the previous eight years combined. That budget now stands at $5,400 million a year—a staggering sum, though somewhat less than we pay for cigarettes and cigars every year. Space expenditures will soon rise some more, from forty cents per person per week to more than fifty cents a week for every man, woman and child in the United States, for we have given this program a high national priority—even though I realize that this is in some measure an act of faith and vision, for we do not now know what benefits await us.

But if I were to say, my fellow citizens, that we shall send to the moon, 240,000 miles away from the control station in Houston, a giant rocket more than 300 feet tall, the length of this football field, made of new metal alloys, some of which have not yet been invented, capable of standing heat and stresses several times more than have ever been experienced, fitted together with a precision better than the finest watch, carrying all the equipment needed for propulsion, guidance, control, communications, food and survival, on an untried mission, to an unknown celestial body, and then return it safely to earth, re-entering the atmosphere at speeds of over 25,000 miles per hour, causing heat about half that of the temperature of the sun—almost as hot as it is here today—and do all this, and do it right, and do it first before this decade is out, then we must be bold.

I'm the one who is doing all the work, so we just want you to stay cool for a minute. [*laughter*]

However, I think we're going to do it, and I think that we must pay what needs to be paid. I don't think we ought to waste any money, but I think we ought to do the job. And this will be done in the decade of the sixties. It may be done while some of you are still here at school at this college and university. It will be done during the term of office of some of the people who sit here on this platform. But it will be done. And it will be done before the end of this decade.

I am delighted that this university is playing a part in putting a man on the moon as part of a great national effort of the United States of America.

Many years ago the great British explorer George Mallory, who was to die on Mount Everest, was asked why did he want to climb it. He said, "Because it is there."

Well, space is here, and we're going to climb it, and the moon and the planets are there, and new hopes for knowledge and peace are there. And, therefore, as we set sail we ask God's blessing on the most hazardous and dangerous and greatest adventure on which man has ever embarked.

Thank you.

Prologue

Realization of a Dream of Ages

Colin Burgess

> Lives of great men all remind us,
> We can make our lives sublime.
> And, departing, leave behind us
> Footprints on the sands of time.
>
> Henry Wadsworth Longfellow (1807–82)

It was the spring of 1961, and the United States was in desperate need of some good news. The nation was experiencing considerable pain and undergoing an inescapable insight, with a mounting number of civil rights protests highlighting a desire for profound attitudinal change. At the heart of this movement was the spreading use of nonviolent "sit-ins," for the most part courageously led by young black college students protesting against enforced segregation in department stores, supermarkets, theaters, libraries, and elsewhere. Over the next few years these demonstrations would escalate in size and turmoil, often marred by violence, deaths, and bloody divisions across the nation.

The Cold War with the Soviet Union also had tensions running high, and a disastrous attempted invasion of Cuba at the Bay of Pigs would prove a monumentally high-profile failure for the United States's ambitious new president, John Fitzgerald Kennedy. The nation's fledgling space program, it was realized, could provide that crucial good news, create a renewed sense of pride and motivation, and offer something around which *all* Americans could rally. But the National Aeronautics and Space Administration (NASA), while still finding its feet as a civilian space agency, was frustratingly guilty of moving too slowly and cautiously and not heeding ominous warnings of pending space activities emanating from the Soviet Union. In this climate of uncertainty the first human spaceflight carried out by cosmonaut Yuri Gagarin stole the thunder of NASA's beloved Project Mercury.

Today there is widespread agreement that President Kennedy launched what became known as the space race primarily to revive the languishing spirits of the American people and to dramatize U.S. technology. The president and his advisors certainly had identified the next great battlefield.

On 25 May, just twenty days after Alan Shepard completed the first U.S. manned space flight, President Kennedy delivered a special State of the Union message to a joint session of Congress. In it he changed forever the course of the United States space program, saying in part, "I believe that this nation should commit itself to achieving the goal, before this decade is out, of landing a man on the moon and returning him safely to the Earth." He then asked for an additional twenty billion dollars from Congress to expedite the program. Kennedy's bold challenge was met with polite but hardly enthusiastic applause from Congress. The president then departed from his prepared remarks, the only time he did so in addressing Congress, by forcefully adding that "unless we are prepared to do the work and bear the burdens to make it successful," it would make no sense in going ahead with the project.

A bold challenge had been laid before the American people; the gauntlet had been thrown down. With only fifteen minutes of manned space flight experience, the United States would set its sights on the moon and try to get there before the Soviet Union. Academic Amitai Etzioni, the so-called guru of the nebulous communitarian movement and later a senior advisor to President Jimmy Carter, would state in bitter criticism, "We are using the space race to escape our painful problems here on Earth."

While NASA was mulling over the president's audacious announcement and examining its options, a little-known but tenacious young engineer from the Langley Research Center in Hampton, Virginia, would champion an entirely new approach to achieving a lunar landing. Dr. John Houbolt had devised a far less expensive but ingenious alternative to the proposed Direct Ascent and Earth-orbit Rendezvous (EOR) plans for setting astronauts on the moon. It was one he called "Lunar Orbit Rendezvous," or LOR. As *Apollo 11* Lunar Module pilot (LMP) Edwin "Buzz" Aldrin would later recall, not only did this innovative approach probably save American taxpayers twenty million dollars, but it helped NASA reach President Kennedy's goal of achieving a manned lunar landing. "LOR, which would require two spacecraft to link up a quarter-million miles from Earth, initially struck many people—me included—as dangerously complex, even bizarre," Aldrin reflected. "But Houbolt stubbornly kept pushing his plan, and the elegant logic of the LOR eventually won over the skeptics."

Houbolt had disliked intensely the Direct Ascent method that had first been proposed, saying it would fail because of the massive rocket required to land the remaining upper stage assembly and manned craft on the moon. "It was a vehicle about the size of an Atlas [rocket]," he once revealed. "Down at the Cape, it takes 3,000 men, a launch pad, and a launch facility to get an Atlas off the ground from the Earth. They were going to land something the size of an Atlas on the moon, backwards, with no help whatsoever. I thought that was preposterous."

Houbolt's plan employed a separate manned vehicle, then known as the Lunar Excursion Module, or LEM, later more simply known as the LM. After two astronauts had transferred into the LEM in lunar orbit, it was planned that they would undock from the Apollo Command Module (CM) spacecraft, descend, and land on the moon. Once the crew had completed their lunar excursions, the LEM would lift off to perform an intricate rendezvous with the *Apollo* craft, piloted by a third astronaut. The two crew members would transfer back into the mother ship, the landing craft would be discarded, and only the *Apollo* spacecraft would finally return to the Earth.

The Houbolt LOR option eventually won the support of influential rocket and aerospace designer Wernher von Braun, director of the George C. Marshall Space Flight Center in Huntsville, Alabama, and Dr. Robert Gilruth, leader of the Space Task Group (STG) located at Langley and director of the new Manned Spacecraft Center in Houston, Texas. Some, however, remained unconvinced, including D. Brainerd Holmes, director of the Office of Manned Space Flight. Despite lingering opposition from some powerful quarters, the lunar rendezvous method was finally approved and adopted by NASA administrator James Webb, who received the president's endorsement for the plan.

In July 1962 NASA formally announced that astronauts would use John Houbolt's Lunar Orbit Rendezvous method to land on the moon. Work could now begin on this colossal undertaking.

On 21 November 1962 President Kennedy would make one of his last speeches at a dedication ceremony for the Aerospace Medical Health Center at Brooks Air Force Base, outside of San Antonio, Texas. In his passionate oratory the president insisted that the United States would not retreat from the conquest of space. "There will be setbacks and frustrations and disappointments," he stated. "There will be pressures for our country to do less, and temptations to do something else. But this research must and will go on."

The young president also threw a stirring metaphor into his speech when he recalled that "Frank O'Connor, the Irish writer, tells in one of his books how, as a boy, he and his friends would make their way across the countryside, and when they came to an orchard wall that seemed too high and too doubtful to permit their voyage to continue, they took off their hats and tossed them over the wall—and then they had no choice but to follow them. . . . This nation has tossed its cap over the wall of space, and we have no choice but to follow it."

In another speech he planned to make the following day in Dallas, the president hoped to inspire his audience by emphasizing, "In today's world, freedom can be lost without a shot being fired, by ballots as well as bullets. The success of our leadership is dependent . . . on a clearer recognition of the virtues of freedom as well as the evils of tyranny . . . we have regained the initiative in the exploration of outer space, making an annual effort greater than the combined total of all space activities undertaken during the '50s . . . and making it clear to all that the United States of America has no intention of finishing second in space."

In a sad irony, given his reference to bullets and ballots, the chilling sound of shots rang out in Dallas just hours after the dedication ceremony, and the popular young president was dead, slain by an assassin. That stirring speech would never be given. The man who had set his nation on a relentless course for the moon was gone, but his extraordinary vision and goal would endure.

Much has been written about the Apollo program and how, ultimately, it evolved into a glorious triumph of human endeavor, persistence, and technology. The flight of *Apollo 11* would doubtless stand as one of the greatest achievements in human history, fulfilling an ambitious pledge made by the president in 1961. It would cost a cumulative twenty-four billion dollars and the lives of eight American astronauts, but on that triumphant day in 1969 the so-called space race to the moon was effectively at an end. The greatest-ever scientific and engineering undertaking had been accomplished—ahead of time—with an unprecedented combination of skill, talent, hard work, and a collective determination, the likes of which we might never see again.

On 16 July 1969 astronauts Neil Armstrong, Buzz Aldrin, and Michael Collins flew to the moon aboard CM *Columbia*, after which Armstrong and Aldrin transferred to LM *Eagle* for the final descent to the lunar surface. On 20 July they became the first human beings on Earth to touch down on another world.

1. The Whole World Was Watching

Rick Houston

> The moon rose above the horizon. Millions of hurrahs greeted her appearance. She was punctual to the rendezvous, and shouts of welcome greeted her on all sides, as her pale beams shone gracefully in the clear heavens. At this moment the three intrepid travelers appeared. This was the signal for renewed cries of still greater intensity. Instantly the vast assemblage, as with one accord, struck up the national hymn of the United States, and "Yankee Doodle," sung by five million of hearty throats, rose like a roaring tempest to the farthest limits of the atmosphere. Then a profound silence reigned throughout the crowd.
>
> Jules Verne (1828–1905)

The world watched and listened, breathless and mesmerized.

"Houston, this is Neil. Radio check."

Could this really be happening? It all just seemed so . . . what's the word for it? Unreal. That's it. Unreal. This stuff happened in comic books, not in real life.

"I'm at the foot of the ladder. The LM footpads are only depressed in the surface about one or two inches—"

The picture was fuzzy, hard to make out. Still, there was no mistaking the ghostly figure as it made its way from the top of the picture to the bottom. That was actually a human being, climbing down a ladder. They were really there, and this was proof positive, believe it or not.

"The surface appears to be very, very fine grained as you get close to it. It's almost like a powder—"

Absolutely amazing. A quarter-million miles from the events taking place, no one dared blink, out of a very real fear that something important might somehow be missed.

"I'm going to step off the LM now."

The grainy black-and-white images broadcast back to Earth during the earliest moments of the *Apollo 11* moon walk instantly became some of the most familiar in the history of mankind. This was the ratification of the Declaration of Independence and da Vinci putting the final touches on the Mona Lisa. This was the end of the American Civil War, World War I, and World War II rolled into one. This was every important event ever recorded, with one very important distinction. It was the first time that anyone other than those directly involved had been able to participate, vicariously at least, in the momentous occasion. Certainly, it was the first truly global media event. On 20 July 1969 hundreds of millions of people were watching, live and in real time, as *Apollo 11* commander Neil Armstrong and Lunar Module pilot (LMP) Edwin "Buzz" Aldrin worked on the flat landscape of the Sea of Tranquility.

The surrenders in 1945 of Axis powers Germany and Japan that signaled the end of World War II carried obvious global implications. The world in which we live was shaped in very large part by the conclusion of that ghastly conflict. Still, only a few hundred people, at the very most, actually saw the signing of those surrender documents take place. If pictures exist of Germany's official surrender, they are obscure at best. Better known are photographs and filmed news footage of the Japanese surrender on board the USS *Missouri* in Tokyo Bay. Radio had never had—and never will have—the same effect, and television was in its infancy. To see what the scene of the surrender looked like, millions had to wait for a newspaper photo or for a newsreel to be played down at the local movie theater in a week or more.

Less than twenty years later the assassination of President John F. Kennedy on 22 November 1963 again brought the world to a standstill. It was Kennedy who set the United States on its course to the moon with his speech to Congress in May 1961, and in the blink of an eye, he was gone. The shocking event changed journalism forever. With television able to fill the news void much more quickly in a dramatic situation like this one, no longer would it be good enough to wait for a newspaper. The public wanted to know the latest developments, and it wanted to know them right now.

2. In this frame from a 16 mm cine film, Buzz Aldrin and Neil Armstrong plant the U.S. flag in the lunar soil. Courtesy NASA.

For all that, coverage of Kennedy's assassination was after the fact, a reaction to the event itself. Dallas businessman Abraham Zapruder was filming Kennedy's motorcade when the shots were fired, but his stunning and rather gruesome work would not be seen by the public at large for more than ten years. There were so few actual eyewitnesses to the shooting itself; assassination buffs can name most, if not all, of them.

The case can be made that only one other event in the history of mankind has received the same kind of live, as-it-actually-happened media attention as *Apollo 11*, and it was under far different, far less peaceful circumstances. The terrorist attacks of 11 September 2001 put singular focus onto New York City and Washington DC. We could not believe what we were seeing, and we were seeing virtually every moment live. The world's news cameras were already trained on the scene when the second plane hit the World Trade Center and would continue to be so through the collapse of both towers and the ensuing cleanup. From Ground Zero to the Pentagon to a lonely field in Pennsylvania, the news was live on network after network, cable news channel after cable news channel. For days the attacks were the sole focus of the entire world. It was, in every sense, 9/11, 24/7.

The *Apollo 11* moon walk was an entirely different proposition. Whether due to its scientific nature or simply a genuine curiosity, there was an attention paid to the flight that went far beyond intense. Of the billions of humans who had ever lived, these were the first steps ever taken on a planetary surface other than our very own Earth. Buck Rogers had come to life, with science fiction no longer completely out of the realm of possibility. How improbable it really did all seem. Less than seventy years earlier, there had never been a powered flight of any kind. Before the Wright brothers flew their short, 120-foot flight on the dunes of Kitty Hawk, North Carolina, on 17 December 1903, those who dreamed of taking to the skies could do so only in hot-air balloons. *Balloons*? Here were Armstrong and Aldrin, three generations later, on the surface of the moon, having begun their 240,000-mile journey at the tip of a rocket more than three times the length of the Wright brothers' first flight.

In typical understated Armstrong fashion, he has tended to downplay the preflight attention paid to *Apollo 11*. In a NASA Oral History interview conducted on 19 September 2001, historian Stephen Ambrose asked Armstrong how he and crewmates Aldrin and Command Module pilot (CMP) Michael Collins dealt with the public's anticipation. Whether intentional or not, Armstrong's answer was a distinct paraphrase of the Serenity Prayer commonly attributed to theologian Reinhold Niebuhr. "We tried to be as focused as we could, work on the things we could do something about and not worry about the things that were beyond our ability to change," Armstrong told Ambrose.

Later in the Oral History, Armstrong added: "I was certainly aware that this was a culmination of the work of 300,000 or 400,000 people over a decade and that the nation's hopes and outward appearance largely rested on how the results came out. With those pressures, it seemed the most important thing to do was focus on our job as best we were able to and try to allow nothing to distract us from doing the very best job we could. And, you know, I have no complaints about the way my colleagues were able to step up to that."

To the untrained eye, not much was distinguishable during the earliest moments of the moon walk, which began, officially, at 10:56 p.m., Eastern Daylight Time. The Capsule Communicator (CapCom) on shift at the time, fellow astronaut Bruce McCandless, told Armstrong and Aldrin, and by extension a worldwide audience, that there was a "great deal" of contrast in the television feed and that a "fair amount" of detail could be seen. "Fair amount" is rather generous. McCandless knew what to look for, but most of the rest of the world did not. The grainy images were a far cry from the pictures that would be broadcast back to Earth during *Apollo 17*, the last lunar landing. In one of those brilliant color scenes mission commander Eugene Cernan uses a brush to clean the Lunar Rover–mounted camera. He lifts his visor; his face, communications cap, and microphone are clearly distinguishable. That kind of amazing clarity had not been available for *Apollo 11*.

Attached to the modular equipment stowage assembly (or in NASA-speak, the MESA) on the side of the Lunar Module (LM), *Apollo 11*'s black-and-white Westinghouse camera was lowered into place by Armstrong as he made his way onto the porch of the LM *Eagle*. The camera was later detached from the MESA and moved, along with a stand, northwest of the LM and pointed southeast to allow for an overall view of the landing site. It was from this vantage point that Armstrong and Aldrin could be seen—in a manner of speaking—during their chat with President Richard Nixon. Once both men were on the surface, it was impossible to tell one from the other. The blurry images served to enhance even further the mystique of what Armstrong and Aldrin were doing. They were a long, long way from home, and these pictures proved it.

For two and a half hours Armstrong and Aldrin bounced, hopped, loped, and skipped across the dusty plain of Tranquility Base. Some of their work was entirely ceremonial, such as Nixon's call and the unveiling of the famous "We came in peace for all mankind" plaque attached to *Eagle*'s descent stage. Still, this was no mere sightseeing excursion. From Armstrong's first step on the surface to his last, every detail of the moon walk had long before been carefully planned. Both moon walkers were allowed time to accustom themselves to moving in the decidedly unfamiliar, one-sixth gravity environment. There were experimental packages to deploy, photographs to take, and, perhaps most important, samples to collect. In one of the moon walk's very few unscripted moments, Armstrong dashed some 200 feet from

Eagle to examine a small crater. It was the farthest either man would venture from the relative safety and security of the LM.

The flight is not famous for featuring the longest, biggest, or most of anything. That was not the mission of *Apollo 11*. The mission of *Apollo 11* was simply this: to land safely, get out, take a look around, grab a few rock samples, set up some experiments, get back in, get the heck out of there, and return safely home to good ol' Mother Earth. If Armstrong, Collins, and Aldrin were able to do that, they would ace their final exam, nothing more, nothing less. *Eagle* wound up farthest by far from its intended landing site and had the shortest overall stay on the lunar surface of the entire program. It was the only flight with just one extravehicular activity (EVA), and its lone moon walk was the shortest of any mission. Lastly, *Apollo 11* wound up with the least amount of lunar material gathered. Depending on the source, Armstrong and Aldrin collected in the neighborhood of forty-seven pounds of lunar material. The pair of explorers snapped a total of 339 photographs while on the surface, including Armstrong's well-known shot of Aldrin. It is an image that has come to symbolize the entire Apollo program, if not manned spaceflight as a whole, nationality notwithstanding. Aldrin is simply standing on the surface in the picture, his left arm slightly raised, and Armstrong the photographer is barely visible in the reflection of Aldrin's helmet visor. For years it was the only known photographic image of the *Apollo 11* commander on the lunar surface. Was it some sort of intentional snub on Aldrin's part or just an oversight? Only Aldrin can say for certain, and he has denied any ill will. Taking pictures of Armstrong on the surface, after all, was not on any checklist.

The sheer enormity of the moment was not lost on the press corps covering the event. Newsman Walter Cronkite, ever the consummate professional at CBS, along with guest commentator and former astronaut Wally Schirra, were all but speechless when Armstrong and Aldrin landed in the Sea of Tranquility. Schirra appeared to wipe a tear from his eye, while Cronkite could add not much more than an enthusiastic "Oh, boy," grinning as he took off his familiar glasses and rubbing his hands together. Finally, Cronkite implored Schirra to say something. Some six and a half hours later, Cronkite yet again had very little to say as Armstrong prepared to make those historic first steps. The veteran commentator was completely immersed in a moment being shared simultaneously by an audience estimated at some six hundred million. When Cronkite did speak, it was in a tone almost literally dripping in hushed awe: "Boy, look at those pictures . . . wow!" While he did regain his composure fairly quickly, in doing so Cronkite very nearly talked over Armstrong's immortal first words, recalling his reaction to the situation in his 1996 autobiography, *A Reporter's Life*:

That first landing on the moon was, indeed, the most extraordinary story of our time and almost as remarkable a feat for television as the space flight itself. To see Neil Armstrong, 240,000 miles out there, as he took that giant step for mankind onto the moon's surface was a thrill beyond all the other thrills of that flight. All those thrills tumbled over each other so quickly that the goose pimples from one merged into the goose pimple from the next. When Neil emerged from the Eagle *I almost had regained my composure, which I'd lost completely when the* Eagle *had settled gently on the moon's surface. I had just as long as* NASA *had to prepare for that moment, and yet, when it came, I was speechless.*

Cronkite was not alone. The BBC's James Burke was almost humorous in his pronounced enthusiasm, his voice rising with virtually every utterance: "Here's the picture!" Then, "There . . . is . . . Armstrong! You can see him moving." As Armstrong paused to describe the lunar surface and to test how difficult it would be to jump back up to the ladder's bottom rung, Burke went completely silent, before finally adding, excitedly: "There he is, putting his foot out. You can see him leaning on it. You can just make out the backpack and the dark circle of the visor in front of it." A split-second later, Armstrong uttered a few simple words that would be remembered through the ages: "That's one small step for man, one giant leap for mankind." Surely, Cronkite and Burke later breathed sighs of relief, thankful they had not walked all over Armstrong.

Note should be made here of the debate that has existed almost from the time Armstrong uttered the famous saying. Did he actually say "One small step for *a* man," with the indefinite article *a* somehow lost in transmission? No, he did not, and to imply otherwise is revisionist history. Granted, it is possible, if not probable, that he intended to say "a man." From the tone and inflection of his voice it seems for all the world that Armstrong caught the mistake immediately. Following "That's one small step for man," he added another *one*, stopped again, then finished the statement with "giant leap for mankind." There's nothing lost in transmission, nothing at all, no matter what any super-scientific studies to the contrary might suggest.

Regardless of what Armstrong did or did not say, most of the world watched with rapt attention as he and Aldrin walked their way into history. It was something unlike anyone had ever experienced before. This was history, yes, but it was at the same time something far more than that. People from virtually every country around the globe could close their eyes and imagine themselves right there alongside Armstrong and Aldrin as they accomplished tasks that had been considered impossible just a few years earlier. The people watching came from every walk of life, and they reacted to the landing in countless ways. Several became astronauts themselves; another would go on to report on man's continued journeys into space and come close to making the fantastic trip himself. They were U.S. soldiers overseas and citizens of countries in the dark shadow cast by the Iron Curtain. They were ordinary people from every walk of life, and these are but a microcosm of mankind's memories of the triumph of *Apollo 11*.

Only a year or so had passed since the assassinations of civil rights leader Martin Luther King Jr. and Robert Kennedy, brother of the slain president, who had all but sewn up the 1968 Democratic presidential nomination. The Democratic National Convention in Chicago triggered violent protests. And then . . . then there was the vicious war in Vietnam. The year 1968 would see the agonizing loss of nearly 16,600 troops, by far the bloodiest year of the conflict. Another 11,616 would die by the end of 1969. There is utterly no way to understate it—the United States in the late 1960s was a nation in conflict, in faraway Vietnam and with itself. Vietnam, the Cold War with the Soviet Union, race relations, violent protests, urban decay, a growing drug problem—name it, and it was taking a toll on the American psyche.

Apollo 11 was not immune from the country's season of discontent. With so many problems facing the country, NBC television commentator Chet Huntley wondered on air about its priorities:

As thrilling as these triumphs and successes are, they sometimes have a tendency to produce a backlash of discouragement over our inability to predict or control our weather, to manage our cities and regulate our traffic, to eliminate poverty and crime and educate all our population and house and feed all of our people. At the same time that there is euphoria and triumph, there may also be a backlash of some disappointment because it may occur to us frequently that if we can correlate 22,000 flight steps and nine million pieces of hardware and make them function as perfectly as all of this, why are we failing with some of our other problems?

During congressional hearings into the *Apollo 1* tragedy, Minnesota senator Walter Mondale tangled repeatedly with NASA management. A member of the Committee on Aeronautical and Space Sciences, he would come to be known as a virulent opponent of the space program. According to Mondale, he actually supported Apollo, to a point.

It was a tough time for a person with my philosophy because I wanted to see progress in education, dealing with poverty and so on at home. I also, at that time, was supporting the war in Vietnam, I regret to say. And we also were moving ahead with the space program. I sat on the space committee and I supported most of that, but as the '60s kept moving along, the war kept costing more and more. The space program was demanding more money all the time and we were . . . slowly beginning to fund some of these so-called Great Society programs. It just got too much and we started getting inflationary pressures.

Not even Bob Hope's annual Christmas visit to American servicemen could avoid the slightest glimmer of controversy concerning the landing. Following his safe return to Earth, Armstrong accompanied Hope's entourage overseas and ultimately visited a number of bases in Vietnam. At Da Nang, Hope brought Armstrong onstage to a standing ovation. He introduced the moon walker as a "very quiet and soft-spoken young man" who had been part of a team that had "provided this world with a thrill they would not soon forget." Those closest to the stage clamored to get nearer Armstrong, all but falling over themselves to shake his hand. At another stop, however, a soldier glared at Armstrong as he bluntly asked why the United States was interested in the moon, rather than the conflict in Vietnam.

"The nature of the American system is that it works on many levels . . . to try to build peace on Earth and good will to men," Armstrong responded. "One of the advantages of the space activity is that it has promoted international understanding and enabled cooperative efforts between countries on many levels and will continue to do so in the future."

It was in this kind of climate that the historic events of *Apollo 11* unfolded. There was discord, of course, but for once in the turbulent 1960s, it was very nearly completely shouted down by overwhelming support both at home and abroad.

The white 1962 Pontiac Bonneville station wagon headed from Maryland to California and then back again, a trip that would eventually take more than three weeks to complete. There were no in-car DVD players—or DVD players of any kind, for that matter—to preoccupy the four children of David and Rosemarie Jones. There were no Game Boys, no PSPs, no Leapsters, and no iPods. No matter. The Jones clan stopped at several National Parks along the way and drove through Texas and other parts of the American Southwest in the depth of summer. An aftermarket Sears air conditioner installed under the dashboard made life fairly tolerable for those fortunate enough to claim spots in the front and middle seats. The Jones children who lost the race to the car were stuck in the back, rear-facing seats, which benefited not in the least from the cool, refreshing air.

Nevertheless, young Tom Jones's mind was on something other than the heat. At fourteen Jones had for years been a space buff. When he was five, his grandmother gave him a copy of the children's book *Spaceflight: The Coming Exploration of the Universe*, which first sparked his interest in reading anything and everything he could get his hands on that dealt with the final frontier. During the two-man Gemini program Jones "sketched like mad" drawings of rockets and launches and astronauts and whatever else he could dream up. He watched *Star Trek* and saw the movie *2001: A Space Odyssey*. Growing up in Essex, Maryland, Jones lived within miles of where Gemini's Titan II boosters were built at the Martin Marietta plant in nearby Middle River. One scrapbook became two, then three, then six, then seven, each crammed with every kind of space-related news clipping he could find. While collecting newspapers and magazines for a Boy Scout recycling project, he kept any and all space-related *Life* magazines for himself. Jones's parents did not have a color television, so for each launch he would make his way next door and plant himself squarely in front of the neighbors' color set, with model rockets in tow.

Tom Jones was, to put it mildly, eaten up with all things space.

There was a tremendous amount of coverage in the daily papers, and that's what I was clipping for my scrapbooks. I would clip out anything that had to do with space, whether it was a satellite launch or something having to do with an actual mission. Then, I found out about NASA Facts. *NASA itself was sending out—to any kid who wanted—reports, pamphlets and posters on the Saturn V, Apollo, and planetary probes. You could get on their mailing list and they just shipped you stuff, just on request. Then, the TV was very dramatic. There would be a big build-up before each mission. If I was in school, we would actually stop class to watch the final half-hour of the countdown. . . . I watched Walter Cronkite the most. He always had some technical experts from North American [Aviation] to talk about Apollo, and then later, after* Apollo 7, *he had Wally Schirra on as a technical consultant, who sat with him on the desk. They would banter back and forth. I think that was intriguing to me, that I got to watch one of the astronauts talk and hear what he said. I'm not sure if the other networks had that or not, but I remember Cronkite did.*

Somewhere between Las Vegas and southern California, Tom Jones convinced his family to pull over to the side of the road so he could listen in on his transistor radio as Armstrong and Aldrin touched down on the lunar surface. Afterward he had to get to a television to see the EVA. He had to. For the first and only time on their trip, the family got a hotel room, near Disneyland, so they could watch history being made.

It was just amazing that you could see these ghostly images of these guys actually walking around on the moon. I wanted it to be crisper than that. The very early steps, when Armstrong was coming down the ladder, were very tough to make out. But as they got out and set up the camera away from the LM, you could actually see them bouncing around and hopping around on the surface. You could follow it pretty well. I think after the first hour, most of my family went out to the pool to splash around, and I was left alone to watch the rest of it. Their attention span wasn't as extended as mine, because I was all wrapped up in this stuff.

Millions of kids dreamed of becoming an astronaut after watching the thrilling triumph of *Apollo 11*. They saw the keen anticipation leading into the countdown and launch, the drama of the landing, EVA, lunar liftoff, reentry, and splashdown, and the pageantry of the postflight ticker tape parades. The astronauts were on magazine covers, on the front page of every newspaper around, and constantly on television. They were the coolest of the cool. For the vast majority of those dreamers reality soon set in. They graduated high school and either got a job or went to college or, in many cases, Vietnam. Tom Jones never veered from his goal of flying in space.

It's not all my doing. There were a lot of lucky breaks along the way. I had the gift of a lot of encouragement from my parents and from teachers who recognized in the '60s that this was an important thing and a kid could aspire to it. I think it's a testimony to the fact that we live in a country where people can dream about almost anything and then find a way to get there and find a way to do it themselves. Certainly, nobody ever told me, "You're dreaming in vain." People thought, "Yeah, that's far out there, kid. But go for it."

When *Apollo 11* flew, Jones was about to enter his freshman year at Stemmers Run Junior High School in Essex. Astronaut requirements in hand, almost as if they were a checklist to be followed at all costs, he made plans to attend one of the service academies following his senior year at Kenwood Senior High School. Check. He graduated from the United States Air Force Academy in Colorado Springs in 1977 and eventually served as pilot and aircraft commander of a B-52D Stratofortress out of Carswell Air Force Base (AFB) in Texas. Check. He received a doctorate in planetary science from the University of Arizona in Tucson in 1988, check, after which he joined the Central Intelligence Agency as a program management engineer. Check. Finally, after a year of training, he officially became an astronaut in July 1991. In April 1994 he flew STS-59, the first of his four space flights, on board the Space Shuttle *Endeavour*. Check, check, check, and double check. Mission accomplished. Jones spent more than fifty-two days in space over the course of his four flights, including three spacewalks that totaled more than nineteen hours. Mission accomplished, indeed.

I don't think I thought about that first moonwalk during my first flight, but I was certainly thinking about the gee-whiz aspects of being a little kid and thinking about going into space one day. The collective memory of the scrapbooks, the TV broadcasts, the movies I watched and the science fiction novels I read—the realization that I was actually in space, some thirty years after starting to think about it for the first time consciously—was very much in my mind.

3. Shuttle astronaut Thomas D. Jones grew up with a cherished dream. Courtesy NASA.

That first day in orbit, I was just very emotionally overwhelmed by the fact that I was actually there after dreaming about it for all those years. The first time I looked out the window and saw the Earth, and could actually have a moment to think about the fact that Tom Jones was in space, it was very moving for me. I wiped away a few tears looking out the window at the beauty of what I was seeing outside.

David Jones, Tom's father, never saw his son fly into space. Jones had been named to the crew of STS-59, but David had a sudden heart attack and passed away in late 1992, a year and a half or so before the flight. Rosemarie, on the other hand, saw all four of her son's flights (and six countdowns). "She never said to me that I'm stupid or 'You shouldn't think about this,'" Jones says. "She said, 'If that's what you want to do, go ahead.'" Remarkably, given the accomplishments of her son, Rosemarie Jones has never flown on an airplane.

At one time or another Jones met all three of the *Apollo 11* crew. When he wrote his book *Sky Walking: An Astronaut's Memoir* (2006), he enlisted Michael Collins to review the manuscript and write a blurb for the book jacket. Armstrong also read and commented on Jones's book before its publication. Jones then served with Armstrong on the NASA Advisory Council beginning in 2006. He had the chance, "as a grown up little kid," to lunch with the man he had seen take the first steps on the lunar surface back in that Southern California hotel room years earlier. As a scientist and former astronaut, he could talk on their own terms with Armstrong and other legendary astronauts such as *Apollo 16* commander and moon walker John Young, who was on the selection board that picked Jones to join NASA. They could compare notes and share common experiences. Finally, at long last, Jones was one of them.

Although *Apollo 11* had its genesis in the cold realities of the tensions that existed between the United States and Soviet Union, there was a sense of ownership in the flight's achievements that transcended nationalities, even within those countries behind the so-called Iron Curtain. When *Apollo 11* landed in July 1969, Laurenc "Lauro" Svitok was a twenty-year-old mechanical engineering student at the Slovak Technical University in the former Czechoslovakia. Living with his parents in his native city of Bratislava, Svitok took time out from preparing for several very challenging exams to watch the landing and moon walk.

Svitok and his compatriots were less than a year removed from the August 1968 Soviet suppression of Prague Spring, a movement in which the country had moved toward a "democratization" of its political system. Some 500,000 troops from Warsaw Pact countries rolled into Czechoslovakia, and twenty-five years would pass until the land would know true freedom again. The January 1993 "Velvet Divorce" separated the former nation into two separate and distinct countries—the Czech Republic and Slovakia, where Svitok now lives. The fact of the matter was that a person did not have to be American to be proud of what had taken place on the flight of *Apollo 11*. Svitok had been a space enthusiast since he heard the official radio announcement of Sputnik in October 1957, captured first by the adventure of

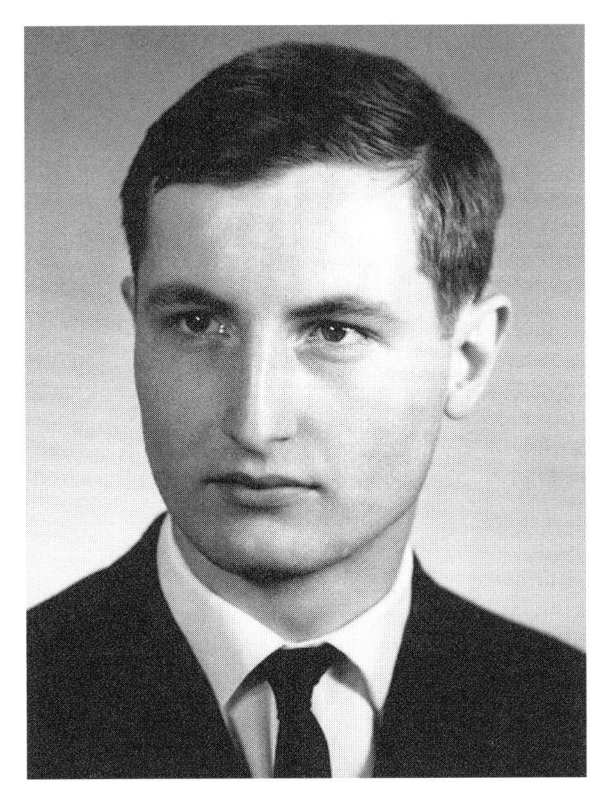

4. Laurenc Svitok around the time of the *Apollo 11* mission. Courtesy Laurenc Svitok.

it all and then later by its science and technology aspects. Spaceflight was, and still is, his biggest hobby.

I really saw Apollo 11 *as a victory of the free society, more advanced technology and much better management. In my opinion, Americans were taken by surprise with Sputnik, Yuri Gagarin and Alexei Leonov. When the United States organized their response and started to beat the Soviets with their Gemini program, it was clear to me that—unless the Soviets had something really big hidden in their sleeves—the Americans would win. Several decades later, knowing now (nearly) the full truth of the moon race, it is the biggest irony that Americans with their free-market economy and competition were able to create one organization—NASA—which covered and controlled the fulfillment of President Kennedy's dream, while strictly centrally organized Soviets allowed the split of their efforts and resources by the quarreling branches of Sergei Korolev, Valentin Glushko and Vladimir Chelomey [the three main architects of the Soviet space effort, otherwise known as the "chief designers"], which brought them to the terminal failure of their manned lunar program. Really amazing.*

Despite being occupied, media in the country was still relatively free and reported on the flight of *Apollo 11* fully and factually and even, Svitok says, enthusiastically. State television carried live coverage of the mission from start to finish, complete with in-studio experts, documentaries, and discussions. Better yet, he lived in the southwest corner of the country, just thirty-one miles or so from Vienna, the capital of Austria, which gave him direct access to Austrian coverage of the flight as well, and he also collected articles from a local publication, *Letectvi + Kosmonautika* (Flying and Cosmonautics).

Even after a crackdown on the last remnants of the Prague Spring a year after *Apollo 11*, living conditions at home were "much, much better than in the Soviet Union":

We were quite independent from the politics. We had a lot of industries, very good agriculture, excellent educational system and living was simple, but good. No luxuries for the ordinary people, but also no hungry and homeless people. Media was another thing. They were fully controlled by the Communist Party and repeated obediently what they were told to say. Our Communist Party was the satellite of the Soviet Communist Party, so we were brainwashed every day for years. People with access to Austrian TV and to the Voice of America had at least that alternative.

The 1970s and early 1980s were years of what the Czechoslovak government called "normalization." State media downplayed American accomplishments in space and was "extremely hostile" against the Space Shuttle program. "I think that many people, having no information but official state media, really believed that the Shuttle was mainly a military project," says Svitok, who still lives in Bratislava. "You know, when you hear something 200 times, you start to believe it."

Lauro Svitok's interest in manned spaceflight today manifests itself in his collection of NASA lithographs, brochures, press kits, DVDs, and books. He still owns a photo of moon walker Gene Cernan, whose grandfather was born in present-day Slovakia, which he received from the Astronaut Office in Houston. "It's really personally written and signed, which was not always the case with other astronauts." Reflecting on *Apollo 11* and what life was once like in his country, Svitok concludes: "I dare to say many people understood *Apollo 11* as a kind of revenge for what the Soviets did to us. Those were really very interesting and complicated times."

That they were, Lauro. That they were.

Dick Stratton was in a world of hurt. The pilot of an A4E Skyhawk launched off the deck of the USS *Ticonderoga*, Stratton was on an armed reconnaissance patrol over Thanh Hoa Province in North Vietnam on 5 January 1967 when a series of unstabilized rockets that he had just fired collided near his plane. At an altitude of 2,200 feet and with shards of the rockets tearing his engine to shreds, it exploded, ripping the tail of his Skyhawk completely off. In effect, he says, he shot himself down. After ejecting, Stratton was immediately captured, and less than a day later he found himself in the infamous Hoa Lo prison, better known by its former inmates as the "Hanoi Hilton." For more than six years—2,251 days, to be exact—Stratton was a prisoner of war, held and tortured in unimaginable circumstances.

Flying his twenty-seventh combat mission and a few months past his thirty-fifth birthday, Stratton had been a naval aviator for nearly ten years when he was captured. A native of Quincy, Massachusetts, Stratton was only "vaguely aware" of what his country was trying to accomplish in space. He had even gone so far as to bet a Stanford classmate that the United States would never make it to the moon because, as he said at the time, "There was no reason to do so." He was a liberal arts major in college, a non-jock who was afraid of heights. Once in the navy, Stratton had his hands full flying a high-performance jet fighter on and off an aircraft carrier. That was reason enough, he says, never to consider joining the space program himself. He was too busy trying to stay ahead of the curve on Earth to worry about anything else, much less becoming an astronaut. Says Stratton, "I wasn't remotely interested in flying anything faster than Mach 1 or at an altitude in excess of 50,000 feet."

Most of the rest of the world was united, for a moment at least, around the happenings of *Apollo 11*. Not Stratton. He knew nothing of it when Armstrong and Aldrin touched down, the day passing just like any other. During the first few years of his imprisonment, Stratton would have been completely and utterly isolated had it not been for clandestine conversations with fellow POWs, which included former vice presidential candidate Admiral James Stockdale and 2008 Republican presidential nominee Senator John McCain, via tapped-out code on the walls of their cells. At best reliable information came at the most infrequent of drips.

Communists always told us, the POWs, the bad news about what was happening in the States in an attempt to demoralize us . . . never anything good. We were subjected to the evening broadcast of the "Voice of Vietnam," known to us as Hanoi Hannah. Each cell had a small bitch box located up in the ceiling where we could not reach it. Nightly, we would hear about all the demonstrations, riots, murders, assassinations, stock market collapses and pronouncements from every anti-war politician, peacenik, agitator and campus loony that bloviated stateside that day. LBJ and [Defense Secretary Robert] McNamara had stopped the bombing. As a result, our ability to get the straight skinny from newly shot down aviators had ceased to exist.

There was little way of knowing that *Apollo 11* had launched, landed on the moon, or returned safely to Earth or, for that matter, that the flight had even existed. Stratton was captured more than a year before the first flight of the Apollo program and was released after the final lunar landing. Three weeks after Stratton's fateful flight, a fire claimed the lives of *Apollo 1* astronauts Gus Grissom, Ed White, and Roger Chaffee. He had not known of the solid, if not somewhat grumpy, return to flight of *Apollo 7* or anything about the risky gamble of sending *Apollo 8* into lunar orbit in December 1968 and its crew's emotional reading from Genesis on Christmas Eve. He was unaware of the warm-up missions of *Apollo 9* and *10*, which tested the LM in Earth and lunar orbit, respectively.

Stratton was in the dark about the historic flight of *Apollo 11* until Trinh Thi Ngo, known as "Hanoi Hannah" to hundreds of thousands of U.S. troops, spilled the beans during a propaganda broadcast over Radio Hanoi. Having been moved from the Hanoi Hilton to a location dubbed "The Plantation," Stratton was all too familiar with the disinformation campaign being waged by the North Vietnamese. While the broadcast was designed to discourage the POWs, it had a completely opposite impact.

One glorious night, locked in solitary confinement, I heard Hanoi Hannah snidely and triumphantly bitching, "The United States might be able to put a man on the moon, but it still cannot defeat the armed forces of the Demo-

cratic Republic of Vietnam." So it was that I and my shipmates in the Plantation learned of the success of Apollo 11. *You can well imagine what a tremendous boost it gave to our morale, what an exhilarating boost of national pride to our battered psyches, and what a great sense of comradeship we felt with our fellow aviators who accomplished such a feat. And the beauty was, the [North Vietnamese] were too stupid to know that they had disclosed the "secret." One for our side.*

In early August 1969, just weeks after the flight of *Apollo 11*, Stratton and other captives took a calculated risk and allowed three of their number—Doug Hegdahl, Bob Frishman, and Wesley Rumble—to accept early release. Designed as yet another propaganda move by the North Vietnamese, the trio instead told the world of Hanoi's inhumane treatment of its American POWs. While torture all but ceased afterward, Stratton continued to live in captivity until he and 590 other American POWs were released as part of Operation Homecoming on 4 March 1973. Stratton says the United States government provided the freed men with no debriefing on world events during their captivity. His main source of information was a "super packet" put together by *Reader's Digest*.

Stratton remained in the navy until retiring in 1986 at the rank of captain. He became a clinical social worker and currently lives in Florida. Two sons, not to mention a daughter-in-law, served in the military as members of the United States Marines Corps. The horrors of the Hanoi Hilton took place long ago, and unlike several other former POWs, Stratton has never returned to that Southeast Asian country. To the rest of the world *Apollo 11* had been a source of golly-gee-whiz amazement. For Stratton, once he found out about it, the landing was one more confirmation of a deep and abiding faith in his country and its abilities.

A couple hundred miles to the south of Stratton's accommodations in the Hanoi Hilton, Jimmy Hatchett was a corporal in the United States Marines on the ground at a small landing zone just outside Khe Sanh, South Vietnam, when Armstrong and Aldrin made their descent into Tranquility Base. An ammunition technician, Hatchett was one month from that magical DEROS—Date Eligible to Return from Overseas. He was going home. Nothing else much mattered, not even *Apollo 11*.

Armstrong and Aldrin could walk on the moon. That was fine. Whatever. Hatchett was far more concerned with going back to The World in one piece. That's what mattered. That's *all* that mattered.

"I don't even remember the first time I heard about it," he says. "When it landed, I was about a month away from coming home and getting out [of the military]. Everything was sort of in a blur because of where I was. We didn't get a lot of news there. . . . My time was taken up with other things, and I just wasn't interested in anything else."

On the other side of the globe, however, Hatchett's sister, Sandy Estep, was consumed with news of the flight at her home in Nashville, Tennessee. Heck, she wanted to make the journey herself. She was ready and more than willing to go. "I was so envious of the 'spacemen,'" Estep remembers. "I was glued to the TV until late that night. I wanted to be up there with them so badly."

Joe Estep Sr., Sandy's husband, never believed that men had actually walked on the moon—it was all special effects, he said. Then again, it might very well be that he was simply trying to tease his wife. Still, he was going to hedge his bet. "He did promise me that if there was any way at all, he would see that I got the first ticket sold to the moon," Sandy continues. "He felt pretty safe that he would not have to keep that promise."

For Sandy, who lost Joe Sr. to a sudden heart attack in October 1994, the dream of flying in space remains to this day. "If I could, I would be on any rocket into outer space," she says. "What a wonderful trip that would be."

Jimmy Hatchett's future wife, the nineteen-year-old Linda Boland, also paid close attention to the journey of *Apollo 11*. Her father, Joseph, worked for General Dynamics Astronautics out of the company's California offices and had helped in the design of the cooling system for the boosters that powered each of NASA's Gemini missions into space. Having just completed her freshman year in college, Linda felt a very real sense of ownership in events taking place nearly a quarter-million miles away. Her dad had helped make it happen.

"My dad was fascinated because he had participated so much," Linda says. "We sat there and watched it live when they got out and walked on the moon for the first time. . . . If my dad hadn't done what he did with the work on the Gemini series, we never would have gone to the moon."

An estimated 400,000 other families of those who contributed in one way or another to the space program could say exactly the same thing.

Waddell Watters, then a twenty-one-year-old enlisted man stationed at Clark AFB in the Philippines, calls man's first landing on the moon one of the most memorable world events of his life, even more so than John F. Kennedy's assassination. Watters arrived in the Philippines just a couple of months earlier and simply was not used to life in a primitive third world country. Everything he had ever known, everything he had ever taken for granted, was being challenged by the lifestyle of native Filipinos, not to mention some of his fellow servicemen.

He had been told not to drink the water. Night buckets filled with bodily waste were emptied in the streets. Surrounded by the jungle, a volcano loomed over the base. Traffic rules? What traffic rules? They seemed, Watters says, optional at best. His sleeping quarters were like something out of a Hemingway story, left over from a bygone era. He lived in open-bay barracks with screened sides and large ceiling fans. He and his fellow servicemen slept on steel-framed bunks hung with mosquito nets that, if not put down early enough each night, would allow a rather large assortment of flying insects to share his accommodations. Watters was in a foreign country disconnected from most activities and many of the ideas around which his life had been formed.

Told by a coworker that the moon landing had taken place, Watters ran to a dayroom that was already standing-room only. Most of the lights were off.

From where I stood just inside the door, I could see the television and its gray "live from the moon" picture. The room was very quiet and we all could hear the commentary on the TV. I remember not talking, just listening and staring at the images. Then, looking out the open door at the bright tropical day with the surrounding coconut palm trees, I was struck with the dichotomy between the modern world from which I came and the technically deprived world I was now in. A society that could put men on the moon versus one which was ruled by a virtual dictator were worlds apart.

Watters came to cherish his time in the Philippines. He loved the people, the food, the places he visited, the scenic wonders that he saw. Yet of all those wondrous scenes, he reported, "the most wonderful was the one I witnessed in that dark day-room."

Miles O'Brien is generally recognized as this generation's Walter Cronkite, the go-to guy in broadcast journalism on all things space. The chief technology and environment correspondent for CNN until leaving the Ted Turner–owned company in late 2008, O'Brien shared a desk with Cronkite during John Glenn's return to flight in 1998 and also anchored the cable network's coverage of the Space Shuttle *Columbia* disaster in February 2003. An instrument-rated pilot himself, plans for O'Brien to fly on the shuttle as a space-going journalist were to have been announced shortly after *Columbia*'s landing.

Growing up in Grosse Pointe Farms, Michigan, O'Brien was ten years old when *Apollo 11* landed on the moon. Little could he have known then the directions his amazing career would take, and on that night he knew only that he could not sleep. The excitement of it all was just too much to overcome.

I was in the basement of my house, because that's where we had the big TV, *one of those old console deals. I remember the landing and being absolutely transfixed. My father's [Miles O'Brien Jr.] birthday is the 21st of July, and according to the original timeline, they were supposed to have landed, take a rest period and then walk on the 21st. My father was pretty happy about that. I stayed up for the entire walk. I wanted to experience that event in real time, it was so exciting. At ten years old, I don't know if I fully grasped the full historical significance of it. Those of us in that era kind of grew up with the notion of space and pushing the envelope in space. I guess I thought it was just sort of a beginning, but of course, it really turns out that it was the beginning of the end of an era.*

In the run-up to Apollo 11, *I was fascinated, like a lot of kids were. I was glued to the tube, transfixed by Walter Cronkite and all that he shared with us. I enjoyed the press' coverage, because it really captured the excitement of the whole thing. As I look back on it, it was extremely supportive coverage. To use a sports analogy, Walter was kind of a "homer" . . . a sportscaster who was pushing for the home team. People of all ages were captured by that event, and couldn't help but be caught up in the excitement. Everybody wanted to know what Walter would say [as the moon walk began], including Walter, I think. After all those years and all that studying he did, all he could do was take his glasses off, mist up and say, "Wow." When I worked with him later, in 1998 covering the Glenn mission, I asked him about that. He just laughed and said, "I didn't have anything planned, obviously. To this day, I can't believe that's all I could come up with." I said, "You know, Walter, that's probably the only word that would've sufficed at that moment."*

Much, but certainly not all, of *Apollo 11*'s coverage had the same kind of enthusiastic feel exhibited by Cronkite, Schirra, and Burke during the landing and moon walk. Few were immune. Not CBS, NBC, or ABC. Not the BBC. Not any major print outlet. Not a fresh-faced rookie in the newsroom or the most grizzled of veterans. If reporting was just the slightest bit biased, well, then, so be it. There had never been another story like this one, so even the hardest of ink-stained wretches could be excused for showing a bit of a rooting interest in the flight. According to O'Brien, the unique nature of the event was one reason at least for the press's generally positive coverage:

I think it's difficult to find an event that's parallel to [the flight of Apollo 11*]. It was the kind of news event that brings us all together. Usually, it's something negative, like the Kennedy assassination or the loss of* Challenger. *But this was an unparalleled news event, in the sense that it brought the world together for something that was really entirely positive. Of course, you could talk a little bit about the Cold War context of it, and how it was perhaps a surrogate battle between the United States and the Soviet Union. But it still was a human achievement. When historians look back on the 20th century, I suspect it will stand out as the pinnacle of what happened in that time frame. I think it's hard to come up with a parallel event that has occurred since. We've had huge news stories, but most of them are bad news, whether it's 9/11 or the loss of* Columbia. *But this event really was unique.*

Nevertheless, it was not *Apollo 11* that gave Miles O'Brien his love for aviation. Instead, it is in his blood. He is a third-generation general aviation pilot and spent much of his youth flying with his father. With CNN O'Brien was able to combine his interests in flying and journalism. After the death of his mentor, John Holliman, O'Brien was thrust into coverage of John Glenn's return to space on board the space shuttle—sharing the desk was none other than Cronkite, the legend. After that assignment O'Brien went on to cover virtually every major space-related event. He reported on the landings of the Mars rovers, and he traveled to Russia and Kazakhstan to be there for the launch of the first multinational crew to the International Space Station. He was on the story when *Columbia* went down, and he was there when NASA began to get back on its feet with the 26 July 2005 launch of *Discovery* and the STS-114 return-to-flight mission. No story, however, would have been bigger than his own venture into space.

The process of getting him there was, he says, "fascinating and ultimately very disappointing." For close to four years O'Brien and CNN went back and forth with NASA. He had talked to people in the Russian Space Agency, trying to get them to come down on their price tag of $20 million or so. It did not work, but when NASA found out that O'Brien had been in negotiations with the Russians, the door in Houston began to open a bit wider. Conversations got more serious, but questions remained. When teacher-in-space Christa McAuliffe died on board *Challenger* in January 1986, it all but ended any kind of NASA-sanctioned space tourism. That was just one of the major hurdles that had to be cleared. Would CNN have to pay for a seat on the shuttle, and if so, how much? Would CNN pay for O'Brien's training? If O'Brien were able to fly, who might be next? Would his getting the chance to fly in space open the door to a virtual lottery system?

It basically came down to this: if O'Brien was to fly, he would have to be fully trained as an astronaut. Training would have lasted a year at the very least, if not two. He knew that it would be a deal breaker. Or would it? His boss told him to go for it, and the deal was done in early 2003. He was going into space. Then came 1 February 2003.

That's when Columbia *happened. That day, I was on the air like sixteen hours. I'd be lying to you if I didn't tell you that I knew my dream of flying was up in smoke, along with the loss of our crew. Obviously, my loss paled by comparison, but it was a very sad day for me on many levels. . . . We pretty much had it locked in place right around the launch of* Columbia. *It was right in that same time frame. We were going to roll the plan out just a couple of weeks after landing and announce it. I was looking at Houston real estate. I was ready to go, move my family and live the life of an astronaut for a couple of years, which would have been amazing.*

O'Brien will most likely never get to fly on the shuttle, but a flight on board some future private space venture such as Sir Richard Branson's Virgin Galactic could be a possibility. Yet being weightless for a few brief minutes is just not the same as a full-blown shuttle mission. He was so close.

If I could scam a ride with Branson or whomever, I wouldn't turn it down, don't get me wrong. But the notion of doing a fifteen-minute suborbital flight versus a ten-day flight to the space station . . . it's a whole different thing. The Branson flights, the suborbital flights, are bungee jumping for the well heeled. I viewed [flying on the shuttle] as a journalistic experience, where I could really share what that notion of exploration was all about in a more meaningful way for people. [Flying with Branson's group] would be fine, all well and good, but the flight I had in mind was really on a whole different level in every respect.

From the first days of his life to his last, Sid Houston had a keen interest in aviation. Raised in the rural eastern Tennessee community of Watauga, he grew up near a small airport. For as long as he could remember, he was fascinated by planes taking off, landing, being serviced, and anything else in between. "I hung out at the airport. If Daddy couldn't find me anywhere else, he knew I'd be over there," Houston said. Years later, while serving as a captain in the United States Army, he hopped a ride on a helicopter every chance he got. It was not necessarily just out of convenience, either. He loved to fly.

Houston obtained his private pilot's license in the early 1990s. More than once, he delighted in taking friends and family for rides in his own personal and decidedly non-NASA-approved "Vomit Comet." When he could no longer fly due to the cancer that would eventually end his life, he became an avid builder of remote control planes, amassing a collection of at least thirty or so, including a Piper Cub with a mammoth wingspan of more than nine feet. It did not matter whether he ever flew the monstrosity—he did not, as it so unfortunately turned out. It was enough for him simply to build the thing and let somebody else take it out for a spin.

Throw in a natural curiosity concerning anything "even slightly mechanical," and *Apollo 11* was just the kind of thing to captivate Houston's attention. He was serving in the U.S. Army at the time of its trip into space, stationed at Camp Zama in Japan. Along with his wife, Betty, and the young

5. Sid Houston with one of his beloved, self-built airplane models. Courtesy Rick Houston Collection.

son they called Ricky, they lived on a nearby base in Sagamihara. Houston did not speak Japanese, so local television coverage was a wash. Instead, he kept up with news of the flight through Armed Forces Radio and by reading the *Army Times* newspaper from cover to cover every day. Interest in all things *Apollo 11* was incredible, and it was not limited to American service personnel. This was a giant leap for *all* mankind, not just those from the United States.

The day they landed on the moon, I think three-fourths of the shops in Japan were closed. Everybody was watching that. I think the whole world was watching that day. After all, that was the first time we'd ever actually touched another heavenly body. We were on the moon. We as a human race had put a man on the moon. That was definitely the feeling in Japan. They took pride in the accomplishment, just like we did in the United States or anywhere else around the world. My Japanese secretary came into work that day because she knew we would have a TV. But her sisters didn't go to work. They stayed home to watch.

For Houston a great deal of *Apollo 11*'s allure centered upon that vicarious experience. For the very first time people could picture themselves on the surface of the moon and not be considered completely insane. If this was possible, then, by gosh, anything was possible, he recalled: "We watched them take off a few days before, and we'd kept up with the news all the way. I guess the pilots I knew wished they were up there, and then a lot of us lay people did, too. If you've got an imagination, you can put yourself in any situation."

A few months after *Apollo 11*, Houston was sent to the Republic of South Vietnam for a year-long tour with the famed First Cavalry Division. Decades later he was diagnosed with lung cancer brought on by the Agent Orange he encountered in that war-ravaged country. He died 30 August 2008, a little more than three weeks after the interview for this book. Sid Houston was sixty-five.

There is no way fully to comprehend the impact *Apollo 11* had on the world. For millions of schoolchildren the names of the flight's lunar voyagers have faded into the pages of their history books. Living in a world where men have walked on the moon is all they—and most of their teachers—have ever known. The impossible seems to happen virtually every other day (can that many songs and videos and movies *really* fit onto a tiny little iPod?), so a certain sense of wonder is lost. That mankind has landed and worked on the moon is, for all intents and purposes, taken for granted by many. Too many.

That is not to say *Apollo 11* has been forgotten. Armstrong's autograph is coveted beyond measure by space enthusiasts. Any mention of an Armstrong signature sighting on eBay elicits prompt and immediate discussion on collectors' message boards. Is it real, the work of an autopen, or an outright forgery? Studies of his exemplars have been conducted to the most minute detail; the distinctive *N* in *Neil* and the *A* and *G* in *Armstrong* face incredible scrutiny in thousands of real and imagined signatures of the first man to walk on the moon. Signatures on his official NASA portraits have been deemed suspect because they happen to cover even the slightest part of the American flag patch on his spacesuit, due to studies that indicated Armstrong refused to allow any part of his signature to overlay Old Glory.

Armstrong was a regular signer for decades after his astronaut selection, even following the moon landing, despite the advice of Charles Lindbergh. The first person to fly a plane solo across the Atlantic Ocean gave the first person to walk on the moon a single piece of advice at a September 1969 banquet, and it was this: Lindbergh told Armstrong never to sign autographs. "Unfortunately I didn't take his advice for thirty years, and I probably should have," Armstrong quipped in *First Man*, an authorized biography for which he did no book signings. After being deluged with requests through the mail and in person for years, Armstrong stopped signing in 1994. He politely but firmly denies virtually all requests, regardless of the circumstance. Whether that has helped or hurt the market for his signature is a matter for debate. Fewer signatures ups their value, which in turn opens the door for a deluge of forgeries. One after another, collectors suggest that Armstrong do a signing for a charity of his choice. He could name his price, they say, and do a world of good. So far, he has not conducted such a signing.

This is the craziness that is the furor over all things Armstrong. Mark Sizemore, a barber in Lebanon, Ohio, was sent a cease-and-desist letter by Armstrong's attorneys in May 2005 after Sizemore allegedly sold a lock of the former astronaut's hair for three thousand dollars. *A lock of his hair*. In 1994 Armstrong also sued Hallmark over the use of his name and a recording of the "one small step" line in a Christmas tree ornament. The matter was settled out of court, and Armstrong later donated its proceeds to his alma mater, Purdue University.

More than forty years on, *Apollo 11* remains firmly entrenched in American pop culture. All three crew members have either written or been the subjects of books. Clips of Armstrong's immortal "one small step" line have been used in countless television shows, news programs, and movies. It was in *Forrest Gump*. It was in *Contact*. It was in *Apollo 13*. The voices of both Armstrong and Aldrin have been featured on the TV cartoon series *The Simpsons*, of all places. Plots of at least three movies have revolved largely around the flight. *Pontiac Moon* from 1994 told the story of a man and his son who drive cross-country as *Apollo 11* makes its way to the moon. In 2001 *The Dish* highlighted the contributions to the flight of an Australian tracking station. More recently, the 2008 film *Fly Me to the Moon* featured the three-dimensional, computer-animated adventures of three young flies as they stow away aboard the Command and Lunar modules *Columbia* and *Eagle*, respectively.

A Walton Easter, the last reunion movie of the American television show *The Waltons*, began with John Boy reporting on Armstrong's first steps, with mom and dad, Olivia and John, watching back home in Virginia. In the series *Cold Case* the murder of a boy whose body was discovered the night after the landing was investigated. Buzz Aldrin has made a slew of guest appearances on television programs, made-for-TV movies, and talk shows, including *The Fall Guy, The Boy in the Plastic Bubble, The Colbert Report, Larry King Live, Numb3rs, The Late Late Show with Craig Ferguson*, and *Punky Brewster*. What's more, the Buzz Lightyear character in the *Toy Story* movies was named in Aldrin's honor.

As much as the events of *Apollo 11* impacted the world, the flight impacted each of its three main participants in a variety of ways. Neil Armstrong has remained outside the spotlight for the most part and Michael Collins a little bit less so, while Buzz Aldrin seems to have embraced it after a series of well-documented problems in the years after the flight.

Whichever approach is best for protecting and even promoting the legacy of *Apollo 11* remains a source of great debate. There are those who vehemently fault Armstrong for not speaking out more on space-related matters, and there are those who just as vehemently cite his right to live however he darn well chooses. Who is right? Who is wrong? Those are questions for the ages to decide.

Ultimately, what did flying in space mean to Michael Collins? It changed him and his perception of himself, without a doubt, even though he says outwardly he seems to be the same person and his habits are about the same. Yet in *Carrying the Fire* he wrote that he does feel different than others:

I've been places and done things you simply would not believe . . . I've dangled from a cord a hundred miles up. I've seen the earth eclipsed by the moon and laughed at it. I've seen the ultimate black of infinity in a stillness undisturbed by any living thing. I've been pierced by cosmic rays on the endless journey from God's place to the limits of the universe, perhaps there to circle back on themselves and on my descendants. If Einstein's special theory is true, my travels have even made me younger by a fraction of a second. Although I have no intention of spending the rest of my life looking backward, I do have these secrets, these precious things that I will always carry with me.

It is virtually impossible to comprehend the pressures to which the three men were subjected. The flight itself, with all its assigned tasks and inherent risks, was demanding enough. The historic nature of what took place during that magical, if not mythical, eight-day journey added yet another unimaginable element of difficulty. When they made it back home, millions upon millions wanted to know what Armstrong and Aldrin felt as they traipsed through Tranquility Base. How could even the most eloquent of poets describe such an experience, let alone a couple of by-the-book fighter jocks?

Their postflight schedule did little to help resolve the matter. From the moment they splashed down in the Pacific Ocean until around 9:00 p.m. Houston time on 10 August, seventeen days later, the crew of *Apollo 11* found itself in quarantine, locked away aboard the USS *Hornet* in the Lunar Receiving Laboratory (LRL), basically a specially modified mobile home. The structure—later airlifted from the recovery ship and flown, eventually, to Houston with the crew and a handful of others tucked safely inside—was as close to peace and solitude as they would find in the months following their return. And really, there did not seem to be much of that. Not surprisingly, given the personalities we think we can understand all these years later, only Armstrong seemed really to have taken to the concept.

We really needed that time to be able to do all of the debriefings and talk to all the various systems guys. The subsequent Apollo crews were very interested in this question and that question that had to do with their own mission planning—what they thought they might reasonably do and whether we had ideas on how they might improve their own flights. Mostly, the discussion revolved around what was doable on the surface; because that affected the planning substantially. So that time was very valuable to us personally, as well as to everyone else. Of course, we would have liked to have been with our families, and we were prevented from that. But we knew they were not far away. All the uncertainty was gone now.

In tune to the relative peace and quiet Armstrong, who celebrated his thirty-ninth birthday in quarantine, was lost to the music of a ukulele his wife, Jan, had stowed inside the LRL. "I couldn't hear what he was playing, but he had his head back, his eyes closed, he was grinning and looking like he was having a whale of a time," said Lee Jones, a NASA photographer. Meanwhile, the "other" precious cargo brought back to Earth—aside from the three astronauts, of course—was treated with extreme care. Still, unbelievably, the rocks had to clear customs formalities in the United States. To get around the problem, each container of lunar rock was carried by an official duly sworn as a United States customs agent.

Early on the morning of 13 September 1969 the three astronauts and their families boarded a plane provided by the White House, bound for New York and a ticker tape parade that surpassed any that had ever been held before or since. An estimated four million people attended the parade, packed fifteen to twenty deep along the route. At times things other than ticker tape were thrown from skyscraper windows high above the chaotic scene. Confetti, shredded telephone books, rice, almost anything people could get their hands on, rained down the motorcade. Armstrong noted in *First Man* that bundles of IBM computer punch cards were also tossed in celebration. When the cards failed to separate properly during their long trajectory downward, they hammered the astronauts' car like bricks, leaving pronounced dents in the hood. The parade route was left carpeted with paper inches deep, if not feet deep in some places. Tugboats and ocean liners greeted the parade with salutes from their horns, and fireboats sprayed water hundreds of feet in the air.

The New York parade was just the beginning. By the time their world tour was over, the crew had been seen in person by an estimated 100 to 150 million people. Of those an incredible 25,000 were said to have received a handshake or autograph directly from members of the crew. They dined in the White House with President Richard M. Nixon and spoke to a joint session of Congress. They had an audience with Pope Paul VI. They met Queen Elizabeth and Prince Philip in London and countless other kings, queens, emperors, dictators, presidents, princes, and paupers around the world.

It was an exhausting itinerary for those who participated. Between the official start of the "Giant Step" global tour on 29 September 1969 until its end on 3 December 1969, the astronauts made no less than thirty or so stops in countries from one side of the globe to the other, in locales as diverse as Las Palmas, Canary Islands; Kinshasa, Zaire (present-day Congo); Belgrade,

6. A global adulation descended upon the crew of *Apollo 11* like the tons of ticker tape, shredded paper, and confetti tossed from office windows during their postflight parade through the streets of New York with Mayor John Lindsay, seated directly in front of Michael Collins. Courtesy NASA.

Yugoslavia (present-day Serbia); and Sydney, Australia. And that's not to mention various side trips that were made to made to satisfy the demands of whatever dignitary, real or imagined, who had some pull with NASA.

To say that it was a whirlwind journey is a wild understatement. The group would tire of hotel rooms and airports and were sometimes—oftentimes—unsure of exactly what city they happened to be visiting. Still, there was the knowledge, if admitted somewhat grudgingly, that it was the rarest of rare opportunities.

The schedule was numbing to both body and spirit. In Mexico City, on the first stop of the tour, the astronauts and their wives were very nearly crushed by a sea of enthusiastic humanity. It would not be the last time such things took place, and the process took a most definite toll. In Bogotá, Aldrin was prescribed medication to ease his anxiety. Of his LMP's onset of depression during the tour, Armstrong was quoted in *First Man* as saying he "wasn't smart enough to recognize the problem." Armstrong continued: "It bothered me then, and bothers me now, that I wasn't up to the job. I've thought to myself, had I been more observant or more attentive, I might have noted something that could have helped Buzz's situation, and I failed to do that. It was sometime after the tour that he started having real problems."

To his credit Aldrin has been open concerning the problems he faced following the flight. In his autobiography, *Return to Earth*, Aldrin described in detail, sometimes rather bluntly, his struggles. His father, Edwin E. Aldrin, was a driven military man, and he drove his only son as hard, if not harder. The elder Aldrin told reporters of his son ranking first in his freshman year at West Point in athletics and academics but seldom mentioned the fact that Buzz would graduate third three years later . . . third. Just third.

When the United States Postal Service unveiled a stamp commemorating the flight, its tagline read, "First Man on the Moon." Buzz Aldrin was disappointed by the perceived slight—it could have read, "First *Men*," after all—but his father was livid. A year or so earlier the elder Aldrin had gotten angry when told that Armstrong would take the first steps on the moon, not Buzz. That Buzz would land on the lunar surface at exactly the same moment as Armstrong, that he would be one of the first *two* men in the history of mankind to walk on the surface of the moon, did not seem to register with Aldrin Sr. According to *Return to Earth*, Buzz had to talk his father out of taking the issue up the chain of command. Later Aldrin Sr. tried to talk his son out of retiring from the air force. He had not, after all, made general yet. Good grief.

It was with this kind of background and these kinds of expectations that Buzz Aldrin entered West Point, served as a fighter pilot in Korea, and became an astronaut and walked on the moon. When chinks in the armor of this astronaut-hero began to show during the world tour, however, none of that seemed to matter. He was Buzz Aldrin, moon walker. He was supposed to be infallible, the perfect husband, the perfect father, the perfect example of what it meant to be a red-blooded American. He was not perfect, and he felt the pressure.

In his book Aldrin described one particularly disheartening episode with his son Andy. Due to some far-flung function, Aldrin and his wife, Joan, missed one of their son's seventh-grade football games. And this was not just another football game either. Previously winless, his team won the contest twelve to seven, with Andy scoring both of the team's touchdowns. His parents were not there to see it. "He was deeply hurt and there seemed no way to make up for his unhappiness," Aldrin mused. "Andy, for the first time, realized that his parents weren't infallible as far as he was concerned. We were becoming human beings and the first chink he spotted in our armor looked to him to be very big indeed. By the time we had been home two hours Joan was in the bathroom crying and I was in my study erasing from my datebook everything scheduled on any Saturday for the remainder of the season and writing in his football games. The team never won again, but eventually we all recovered."

Drinking became a problem, as did a deep depression. Aldrin broke down in tears following a public function. There were days he could barely force himself to get out of bed. He had an affair. He went through two divorces. After leaving NASA in July 1971, his problems came to a head. "The rule of my emotions was absolute and ruthless," he wrote. "In no way could I stop what I felt, but I hoped somehow to stop feeling anything at all. I yearned for a brightly lit oblivion—wept for it." On 28 October 1971 Aldrin was admitted to Wilford Hall, a hospital near Brooks AFB in Texas. Through intense therapy he began to come to grips with what ailed him. Basically, what it amounted to was that he had always pushed—and been pushed—toward increasingly lofty goals. "It soon emerged that my life was highly structured and that there had always existed a major goal of one sort or another," he explained. "I had excelled academically, being at the top of the schools and classes I had attended during my life. Finally, there had been the most important goal of all, and it had been realized—I had gone to the moon. What to do next? What possible goal could I add now? There simply wasn't one, and without a goal I was like an inert ping-pong ball being batted about by the whims and motivations of others. I was suffering from what poets have described as the melancholy of all things done."

Since coming to grips with his alcoholism and depression, Aldrin has poured himself into other pursuits, such as the writing of his memoir and a number of other science fiction and children's titles. In 2002 he had a well-publicized encounter with a hostile conspiracy theorist who tried badgering him into stating the moon landings were a hoax, calling the celebrated astronaut a liar and a coward in the process. For his troubles the theorist—who had rudely confronted other Apollo astronauts and their families in the past, also calling them liars to their faces—received a swift left hook to the side of the head from Aldrin. The incident had been caught on film, and it quickly became one of the most viewed and popular items on the Internet. To everyone's delight and relief, apart from the publicity-seeking theorist, the police and city prosecutor subsequently dismissed all claims of assault and refused to file any charges against the former astronaut.

Buzz Aldrin is anything but a coward. Far from it. He has become a leading advocate for space travel, an issue he and the rest of the *Apollo 11* crew addressed during a NASA press conference held in conjunction with the twentieth anniversary of the flight. Asked by CNN's John Holliman the best and worst things that had happened in the two decades following the first lunar landing, Aldrin responded: "Well, I certainly feel the best achievement . . . was the achievement of Apollo and the continuing to expand on that while we had the wherewithal. And the worst part was the gradual demise of that." So what happened to that drive, that all-encompassing, ironclad will to make it to the moon? Collins answered that question in the same press conference:

The world in the '80s is a lot more complicated than the world of the '60s, at least in this country. President Kennedy said that we were going to land a man on the moon and return him safely to the Earth—that was his goal. President [George H. W.] Bush, whom I consider to be a president as dynamic as Kennedy, I think, in today's climate, would have to say, "I think we ought to dedicate ourselves to the goal of perhaps considering appointing a commission, after due deliberation with the Congress, of investigating the feasibility of certain long-range goals for the space program, perhaps even including a mission to Mars." It's just a sign of the times. The times are a lot more complicated today, plus there were a couple of precipitating events that caused Apollo to be launched that do not exist today.

Finally, the question was asked: why continue to explore? Why should we go to the trouble, the expense, the danger? Why explore? Because man is compelled to find out what is next, a point that Armstrong drove home during the interview: "It's just inherent in the human condition," he explained. "We are a nation of explorers. We've started on the East Coast, we went to the West Coast and then vertically. Starting with the Wright brothers, [Chuck] Yeager through the sound barrier, Armstrong and Aldrin on the moon . . . it's our tradition. It's in our culture. It's a fundamental thing to want to go, to touch, to see, to smell, to learn. I think [that tradition of exploration] will continue in the future."

Such questions were more than likely asked of Ferdinand Magellan, Francis Drake, Richard E. Byrd, Meriwether Lewis, and William Clark and certainly of men like Charles Lindbergh, Edmund Hillary, and Tenzing Norgay. They circumnavigated the globe, explored the South Pole, flew solo over the Atlantic, and climbed Mount Everest, all to the amazement and adoration of their fellow man. And on 20 July 1969, with very nearly all of humanity along for the ride, the names of Neil Armstrong and Buzz Aldrin, the first human beings to step foot on the moon, along with their ticket home, their Lunar Module pilot, Michael Collins, were added to that list of immortal explorers.

2. The Eagle and the Bear

Dominic Phelan

We hope that, before long, man's foot
will step on the moon's surface.

Soviet cosmonaut Lt. Col. Yuri Gagarin (1934–68)

In Europe only citizens of the Soviet Union and isolated Albania missed one of history's most inspiring moments as most of the Eastern bloc had decided to ignore the Kremlin's wishes and broadcast live coverage of Neil Armstrong and Edwin "Buzz" Aldrin landing on the moon. A privileged few in Moscow did get caught up in this "Apollo fever" as the military and space elite assembled at a secret monitoring center on Komsomolsky Avenue to follow the mission.

Unlike others in attendance, however, cosmonaut Alexei Leonov's reactions were tempered by the knowledge that if things had worked out a little differently, it could have been him and not Neil Armstrong who became the first person to leave a footprint in the lunar dust. Four years earlier he had become famous throughout the world as the first person in history to walk in space, but since then he had been working toward a new and even more impressive goal. As he witnessed those flickering images emanating from the surface of the moon, he would undoubtedly have experienced some frustration. And yet, at the same time, he felt privileged to be seeing history unfold before his eyes.

"It was with mixed emotions that I stood watching events unfold on our television monitors that July morning," Leonov admitted years later. "If it couldn't be me, let it be this crew, I thought, with what we in Russia call 'white envy'—envy mixed with admiration."

Although Leonov realized he and his nation had lost the race to be first, he still believed—like many in the room—that Soviet cosmonauts would follow the Americans to the moon. As their conversation turned to professional talk about the mission, one of the military officers present sympathetically patted Leonov on the back with the words, "That's how it's done . . . that's the task that lies ahead of you."

Although the moon race might have been important in the early 1960s, by the end of the decade the White House and the Kremlin had other things on their minds. The Nixon administration was preoccupied with ending the Vietnam War, while Moscow was busy trying to reestablish Soviet authority over an increasingly rebellious Eastern Europe as well as trying to prevent a war with China over a disputed border.

After the *Apollo 11* moon walk the Kremlin grudgingly acknowledged the United States's feat by devoting seven minutes to the mission on the main evening news, while the *Pravda* newspaper relented on its previous indifference in providing a six-column account of the successful landing. Unfortunately for Alexei Leonov, this recognition, however restrained, also signaled the Kremlin's own loss of interest in the moon race, having been so convincingly beaten in the eyes of the world. The party line now was that there had never been any thoughts of trying to send Soviet cosmonauts to the moon, not when unmanned probes could do a better job with far less risk.

In retrospect any chance of a Soviet cosmonaut reaching the moon first probably died along with the legendary chief designer Sergei Korolev on an operating table in January 1966. His OKB-1 (Special Design Bureau Number 1) in Kaliningrad, northwest of Moscow, had played a pivotal role in the space program's success up to that point. Brilliant at his work, cautious, and an excellent manager, Korolev had designed and developed the world's first intercontinental ballistic missile (ICBM), the R-7, or *Zemyorka* (Little Seven). This workhorse rocket had been responsible for launching all major Soviet space payloads to that time, including the Sputnik, Luna, and Venera probes and the manned Vostok and Voskhod spacecraft. Korolev's success in his early achievements, particularly following the propaganda that ensued from the successful launch of the first Sputnik satellite in October 1957, quickly brought him the full support of Soviet leader Nikita Khrushchev. Khrushchev's desire for ever more daring space feats actually fueled Korolev's own ambitions.

According to Belgian space researcher Bart Hendrickx, it was Korolev, not Khrushchev, who masterminded most of the spectacular space firsts, "often skipping steps that his colleagues deemed necessary. In most cases, Korolev's sound technical intuition made these leaps justifiable, but in others, especially the Voskhod program, the safety of space crews appears to have taken a back seat to the relentless drive to beat the Americans, although . . . the crews' safety was always Korolev's foremost concern." By taking these calculated risks, he was able to place the first three-man crew in orbit on board *Voskhod 1* and conduct the first spacewalk with *Voskhod 2*, months ahead of similar American Gemini missions.

One of Korolev's most enduring frustrations lay in the fact that, in complete contrast to NASA, the Soviet space program had no truly centralized organization or long-term plans. Khrushchev would repeatedly ignore any of Korolev's proposals for a reorganization, and as a consequence the Soviet space program remained reliant upon mostly nonspecialized design bureaus, many of which were administered by different ministries. Korolev, with nowhere left to turn, was forced to delegate most of the design work on unmanned spacecraft to his associates, but the strain of overseeing such a vast number of projects, and concurrently, imposed a tremendous burden on him and his health, which had already suffered badly during his wrongful internment in one of Stalin's notorious gulags.

Unfortunately, the man selected to replace Korolev as head of OKB-1's staff of sixty thousand workers, his loyal deputy Vasily Mishin, simply was not up to the job. An accomplished scientist-engineer who had studied applied mathematics at the Moscow Aviation Institute, Mishin began his career in rocket science at the NII-1 research institute. At the end of World War II he was dispatched to Germany to study the V-2 rocket program and there worked alongside and befriended Sergei Korolev. In 1946, following his return to the Soviet Union, Mishin willingly took on the position of Korolev's deputy designer. He subsequently became deeply involved in the Sputnik program but sadly displayed little of the chief designer's uncompromising drive, initiative, and foresight. In fact he was once described in a *New York Times* article as "a solid engineer utterly lacking in his boss's charisma or political gifts." Following Korolev's sudden death in a Moscow hospital, it seems that Mishin had only gained the title of chief designer by default, perceived as a compromise candidate who was generally acceptable to all within an increasingly fractured space program.

Although many imagined the Soviet space program to be one vast monolithic enterprise bringing communist-style central planning to the cosmos, we now know that in reality there were a number of design bureaus literally fighting among themselves for the patronage of the military and the Kremlin. Any miscalculation was quickly used by a rival, resulting in a bitter game of personal attacks that seriously hindered the manned lunar effort.

"Without Korolev, Mishin was lost," Leonov declared, offering a stinging indictment of the man tasked with getting a Soviet cosmonaut to the moon before the Americans. "He was a very good engineer, but he had his weak points; one of which was that he drank. He was also hesitant, poor at making decisions and reluctant to take risks. This was to cost us dear."

Shortly after Mishin took over OKB-1 in 1966, its name was changed to the rather austere Central Design Bureau of Experimental Machine Building (TSKBEM the Russian acronym). TSKBEM's main serious rival was the smaller OKB-52 bureau based in the Moscow suburb of Fili, established by the sophisticated, well-dressed Vladimir Chelomey, who had cultivated excellent relations with the military by supplying it with cruise missiles. Although he only had a staff of eight thousand, the fact that Chelomey also hired Khrushchev's son Sergei to work on missile designs opened many doors for him in the early 1960s. Like OKB-1, the name of Chelomey's OKB-52 bureau underwent a name change, to the nearly identical Central Design Bureau of Machine Building, or TSKBM.

Mishin also had a tense relationship with the head of cosmonaut training, Gen. Nikolai Kamanin. Cosmonaut training had been an all-military matter since 1959, but by the mid-1960s Mishin wanted his own civilian engineers flying on his design bureau's spacecraft. Unfortunately, Kamanin was a strict Stalinist puritan who ran the cosmonaut center like his own private fiefdom, and not surprisingly the two soon argued over whether these new cosmonauts should be trained directly at TSKBEM or in the existing cosmonaut training facility that later came to be known as Zvezdny Gorodok (Star City).

To Kamanin the fact that Mishin was also a heavy drinker compounded the problem, as he often had to censure or even ground cosmonaut candidates when they got into alcohol-fueled trouble. "Mishin wants us to regard him as a resolute leader as well and tries to copy Korolev but not having the authority, the experience and the knowledge of Korolev, he often ends up in unpleasant situations," Kamanin confided in his private diary. His once-secret journals offer a unique insider account of the period. Although published in the late 1990s following the collapse of the USSR, these journals only gained a wider audience in the West after a painstaking translation by Russian speaker Bart Hendrickx became available. All these confrontations only added to Mishin's stress levels as he tried to get Korolev's moon project off the ground.

Sergei Korolev had originally considered sending a Vostok capsule around the moon, but he soon realized that such a complex mission was beyond the capabilities of both its simple design and a solo pilot. He then started work on what we now know as the Soyuz spacecraft, intended to carry three cosmonauts on a circumlunar mission by the middle of the 1960s. It was only after the full, ambitious details of the NASA lunar landing plan became public that he decided to scrap this small-scale thinking and match the grand scale of Apollo.

The original concept for a "universal rocket" called the N1 had been on the drawing board since 1956, envisaged as a massive booster capable of launching military satellites, interplanetary probes, and even a large space station. To do so, Korolev would have to rely on Valentin Glushko, the only man who could have provided the powerful rocket engines he needed for the proposed super-booster.

Glushko and Korolev had both come of age during the early days of amateur rocketry before they were swept up in the brutal Stalinist purges that nearly destroyed the country in the late 1930s. Both men managed to survive, although they remained deeply suspicious of each other's involvement in their own, unwarranted arrest by the brutal NKVD secret police. Following World War II they were both sent to Germany to learn the secrets of Wernher von Braun's V-2 rockets, returning to establish their own missile design bureaus on Stalin's orders. Their relationship was mired in a longstanding mistrust of each other. A further feud developed when Korolev was selected as the chief designer of the main OKB-1 design bureau outside of Moscow, with Glushko's own OKB-456 factory in a subservient role, supplying rocket engines.

Glushko's ill feeling centered on a disagreement with Korolev about the significance of the engines. Glushko insisted that they were the most important part of any rocket, once boasting that a wooden stick would fly with one bolted on. Korolev, meanwhile, believed the engines were only a part of his overall creation. In years to come the failure by the two men to reach a compromise would doom the Soviet moon project.

By the early 1960s Glushko's renamed Energomash design bureau was championing the hypergolic nitrogen tetroxide / UDMH propellant mix, originally developed for use on silo-stored ballistic missiles that needed to be launched at a moment's notice. Korolev had an aversion to using these highly toxic propellants for flights into space, fearing that any accident would result in a catastrophic explosion similar to an R-16 missile at Baikonur in October 1960, which had killed over 120 army personnel and rocket engineers when it exploded on the pad. The fuel had subsequently become known as the "Devil's Venom," and Korolev was determined to prevent its use on any manned spaceflight. Instead, he insisted on the more familiar liquid oxygen and kerosene engines.

Although that devastating 1960 explosion had strengthened Korolev against rival Mikhail Yangel's OKB-586 design bureau, now it was Glushko who had the upper hand, and he simply walked away from the N1 project altogether. This move effectively called Korolev's bluff and made life exceedingly difficult for the OKB chief designer.

When Soviet rocket engineers first learned that NASA wanted to use five massive first-stage engines and liquid hydrogen / liquid oxygen upper stages on their Saturn V moon rocket, they had serious doubts that the Americans could perfect the technology by the end of the decade. But Korolev soon realized he had been wrong to underestimate the Americans. That recognition coupled with Glushko's refusal to develop the needed N1 engines forced Korolev to turn to aircraft jet engine designer Nikolai Kuznetsov to develop a new liquid oxygen / kerosene engine for the moon rocket. Although Kuznetsov lacked experience with rocket engines, Korolev gambled that by clustering enough of them together, he could create a giant booster equal to the Saturn V. The overall thrust would have to be a fifth more than that of its American counterpart, making it the most powerful booster in the world, and its first-stage engines alone would consume some forty tons of propellant just to get the 345-foot booster off the pad.

Unlike the three-stage Saturn V, the N1 used liquid oxygen and kerosene in all its five stages. After its first three conical stages had placed the manned lunar complex in low Earth orbit, the fourth stage would begin the Trans-Lunar Injection burn to send the cosmonauts on their way to the moon. A fifth stage would then perform midcourse corrections, the Lunar Orbit Insertion and the final burn dropping the lunar lander out of its orbit at the beginning of the moon-landing descent.

Although it could only place about 95 tons into a fifty-one-degree inclination Earth orbit when launched from Baikonur, it is still startling to realize that if the N1 had been launched from the better twenty-eight-degree latitude geographic location of Florida, it could have placed nearly 105 tons into orbit. It was a most respectable figure when compared with its far more sophisticated American rival.

Unfortunately, this brute force approach would need a total of thirty NK-15 engines on the first stage, eight NK-15s on the second stage, and four NK-19s on the third stage just to get it into Earth orbit. In fact, so many engines were being used that Kuznetsov's engineers decided to downgrade the performance of each engine in order to lower the chances of failures during a launch. To save time only two out of every six engines leaving the factory could be tested, so by extension just ten were guaranteed to work at ignition. Somehow they hoped the mighty booster could be "tamed" in flight and were resting their hopes on the primitive KORD computer system to shut down any faulty engines during the launch.

This amazingly cavalier approach was compounded by a decision to ignore testing of a complete first stage on the ground in order to concentrate scarce resources on the actual flight hardware. In contrast to the Americans, who had conducted their first full-on testing of the Saturn V's first stage engines in mid-1965, the Soviets would not get to do the same until they ignited all the engines of an actual N1 during its maiden flight. It was an appallingly risky undertaking, and even some of its own designers expected around ten N1 rocket failures before a successful flight was achieved. As senior N1 engineer Vladimir Vakhnichenko later admitted: "The underestimation of the scale factor—the immense size of the launch vehicle, each launch of which was an event in the life of the country—played a fatal role. It was no big deal that for some rockets it was necessary to carry out forty to fifty launchings before they 'learned' to fly, but that approach was unsuitable for the NI."

Not surprisingly, some of Korolev's own OKB-1 staff had been openly skeptical of the whole project even before it left the drawing board. One of the most vocal critics was a highly experienced engineer named Konstantin Feoktistov. "The work started in a more or less serious way in the spring of 1964," he recalled in an interview with Bert Vis in Washington DC in 1992. "This was the beginning of some very severe contradictions between me and Korolev. I felt the lift capacity of the NI rocket was not sufficient for a lunar expedition and I thought that it wouldn't work." Korolev appears to have gotten around Feoktistov's objections by the simple expediency of offering the ambitious young man a much-coveted seat on the newly built, three-man spacecraft called Voskhod. "This enabled him to get me away from this program for a couple of months so that he could use my engineers by getting me out of the picture," Feoktistov reflected. "After coming back from the flight I went back to the design bureau and had to keep working on the lunar program but . . . I kept saying that it wasn't going to work and that the project would die sooner or later, which is what happened."

Soviet leader Khrushchev himself clearly was also losing his early fascination with Korolev by 1964 and decided to allow Chelomey into this bitter moon mix after the designer had proposed a two-man circumlunar spaceship called LK-1 (which resembled a cross between a Gemini and Apollo capsule) for launch on the new UR-500 booster, which later came to be known as the Proton rocket. That August Chelomey took over the circumlunar project altogether so that Korolev could concentrate on the troubled NI-L3 moon landing plan. Unfortunately, Korolev was intensely jealous of Chelomey, so when Khrushchev was unexpectedly deposed by the Leonid Brezhnev faction in the Kremlin in October 1964, the chief designer put pressure on the new leaders to allow him to resume control of both projects. Chelomey was ordered to scrap his LK-1 plan and instead launch a deconstructed Soyuz on a "lunar loop" using his UR-500 booster. He was also given a timetable: the flight would be ready in time for the fiftieth anniversary of the October Revolution in 1967. Even General Kamanin was surprised by all this political maneuvering, complaining that "Korolev scored an easy victory over Chelomey, but this easy victory cost us almost two years."

The new LI craft was basically a Soyuz minus its spacious front orbital module. It would carry a crew of two cosmonauts without spacesuits on a week-long, free-return loop around the moon, followed by a safe touchdown back in Russia. As it had a new, thicker heat shield to protect it from the higher lunar velocity reentry, the craft's designers insisted on several unmanned test flights before any cosmonauts were permitted to occupy the spacecraft on an actual mission. Eventually, fourteen complete LIs were built, of which three could have been used for crewed missions.

Considering his earlier bitter opposition to hypergolic propellants, Korolev's agreement to use a Proton vehicle to launch the lunar Soyuz was surprising, as the UR-500 was virtually an old ICBM design using the very same toxic storable propellants over which he had fallen out with Glushko.

When it came to the NI-L3 moon-landing plan, the Soviets decided to copy the American Lunar Orbital Rendezvous (LOR) approach. Ironically, this plan—grudgingly accepted by NASA after the agency realized it was more economical than its favored options of Earth Orbital Rendezvous (EOR) or Direct Ascent—was inspired by the writings of Ukrainian space theorist Alexander Shargey, who had been forced to live under the pseudonym of Yuri Kondratyuk to avoid the attention of the NKVD during the 1930s, before being killed fighting the invading Germans in 1942.

After the disappointment of losing the circumlunar project, Chelomey had asked the Kremlin to adopt his own moon-landing plan to replace the troubled NI-L3 at the time of Korolev's death. This LK-700 plan envisioned a forty-five-ton Direct Ascent lunar lander launched on a UR-700 booster—a new giant booster made up of clusters of Proton propellant tanks fitted with nitrogen tetroxide / UDMH rocket engines, which Glushko had earlier offered Korolev for use on the NI. One of its strongest supporters, General Kamanin, even claimed the UR-700 would have cost one-tenth as much to develop as it was based on Proton components and could be launched from the existing NI launchpad at the Baikonur Cosmodrome.

The seriousness of his proposal led to the formation of the Keldysh Commission in late 1966, headed by the respected Academy of Sciences president, Mstislav Keldysh, which examined the merits of both moon-landing concepts. Predictably, given that many of the commission's thirty-six members were still sympathetic to the memory of the recently deceased chief designer Korolev, they decided to continue with NI-L3.

Because work had visibly begun on the launchpads for the N1 moon rocket, American spy satellites had been keeping watch for any concrete evidence that the Soviets were still competitors in the moon race. Their concerns were somewhat premature; even the launchpad was behind schedule. Vladimir Barmin, the man who had designed the pads for all of Korolev's previous rockets, initially refused to work on the N1 before being reluctantly persuaded.

The new assembly building for the booster measured some two hundred feet tall by over six hundred feet wide and close to eight hundred feet long and would be the largest concrete hangar in Europe or Asia. When the CIA first saw satellite imagery of the construction, the agency knew it was for something special, as it closely resembled NASA's Vehicle Assembly Building (VAB) at the Kennedy Space Center, albeit laid on its side. Barmin's team had decided to scale up the horizontal assembly methods used on smaller rockets so they would not waste concrete constructing a tall building like the one in Florida. Unlike the Saturn V, which would exit the VAB mounted vertically above a gigantic tractor crawler, the N1 was designed to be loaded horizontally onto flatbeds and hauled by locomotives to the launchpad. The decision had also been made to build the two N1 launchpads less than two thousand feet apart to save money, an economy measure they would later regret.

Construction work on the N1 itself started in mid-1965, when two hundred factories in Samara, a vast city in the southeastern part of European Russia, began making components. These parts were then taken by rail to the Baikonur Cosmodrome in Kazakhstan for assembly. While the Saturn V used cylindrical propellant tanks, the N1's conical design was the result of a more utilitarian approach using progressively smaller internal spherical fuel tanks in each stage to maximize the volume while minimizing construction materials.

Although a Saturn V mock-up was taken to a Florida launchpad as early as May 1966, it would be another year and a half before an American spy satellite managed to photograph its counterpart at Baikonur. By then the CIA was perhaps overconfidently predicting that a Soviet moon landing could not take place before 1972 but that NASA could still be upstaged by a cosmonaut looping around the moon before Kennedy's deadline could be achieved.

In the years before the CIA offered its pessimistic outlook for its opponents' lunar program, a growing optimism had prevailed among many involved in the Soviet space effort. According to Alexei Leonov, who had been selected to command the L1 group, "Vasily Mishin's cautious plan called for three circumlunar missions to be carried out with three different two-man crews, one of which would then be chosen to make the first lunar landing," he revealed. "The initial plan was for me to command the first circumlunar mission, together with Oleg Makarov, in June or July 1967. We then expected to be able to accomplish the first moon landing—ahead of the Americans—in September 1968." The race was getting tight.

The number of cosmonauts assigned to the two lunar programs was often fluid, with cosmonauts taken from the group to train for Earth-orbiting Soyuz missions as they were needed. From early 1967 regulars training alongside Leonov and Makarov in the L1 lunar group included veteran cosmonauts Valery Bykovsky, Georgi Grechko, Pavel Popovich, Nikolai Rukavishnikov, Vitaly Sevastyanov, and Boris Volynov. As the L1 ground simulator would not be ready for use by the cosmonauts at Star City until early 1968, they often found themselves spending hours in the Moscow planetarium familiarizing themselves with the constellations. Some of the lesser-known faces among them were even sent on clandestine trips to Somalia to see the stars of the southern sky for real.

Oleg Makarov was a civilian engineer, a person Vasily Mishin had selected from his own design bureau for space training. At the time this move away from selecting only air force candidates had caused a great deal of resentment among the military cosmonauts, but Makarov was now slated to spend seven days sitting beside Leonov inside a cramped L1. Buoyed by the apparent success of their Voskhod flights, the Soviets had considered their manned capsules safe and airtight, safe enough in fact that they could launch crews without their bulky protective spacesuits. As a result, cosmonauts on L1 missions were expected to undertake their assignment dressed simply in lightweight woolen tracksuits. Only the N1-L3 crew would take along spacesuits, designed for the transfer spacewalks in lunar orbit and the all-important moon walk.

Although the loss of the three *Apollo 1* astronauts in a spacecraft fire during a countdown test in January 1967 was a terrible tragedy, the Soviets must have secretly seen it as offering some much-needed time to catch up with the Americans. A month before their own *Soyuz 1* disaster leveled the playing field once again, they were ready with the first test flight of their moonship.

The L1 version of the UR-500 (now officially called Proton) made for an impressive sight as it stood some two hundred feet tall on its launchpad. The mighty booster consisted of a new third stage and a Block-D fourth stage designed to rocket the L1 to lunar velocity. Although this first four-stage configuration worked perfectly when *Cosmos 146* (as it was misleadingly designated) lifted off on 10 March 1967 in a first test of the L1 hardware booster, many within TSKBEM still wanted manned missions to be dual launched. This concept would see the L1 placed in orbit unmanned, with the crew launched separately on a safer Soyuz rocket before transferring to the L1/Block-D vehicle during a spacewalk. This plan would have been very uncomfortable for the two space-suited cosmonauts, as the L1 only had one cramped eight-cubic-foot cabin. By contrast the Soyuz, with its spacious frontal orbital module, had twenty cubic feet of living space.

General Kamanin missed this first L1 launch, so he made sure he took the L1 cosmonaut group to see the next liftoff from Baikonur on 8 April 1967. For most of them it was the first time they had seen a Proton rocket up close. Their amazement was only marginally tainted when the L1/Block-D mock-up payload became stranded in Earth orbit after a malfunction and given the nondescript cover designation of *Cosmos 154*. More serious was the tragic death two weeks later of veteran cosmonaut Vladimir Komarov on *Soyuz 1*. The tragedy had major repercussions for the L1, as they were essentially the same spacecraft. Much soul-searching must have been done because the backup parachute on the L1 had been removed in order to save precious weight.

Any hopes for a manned flyby of the moon to celebrate the October Revolution in 1967 had now vanished, but while the Americans could only guess at their space rival's intentions, the Soviets had the advantage of full access to American documentation. One surprising example of their intimacy with the American technology came when cosmonaut Pavel Belyayev paid a courtesy call on the American pavilion at the Paris Air Show in May 1967. He jumped into the Apollo mock-up on display and amazed his hosts by seeming to know his way around the interior and displaying an unexpected familiarity with its systems. At the same show

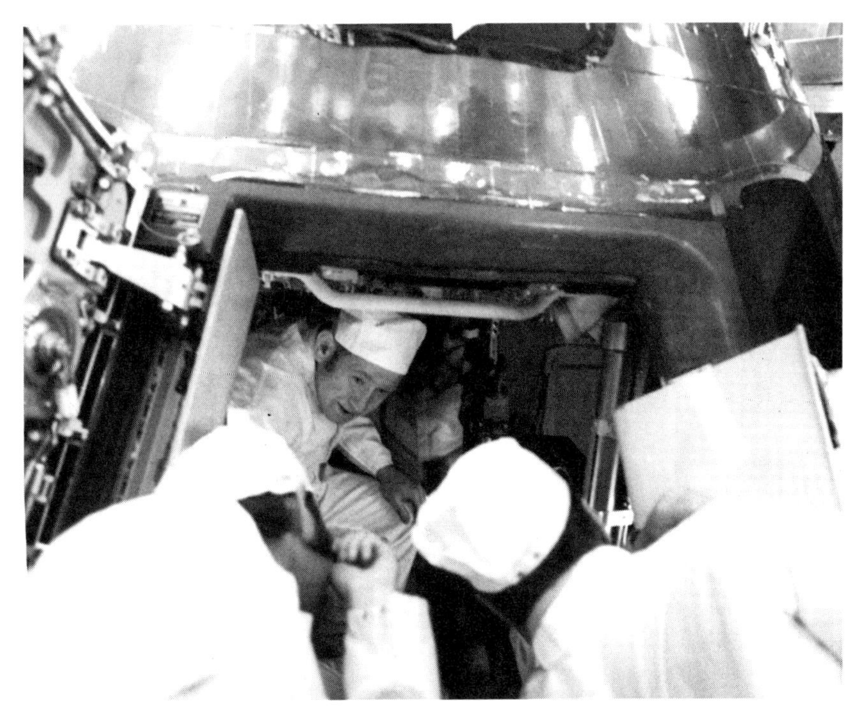

7. Soviet cosmonaut Alexei Leonov is shown inside an Apollo spacecraft in the early 1970s. Courtesy NASA.

Belyayev also let slip to *Gemini 10* astronaut Michael Collins that he expected his next flight to be to the moon and that he was practicing unpowered helicopter landings—something only useful for aspiring lunar lander pilots who wanted to see what a steep moon-landing descent was like. We now know several cosmonauts completed these dangerous flights using a helicopter fitted with the same type of digital computer that would have been installed on the Soviet lunar lander.

As if to confirm the stalling of both moon efforts, during the next Proton launch on 28 September 1967 one of the first-stage engines failed to ignite, and the rocket flew off-course for about one minute before the emergency escape system dragged the LI to safety. Although the capsule landed some forty miles downrange, the recovery team, which included Alexei Leonov, was hindered by fumes from the crashed rocket tanks nearby. The "Devil's Venom" was living up to its toxic reputation. The recovery team had to drive around the flat terrain of the steppe searching for some high ground in a frantic attempt to escape a creeping cloud of toxic gas. The cosmonauts must have wondered if they would face the same problem if they ever had to survive a malfunctioning launch. On 22 November 1967 the next unmanned LI launch saw the second-stage malfunction, but this time the capsule was heavily damaged when it plowed into the ground at high speed 177 miles downrange.

In a bid to inject some urgency into the project during 1968, Mishin planned to launch one unmanned LI a month starting in March with the aim of sending a cosmonaut to the moon by the end of the year. The first launch on 2 March saw the ship, designated *Zond 4*, successfully boosted to a lunar distance by its Block-D stage. It had been deliberately launched in the opposite direction to a lunar trajectory in order to avoid any complications caused by actually going anywhere near the moon, and its true purpose as a test of a spacecraft intended for transporting cosmonauts was disguised by recording it as one of a series of scientific biosatellites. Some of the LI crews who had been at Baikonur for the launch of *Zond 4* then flew to the Yevpatoriya tracking center for communications tests with the craft.

As the capsule looped out to lunar distance, the cosmonauts decided to toast the health of the craft with champagne. They also took time to toast Yuri Gagarin's thirty-fourth birthday, occurring just a few days later. Vitaly Sevastyanov remembered that it was only "a tiny glass of champagne, because we had to do some work," but even these modest celebrations were premature. What initially seemed like success over lunch turned to failure at the last possible moment, when Dmitri Ustinov of the Soviet Defense Ministry ordered the capsule's destruction during reentry after a sensor indicated it was off-course and could end up falling into the ocean. A sophisticated "double skip" reentry into Earth's atmosphere on the way back from the moon was essential to guarantee a landing in the USSR. Unfortunately, the Soviet military had balked at the costs involved in any such ocean recovery operation, realizing it would involve over sixteen thousand navy personnel, and there were no ships available to recover the Zond craft quickly if it had been allowed to splash down well off-target. Traditional Soviet paranoia now dictated that it be blown up in case it fell into the eager hands of a foreign power. This was especially frustrating for the cosmonauts and its designers as they knew any crew aboard would have survived the 20-g ballistic reentry.

If the destructive loss of the Zond spacecraft was not depressing enough, the cosmonaut corps was deeply shocked only weeks later when Yuri Gagarin was killed in a fiery jet crash. Although the lunar pairings had backup pilots, able to step in at the last minute to replace the commander, no one had seriously considered this option until Gagarin died. As a result, the Kremlin grew increasingly wary about risking any more heroic household names, especially the world's renowned first space walker. Leonov's reserve, Anatoli Kuklin, suddenly found himself being prepared for a manned Zond mission during 1968.

"On 27 March Yuri Gagarin died. Our chief started to think things over," Kuklin revealed in a 2001 interview with researcher Rex Hall. "The first cosmonaut had died, and now, there was the first cosmonaut to walk in space, Alexei Leonov . . . [our chief] said 'Let's assign Kuklin,' so I trained in March and April. One flight-engineer and two commanders: poor Makarov underwent two training cycles." Luckily for Leonov, continued problems with both the Proton and the L1 craft meant that any decision to continue was held over in the corridors of power. That June a list of cosmonauts selected for the N1-L3 landing missions was finalized. Along with cosmonauts already training for L1 flights, pilots selected included Viktor Gorbatko, Yevgeny Khrunov, Viktor Patsayev, Anatoli Voronov, and Alexei Yeliseyev.

The lessons of the *Zond 4* fiasco were well learned: when the next unmanned L1 lifted off on 23 April 1968, ten Soviet navy ships were dispatched to the Pacific Ocean. A troublesome second stage on this unmanned test flight, however, meant that the payload failed to reach Earth orbit. On a more positive note, but with classic irony, the escape system worked perfectly once again, and the capsule touched down safely on the steppe, not far from Baikonur. Someone, at least, had done his job well.

The next Proton launch attempt, set for late July 1968, almost caused a catastrophe on the scale of the R-16 explosion back in 1960. With more than 150 technicians working on the launchpad, the rocket's Block-D liquid oxygen tank burst after becoming overpressurized and the top part of the rocket containing the spacecraft began to topple over. Momentarily, it looked as if the whole booster might explode around the pad workers, but amazingly the L1 escape tower became entangled in the pad gantry, preventing it from falling onto the fully fueled booster below. Although it would take several weeks to dismantle and remove the twisted rocket, it could have had a far worse outcome.

By the time *Zond 5* lifted off successfully from Baikonur on 15 September 1968, Mishin's schedule was in dire trouble. Up to that time the unmanned tests had mostly proven to be failures, so much so that the Soviet navy decided once again to cut back on its ocean recovery fleet in order to save money, declaring that the odds on the unmanned LI actually making it back to Earth looked very slim. But the craft—which this time included a "crew" of two tortoises—successfully achieved Earth orbit before being rocketed toward the moon by the Block-D stage. It was the first time one of the unmanned LIs had progressed to this stage, and cosmonauts such as Bykovsky and Popovich were again on hand at the tracking station to test voice communications with the spacecraft. These transmissions, with Popovich playfully pretending to be in communication with an onboard cosmonaut, caused great confusion in the West when they were detected. Intercepting these communications from deep space did at least confirm one worrying, growing suspicion—that the Soviets were preparing for their first manned circumlunar flight.

The Soviet navy's decision to scale back its recovery fleet would backfire badly when *Zond 5*'s reentry guidance system malfunctioned on the return journey and it had to make a ballistic descent into the atmosphere. When it splashed down in the Indian Ocean at night, the nearest Soviet ship was some sixty-five miles away. As a result, it was several hours before the bobbing capsule could be retrieved from the ocean. Embarrassingly, this recovery took place in front of a group of U.S. Navy ships that had also been alerted and had arrived at the scene. Apart from giving the Americans a good look, this mission was the first triumph for the LI program, as a spacecraft had not only traveled to the moon and back, but it had also captured good-quality, close-up images of the moon. Furthermore, it had proved that any technical problems could be overcome to bring the ship home safely. Adding to the elation of the mission scientists was the fact that when the spacecraft was opened up, the two tortoises were recovered alive, having survived their lunar loop.

Soon after, when Nikolai Kamanin heard rumors that NASA was examining an audacious plan to send its second manned Apollo spacecraft to the moon, he was shocked at its sheer adventurism. "One can understand the

Americans," he recorded in his diaries. "Their Apollo cannot fly around the moon unmanned. In order to test their ships the Americans are forced to risk the lives of the crews, but we, having ships like *Zond 6*, can test them without such risk." Even Kamanin must have realized, however, it was the Soviets' own insistence on conducting several unmanned test flights before putting men aboard the L1 that had cost them dearly in the race to the moon. If only they had shown less caution and allowed cosmonauts aboard earlier, even at some risk, they might very well have been the first to send men around the moon in 1968. They certainly had that capability, as the *Zond 5* mission had proven. The only forlorn hope they could now cling to was that *Apollo 8* might somehow fail in its attempt to orbit the moon, leaving the way clear for a Soviet manned mission in early 1969.

Then there was another stumble in Soviet plans. Immediately following the launch of *Zond 6* on 10 November 1968, the craft's high-gain antenna failed to deploy as the Block-D blasted it toward the moon. Although this did not prevent Leonov, stationed at Yevpatoriya, from conducting communications tests with the lunar craft, it made life for the engineers on the ground particularly difficult.

On 12 November NASA finally confirmed that *Apollo 8* would head for the moon in December. Despite coming nowhere near to matching this bold venture, the Soviets stoically pressed on with their own lunar program. *Zond 6* passed around the moon just two days later, recording impressive images of the lunar landscape below.

Kamanin openly discussed the situation with the anxious cosmonauts, telling them that although they were prepared for their mission, a flight-ready version of the L1 would not be available until January 1969. In an attempt to boost morale he asked them to provide a new name for the manned L1. Normally, the superstitious cosmonauts would not have been involved in this process, believing it could be some sort of bad omen. As Oleg Makarov explained, "We had a tradition that the names are chosen only a few days before the launching. Before that, privately, everyone used the factory codes L1 and L3." Now for the first time more romantic names were being openly discussed. Suggestions included *Rodina* (Motherland), *Ural*, and *Akademik Korolev*.

Even as names for the first ship were being debated, things began to go wrong aboard *Zond 6*, when the hydrogen peroxide supply for the control thrusters froze. This situation, if not resolved, would make the spacecraft uncontrollable during reentry. In an effort to thaw out the hydrogen peroxide, controllers transmitted instructions that turned the craft's tanks toward the sun. This innovative measure, however, also exposed a faulty rubber seal around the main hatch to the heat, causing the air to begin to leak out of the cabin. Although the cured thrusters initiated the sophisticated double-skip reentry needed to land inside Soviet territory, an electrical short-circuit during the landing caused the parachutes to be cut free automatically while it was still several miles above the ground. The free falling spacecraft slammed into the ground at high speed. Fortunately, the retrorockets had also fired early, so unlike *Soyuz 1*, the *Zond 6* capsule did not explode on impact, which meant that some of the scientific experiments were safely recovered. Nevertheless, it was abundantly clear and deeply troublesome that had there been cosmonauts aboard, they would have perished. Later on, photographic film recovered from the wreckage was developed and shown around the world to give the impression that the mission had been a complete success.

By the end of 1968 NASA was ready to send *Apollo 8* to the moon. It was a devastating time for Alexei Leonov, as he later recalled in *Two Sides of the Moon*. "I was in Moscow, busy working on the L1 circumlunar program, when news came through that the Americans had sent a manned spacecraft into orbit around the Moon," he recalled. "I suddenly had the feeling that everything was slipping through my fingers. I could see my dreams going up in smoke."

Others who had flown to Baikonur to take part in an upcoming Soyuz mission could not help but stare longingly at the moon as three American astronauts slipped into orbit around our nearest celestial neighbor. "We saw the bright crescent of the moon and for a minute we were all silent," Kamanin wrote in his diary on 24 December. "We were filled with contradictory feelings: it hurt that not our guys were first around the moon but [nevertheless] we all admired the courage of the American astronauts and silently wished them success."

An official meeting was held on 30 December to suggest an immediate Soviet response to *Apollo 8*, but optimism was at a low ebb. Many still dreamed of a Soviet moon-landing mission in the 1970–71 period, believing that the Americans might not enjoy a successful manned landing on their first attempt. But even at a time when a circumlunar flyby was seen to be irrelevant, scarce resources continued to be wasted on unmanned Zond flights, as spacecraft designer Igor Afanasiev discussed in a 1991 article. "Immediately stopping a flywheel once it has been started is virtually impossible," he reasoned. "A program that had produced very satisfactory results couldn't be cancelled. Besides, the spacecraft were built, the launch vehicles were waiting. The schedule of flights had to be observed."

With plans for a manned L1 flight now abandoned, Kamanin and the L1 cosmonauts had to be content with watching the next Zond spacecraft taking off from Baikonur on 20 January 1969. This flight used the refurbished descent capsule salvaged from the previous (23 April 1968) launch failure. It was with some irony that this capsule had to be pulled free, once again, by the escape system from an errant Proton booster eight and a half minutes into the flight. This time it landed in a valley in Mongolia and was recovered three hours later.

General Kamanin confided to his secret diary his growing frustration that the cosmonauts were being denied their chance to fly around the moon. "I dreamed of seeing the day when I would fly to Moscow with our guys after their return from the moon," he wrote. "These were completely realistic dreams but major mistakes by the leaders of our space program and excessive automation of spaceships led to the Americans jumping ahead and flying around the Moon on *Apollo 8*."

To have any rapidly dissipating hope of matching Apollo, the N1 had to be launched in early 1969, but many of its designers still did not believe it was ready. Only enormous pressure from Mishin and his supporters resulted in a February launch date being set. Movie footage of the roll-out of the booster on its huge horizontal transporter, mounted on twin railway lines, shows onlookers with nervous smiles on their faces. Perhaps they were simply relieved the day had finally come to see if it could actually fly.

This first N1 to fly was codenamed "3L," and its three painted green-and-white conical stages made it resemble a military missile as it took off into the early afternoon sky on 21 February, carrying a modified L1 and Block-D stage under a 140-foot-long white fairing. An explosion in the first-stage engine compartment at sixty-nine seconds into the flight resulted in the loss of the whole vehicle. As on numerous previous Proton failures, the escape system worked perfectly, dragging the spacecraft mock-up safely away as the remaining N1 debris impacted the ground thirty-one miles downrange. Amazingly, in light of all the development problems, only two of its thirty first-stage engines had failed to work on liftoff.

After years of waiting for this very moment, the CIA managed to miss the entire show as this short flight literally flew under the radar of American monitoring stations in Turkey. It would only be later, when old satellite photographs were closely examined, that the large crater made by this crash was discovered.

Optimistically, in June 1969 Kamanin and Mishin selected a final group of eight cosmonauts for moon-landing missions. They were Bykovsky, Khrunov, Leonov, Makarov, Patsayev, Rukavishnikov, Voronov, and Yeliseyev. By then, surprisingly, General Kamanin was also expressing rage at what he saw as the downplaying of historic American achievements by the Soviet press. A prime example came, he wrote, when *Apollo 10* was unfavorably compared to unmanned Soviet probes sent to the planet Venus. "While little is written here about this outstanding mission, the successes of Venera are being drummed into us," he observed. "Although even to Boy Scouts it's clear that the importance of the Apollo flight is ten times greater than the flights of all our Veneras."

After studying the first N1 failure, the Soviets decided it needed extensive modifications to the 4L booster, so it was replaced with the newer, 5L for the second launch. This booster carried another L1 mock-up as it lit up the night sky on the evening of 3 July 1969. Then, less than ten seconds into the flight, a shard of metal entered the oxygen pump of one of the engines, creating an explosive chain reaction that forced the KORD computer to shut down all the remaining engines. When this occurred, the booster was only 650 feet above the launchpad. Slowly, inevitably, gravity took hold, and the giant booster slumped back to the ground before a massive detonation took place that knocked stunned observers a dozen miles away off their feet. The flight had only lasted twenty-three seconds, but when it was all over, so was any last hope of a Soviet cosmonaut being on the moon in the early 1970s.

That accident effectively destroyed the main launchpad, in addition to damaging the assembly building and an N1 mock-up sitting on the backup pad nearby. For some of the cosmonauts present—including Khrunov, Leonov, Makarov, and Rukavishnikov—it was painfully obvious to them that they would probably never walk on the moon. Years later Khrunov admitted that he had wept after the accident as he realized it was all over for the project.

At that same time, less than two weeks before the Americans would attempt their first manned moon landing, astronaut Frank Borman was in the Soviet Union on a nine-day semiofficial tour organized by the Institute of Soviet-American Relations. His scheduled visit to the cosmonaut training center near Moscow could not have been more poorly timed, as it was due to take place less than thirty-six hours after the launch of the latest N1. General Kamanin, who had refused to attend the launch, was on hand to greet Borman personally, but cosmonauts who had witnessed the explosion kept what they had just seen at Baikonur to themselves as they sat through Borman's presentation of his *Apollo 8* "home movies." Borman must have wondered about the sea of gloomy faces surrounding him. Alexei Leonov arrived later and managed to catch up with Borman at a dinner in Moscow's plush Metropole restaurant. He noted that it was packed full of uniformed military personnel, there to see the famous American. "Everyone wanted to stand near him. To touch him," Leonov remembered. "When I eventually met him I congratulated him on his mission. I of all people, I said, knew how hard it must have been. I did not tell him I, too, had been training for a circumlunar mission. But I felt as if he knew [and] we discussed good locations for lunar landing."

Ironically, the favored landing site for the first Soviet mission would have been the Ocean of Storms, the eventual landing site selected for *Apollo 12*, while the Sea of Tranquility was actually the third choice. These locations had been independently selected by scientists at the Vernadsky Institute in 1968 using data gathered from several early Luna probes.

In many ways the N1-L3 landing would have been very similar to that of Apollo, consisting of the *Luniy Orbitalny Korabl* (LOK) mother ship and a *Luniy Korabl* (LK) lunar lander. The LOK looked to the untrained eye like a longer version of the standard Soyuz without its solar panels, but it was almost totally different internally. The mother ship was designed by a team headed by Yuri Semenyov (who later took over as the head of the space program following the death of Glushko) and included a roomy orbital module. This space was needed to store the bulky moon suits and also act as an airlock for a cosmonaut transferring to the lunar lander during a spacewalk. If one had taken place, it would certainly have been one of the most awe-inspiring moments in human spaceflight, as it would have occurred in lunar orbit only thirty-seven miles above the surface.

The LK lander itself was about as tall as its Apollo counterpart but was only about a third of its mass, as most of its powered descent was performed by the attached Block-D stage. This gave the Soviet moon lander a lean, top-heavy appearance, but with designers conscious of the safety features needed to protect its solo pilot, it incorporated a few extras when compared to NASA's LM. The LK's final descent itself would have been very short. Once the Block-D had separated just above the lunar surface, after firing to brake the lander out of its lunar orbit, the cosmonaut would only have somewhere between twenty-five seconds and a minute to find a safe landing site for touchdown. Up to that point most of the landing was highly automated, using the same computer tested by the pilots on helicopter training flights. In reality, however, it has been calculated that the cosmonaut only had about three seconds to make the decision about where to set his craft down on the lunar surface. Thankfully, he could land on up to a twenty-degree slope, as the LK was fitted with four upward-facing solid propellant thrusters that fired on touchdown to plant the lunar lander legs firmly onto the surface of the moon, keeping the lander in an upright position.

As an added safety feature, the designers at TSKBEM had also fitted the LK with a single descent/ascent engine (Block-E), so the cosmonaut did not have to waste valuable seconds starting a second engine if he had to abort. All he basically had to do was throttle up once again to make it back to the LOK waiting in lunar orbit. The Soviet lander was also fitted with a second, backup ascent engine in the event the main engine failed to ignite.

Had it taken place, the first Soviet moon walk would have lasted about ninety minutes. As on *Apollo 11*, a television camera fixed atop the exit hatch of the LK lander would have recorded the historic moment as the lone cosmonaut stepped onto the surface, planted the red flag of the USSR, took a congratulatory call from Leonid Brezhnev, installed scientific instruments, and gathered some lunar rocks for the return to Earth.

As it happened, the only piece of moon-landing hardware ready to fly by 1969 was the lunar spacesuit. Designated *Kretchet* after a large arctic bird of prey, this was a radical semi-rigid torso design, which allowed the cosmo-

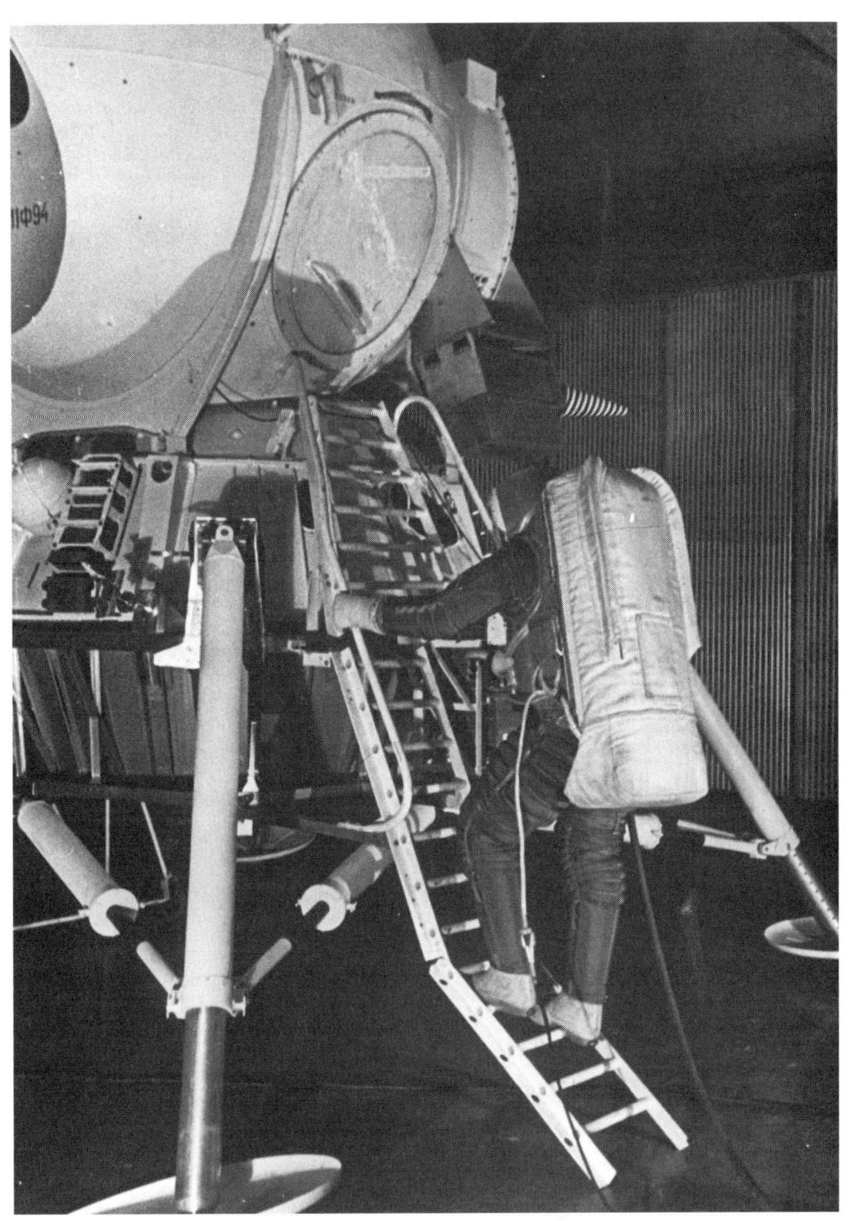

8. Cosmonauts practiced egress and moon-walking techniques on a full-scale LK lander training mock-up. Courtesy JSC RD&PE Zvezda and S. P. Korolev RSC Energia.

9. The *Kretchet* moon suit was ahead of its time. Courtesy Sotheby's.

naut to step into the suit through the backpack and shut it like a door behind him. It was an elegant concept that overcame the lack of room inside the lander. Before his moon walk the cosmonaut attached a bizarre Hula Hoop–like tube around his waist, preventing him from becoming trapped on his back like a stranded turtle if he fell over on the moon's surface. The second member of the L3 crew, who remained in lunar orbit, had a lighter spacesuit called *Orlan*, which he could use in an emergency to help the lunar surface cosmonaut if he got into trouble during the spacewalk between the LOK and the LK.

By late 1968 the cosmonauts, wearing the *Kretchet* suits, were training under lunar conditions on one-sixth gravity simulations aboard Tupolev 104 aircraft. Some of them had even donned the spacesuit at the Kamchatka Peninsula in the Russian Far East in order to test its effectiveness while clambering around the area's rugged volcanic slopes. The *Kretchet* could have been used on the moon in 1969 if only there had been a spaceship to take the cosmonauts there.

Realizing that a space station was the most likely follow-on project, from 1970 the spacesuit's manufacturer, Zvezda, worked on modifying the design for use in Earth orbit. It would become the basis of the EVA suit successfully used since 1977. Recently, it has become the standard Russian spacesuit aboard the International Space Station and has even been used by American astronauts from time to time. Many consider it one of the best-ever spacesuit designs.

As the world watched the *Apollo 11* saga unfold in July 1969, the often forgotten Soviet response, *Luna 15*, was staged as a last-gasp effort to land, scoop up, and return some lunar soil to Earth ahead of the manned landing. Although it was an unmanned probe, a successful mission by *Luna 15* would—to a degree—have upstaged the historic American effort. To achieve its mission, its NPO Lavochkin designers had taken the landing stage of a *Lunokhod* rover and hastily designed a drill and an Earth-return vehicle fitted with a basketball-sized recovery capsule to carry back a small soil sample. Launched on 13 July, just three days ahead of *Apollo 11*, celestial mechanics dictated that their sample of moondust would arrive back on Earth a few hours after the Apollo astronauts had splashed down in the Pacific.

If successful, it was felt, the flight of *Luna 15* might draw some attention from the simultaneous American effort to place Armstrong and Aldrin on the moon by demonstrating the effectiveness of sending robotic machines instead of risking human lives. Space historian Brian Harvey discovered details of the bizarre propaganda show waiting back on Earth if the spacecraft returned safely: "Once the launch was successful, preparations were put in train for a triumphant parade through Moscow, probably for 26th or 27 July. An armored car, covered in the Soviet flag and bedecked with flowers, would bring the rock samples from Vnukuvo Airport into Moscow, past the west gate of the Kremlin and on to the Vernadsky Institute where they would be displayed to a frenzy of the world's press before being brought inside for analysis."

The launch of *Luna 15* caused immediate anxiety in the United States once its objectives had been revealed, and assurances were sought from the Soviet Union that it would not interfere with or even endanger the manned mission. It was with some relief that these assurances were issued by the Kremlin. But any plans and propaganda strategies went awry when problems were encountered with *Luna 15*. It had entered into an elliptical orbit around the moon that brought it to within ten miles of the lunar surface at its lowest point, at one stage practically grazing the area around the Sea of Tranquility. Then, within hours of Armstrong and Aldrin setting foot upon the moon, *Luna 15* began its powered descent but slammed into the aptly named Sea of Crises at high speed and was destroyed.

A month after the triumph of *Apollo 11*, Leonov and Makarov found themselves at Mission Control acting as operators for the unmanned *Zond 7*, which had been launched on 8 August 1969. Frustratingly, its living occupants consisted of four tortoises. Although two years late, this endeavor turned out to be the first totally successful LI mission and energized its supporters enough to lobby the Kremlin to fly a manned version around the moon to commemorate the centenary of Lenin's birth in April 1970. The idea was quietly shelved.

Deprived of the opportunity to impress the world with anything comparable to *Apollo 11*, Leonid Brezhnev was reduced to using the unflown hardware as vast props for a piece of political theater he staged to impress a visiting delegation of Czech Communists touring the Baikonur Cosmodrome that October. They were shown a *Lunokhod* lunar rover and an LI before being taken to see the repaired NI mock-up mounted on the surviving launchpad. As this display of Soviet technology could not be spoiled by failure, the twisted remains of the other launchpad destroyed by the July explosion were skillfully hidden using strategically placed fencing. They would be the last foreigners to be shown anything related to the Soviet manned moon program for the next two decades.

Around this time Nikolai Kamanin confided once again to his diary that the scarce coverage of Apollo in the Soviet press had spoiled his mood. "The fulfillment of a dream that has captured the imagination of the whole planet for the past ten years has ceased to interest our leaders only because the ones flying to the moon are not we, but the Americans," he wrote. "I cannot join this conspiracy to remain silent about this great achievement of mankind and with all my heart welcome the successes of our American colleagues, although I know the Soviet cosmonauts have a tough time coping with our defeat."

Kamanin's emotional feelings combined with a sadness for his own cosmonaut team. Like him, they now realized that their slim chances of going to the moon were effectively blocked by a disinterested Kremlin. As if to confirm this fear, the cosmonaut corps in training for lunar missions was disbanded and the men transferred to the space station project by May 1970. Leonov was particularly bitter. "I argued very hard that we should continue with our work but the higher powers were adamant," he recalled. "The lunar groups which I had commanded and trained with for three long, hard years were disbanded. . . . There was no use, as the Russian saying goes, rubbing ashes on my forehead—crying over spilled milk. I knew I had to accept the decision. If I did not, there was a risk that I would never fly again."

Any slim chance of a future manned lunar mission all but disappeared on 24 September 1970, when the *Luna 16* probe successfully returned some lunar soil to Earth, confirming the Kremlin's proclaimed stance that the Soviet Union's only interest in the moon had been in planning and designing a cheaper unmanned alternative to Apollo. There had never been any Soviet plans, it declared emphatically, to send cosmonauts to the moon.

In that atmosphere the flight of *Zond 8* on 20 October 1970 can be seen as a final sad footnote, but at least the project went down fighting. Its biological cargo of tortoises, flies, and plants ended their lunar flight with a planned splashdown in the Indian Ocean, demonstrating that the Soviets could have successfully sent men around the moon in 1970, if only the political will had been there. Years later pictures of what might have been the manned ship emerged from the archives, showing an LI with the name *Zond 9* boldly emblazoned on the side in large red letters.

The poor performance of the Proton rocket arguably cost the Soviets the honor of flying the first men around the moon. Despite spending the equivalent of more than two billion dollars on the first nineteen Proton rockets for the lunar project up to February 1970, only six of them could be regarded as totally successful. It was a miserable record that cost them dearly. But in the meantime Vladimir Chelomey, who had been concentrating on a military space station project called Almaz, had the last laugh when this design and the Proton booster were used by the Feoktistov faction within TSKBEM as the basis for the new Salyut space station.

Vasily Mishin, for his part, had been opposed to any Proton-launched space station project from the start, knowing it would deflect attention away from his faltering N1. Following the launch of the first Salyut station on 19 April 1971, he had shown a distinct lack of interest in the station program, which some later saw as a contributing factor to the subsequent *Soyuz 11* disaster. The inexperienced, under-trained crew of *Soyuz 11*, which consisted of former LI candidate Georgi Dobrovolsky, together with Vladislav Volkov and Viktor Patsayev, had surprisingly replaced Leonov's prime crew in its entirety only three days before the launch because of a possible medical problem with crew member Valery Kubasov. It was a surprise to many when Mishin took some of his senior engineers with him to Baikonur for the launch of the third N1 when they should really have been on hand at Mission Control to help support the first Salyut space station crew, who were far less familiar with the station's systems and hardware than the original prime crew.

Mishin had openly demonstrated his first love when he ordered the cosmonauts on board *Salyut 1* to watch the launch of the 6L booster, carrying an LOK capsule, as they passed over Baikonur on 20 June 1971. Perhaps symbolically, the giant booster refused to cooperate, and Mishin had to postpone the N1 liftoff. When it finally left the pad on 27 June 1971, it did not have an audience above, but the booster, now painted pure white like the Saturn V, was a spectacular sight as it lit up the night sky on a pillar of flame nearly three times its own length. In a bizarre twist this launch failed because all of the engines worked perfectly, creating a previously unforeseen roll motion that could not be counteracted. Eventually, rotational dynamic forces on the whole booster grew so strong that the top stages literally sheared off and crashed near the launchpad. The remaining section thundered away in a giant arc, eventually impacting and exploding downrange.

Mishin finally took charge of the *Soyuz 11* mission at that point when just three days after the N1 failure, he flew directly to the Yevpatoriya tracking center to supervise the crew's return from their highly successful three-week mission. Subsequent commentators have linked Mishin's obsessive focus on the failing moon program, when he should have been concentrating on the new space station, as one of the reasons for the safety lapses that killed the *Soyuz 11* cosmonauts when their cabin depressurized on reentry.

After these two disastrous events, the loss of the third N1 and the deaths of the *Soyuz 11* crew, Mishin's authority diminished rapidly. With the current N1-L3 project looking increasing irrelevant, he had to start promoting a follow-on "lunar outpost" project called N1-L3M, which utilized two N1 launches and a larger, multi-person lunar lander capable of longer stays on the moon. But even he admitted that none of these endeavors could be ready until the late 1970s.

Mishin was in the hospital and missed what would turn out to be the last and best N1 launch on 23 November 1972. Less than two weeks before the United States's final manned lunar mission, stand-in launch director Boris Chertok must have felt that success was at last possible with the improved 7L. This latest rocket was fitted with new roll control motors to prevent a repeat of the previous failure and over thirteen thousand sensors to monitor the progress of the flight. After liftoff the N1 climbed smoothly for 107 seconds before an unexpected vibration in the propellant lines for engine no. 4 caused them to rupture, and the ascending booster disintegrated in a massive fireball. The frustration of everyone involved must have been intense as the explosion took place only 7 seconds before the scheduled first-stage cutoff. As the escape tower once again pulled its payload to safety, the contrast with the Americans could not have been more evident. As NASA was winding down Apollo as a technological and political triumph, the Soviets had not managed to fly their own moon rocket successfully even once.

During a trip to Moscow by moon walker David Scott in 1973, in preparation for the Apollo-Soyuz Test Project (ASTP), Leonov could not contain himself any longer and over a quiet drink at his apartment told the astonished astronaut of his own lunar training. Although NASA's senior management had been given CIA briefings on the subject, the information had never filtered down to the astronauts, and much of what Leonov confided to Scott was news to him, as he later wrote in *Two Sides of the Moon* (co-authored by Leonov). "It was fascinating to learn that the Russians had been that far along the path towards a lunar landing and to learn that Alexei was their key man," he revealed. "Our mindset at the time was that the Russians did not tell anyone anything, so the openness with which Alexei and I talked that night was, to me, quite fascinating."

The improved 8L and 9L boosters, using the modernized NK-33 Kuznetsov engine, were being readied in 1974. But just as the 8L booster, the fifth N1, was being prepared for launch sometime in August 1974, Mishin's enemies finally pounced. "Many mistakes were made in the L3 project," Nikolai Kamanin wrote in his diary. "They were visible even before the appearance of Apollo and after the first successful flights of it later it became clear to everyone that our lunar ships cannot compete with it. The sad story of the N1 and L3 is not finished yet: Mishin and his high patrons are trying to 'cure' a bad rocket and a bad ship and at the same time we are continuing to lag ever further behind the USA."

Kamanin and many others wanted to kill off the N1 project before it successfully placed something into orbit. They knew that once that happened, they would be stuck with a rocket they all knew was flawed. With the tragic deaths of the returning *Soyuz 11* crew on his hands and the failure to match the new American space station Skylab, Mishin's enemies now had enough ammunition finally to get him sacked by the Kremlin.

"There was no political advantage in continuing to go to the moon because the Americans had already been there six times," rationalized space designer Boris Chertok. "If we had stepped on the Moon even once after that, nobody would have been particularly happy. So it would have been necessary to plan a new program—build a real lunar outpost—and that would have been very costly. The political leaders then in power were not prepared to go along with that."

By this time most problems with the four-stage Proton had been overcome. Three LK lunar landers launched by Soyuz boosters had been tested successfully in Earth orbit during 1970–71. If only the Soviets had begun

10. Four failures of the N1 ultimately prevented a Soviet cosmonaut from walking on the moon. Courtesy Edwin Cameron.

building all these spacecraft a few years earlier, then the Americans might have had a real challenge when it came to Kennedy's lunar goal of a safe landing and return by the end of the decade. Now, suddenly, the decision had been made to suspend N1 flights, followed by an order to destroy the two remaining flight-worthy N1 rockets.

One authoritative estimate quotes a figure of four billion rubles for the construction of ten N1 rockets, so each one lost would have cost an unsustainable four hundred million rubles. Considering the fact that the gross domestic product of the USSR at that time was between one-quarter and one-half that of the U.S. economy, this was a serious expense the Soviet Union could ill afford. Costs aside, terminating the N1 program was a massive blow to those involved. "Six rockets, two of them fully assembled, were dismantled," Vasily Mishin commented with obvious bitterness. "The people who devoted the best years of their lives to them cried." He would later say that he understood everything—except Valentin Glushko taking over as the new chief designer.

Although Glushko is often portrayed in the media as the villain of the piece, the race for the moon was over, and he, more than anyone, realized that a new booster was needed for a new era of space shuttles and manned missions to Mars. One of Glushko's subsequent orders can be viewed as totally vindictive: his instruction to Nikolai Kuznetsov to destroy all the remaining N1 engines. There was little sense in doing it; by that time the NK-33 had developed into a superb engine. Engineers had even managed to run one continuously on a test stand for an astonishing five and a half hours. Perhaps Glushko realized he had a potential rival in the making.

Not surprisingly, Kuznetsov was outraged by this order, which he decided to ignore. He then covertly hid over one hundred of his beloved NK-33s in a disused section of his factory under large NUCLEAR DANGER warning signs. That pride in his engines would turn out to be a lifeline for the cash-strapped factory in the 1990s, when American engineers from Aerojet declared it one of the best rocket engines ever and purchased dozens of them for a million dollars apiece. These NK-33 engines were even considered for use on the latest version of the Atlas launch vehicle—the same American former ICBM used to launch John Glenn into orbit at the height of the Cold War back in 1962.

Soon after achieving the job of chief designer, Glushko merged TSKBEM with Energomash to form one giant new organization that became known as NPO Energia, which he kept under tight control. This enabled him to start again from scratch on a super booster design called *Energia*. Using high-tech liquid hydrogen / liquid oxygen propellants like the Saturn V, the booster made its debut flight in 1987, before successfully launching the Soviet space shuttle *Buran* on a fully automated, two-orbit mission in November 1988. This superb rocket—which flew less times than the N1—sadly became one of the first casualties when the Russian space budget collapsed following the demise of the Soviet Union.

Although the twentieth anniversary of the *Apollo 11* landing is often regarded as the main spur for an increasingly open Soviet society finally to reveal its past lunar ambitions, it is no coincidence that the controlling influence of Glushko had just ended with his death in January 1989. Others could finally challenge his "official history" of the late 1960s—something he had easily managed as editor of the official *Soviet Encyclopedia of Cosmonautics*.

At first many were still reticent to reveal details about plans to send cosmonauts to the moon ahead of the Americans—a sensitive blank page in space history. Then, in the same month in which Glushko passed away after a prolonged illness, Vasily Lazarev became the first cosmonaut to confirm publicly, to a lecture audience, the existence of a manned lunar program. Lazarev, who had flown twice with Oleg Makarov, may have been the first to reveal the existence of this colossal Soviet program, but he still refused to divulge more details because it was "a matter of the past."

The first solid facts came in a book about cosmonaut Valery Bykovsky published that summer, which told the story of his training for a manned Zond flight in the 1960s. Once the secret was out, Mishin used a series of newspaper interviews as an apologia for his own failures. "We came just one step from success but we were not allowed to take that step," he told *Pravda* newspaper in October 1989. "The Americans had already invested $25 billion in their program and eventually reached the moon. We invested ten times less and had to fight for every million rubles."

In December 1989 a visiting group of professors from the Massachusetts Institute of Technology (MIT) were shown a Soviet lunar lander during a tour of the Moscow Aviation Institute museum, which had become Vasily Mishin's second home since his forced retirement in 1974. The surprised Americans took some pictures and passed them onto the *New York Times* when they got home. The pictures were published on the newspaper's front page under the dramatic headline "Russians Finally Admit They Lost Race to Moon."

The moon project might have started out as a dream, but it soon turned into a nightmare that not only cost billions of rubles but failed to give anything back in return to either the Kremlin or the Soviet people. Unlike NASA, everyone at the heart of the Soviet space effort had been reluctant to cooperate with other agencies because they were too busy protecting their own interests and trying to win favor with the Kremlin. Any advances they might have gained by Korolev's series of spectacular space firsts in the early 1960s were quickly wasted by infighting over precious funds and who should get them for their own space projects.

Although the cosmonauts would also have given anything to try and make a landing on the lunar surface, their superiors eventually believed the odds were stacked against them and cut their losses after the demoralizing success of the rival Apollo program. It is nevertheless a tribute to their skill that they were still within sight of NASA when it came to flying cosmonauts around the moon with the LI.

Ultimately, it was the Soviet people who missed out. Denied the truth for over twenty years about how their nation's engineers and cosmonauts had planned to fly to the moon ahead of the Americans, they were also prevented from watching one of humanity's most defining moments as it took place in July 1969. As Leonov was to later lament: "Not showing live coverage of the *Apollo 11* moon landing was a most stupid and short-sighted political decision, stemming from pride and envy. The Soviet Union had been working for so long that there were those who felt they could not show another achieving our goal. But our country robbed its own citizens by allowing political considerations to prevail over genuine human happiness at such events."

Boris Chertok had also been one of that small group of military men, scientists and cosmonauts who had secretly gathered in the Komsomolsky Avenue facility to watch live pictures of Armstrong and Aldrin walking on the moon. Only able to witness one of the most monumental events in human history because their small television monitors were hooked up to a bootleg cable system coming in from Europe, he shook his head in wonder as he watched events unfolding on the Sea of Tranquility. "We were delighted as engineers as they had done a wonderful job," he would state many years later. "But on the other hand we felt disappointment. Why them and not us? It was bitter."

"It was not a fair race," was Vasily Mishin's final conclusion. "First of all, America was richer than we were, especially then, and Russia was weakened by the fight against German fascism and weakened by the cost of the arms race. As soon as America began the moon race, we understood we could not win." Although Mishin presided over the Soviet Union's failed efforts to beat the United States to the moon, it is his unfortunate legacy to be widely perceived as a failure after the stunning achievements of his predecessor. Despite this reputation, he still attracts sympathy for trying to achieve the near-impossible in the face of bitter rivalry and bureaucratic obstinacy. James Harford, who interviewed Mishin several times about his mentor for his biography *Korolev: How One Man Masterminded the Soviet Drive to Beat America to the Moon*, views Mishin's unfulfilled dreams of a Soviet moon landing in the context of those times. "In my opinion he was a very competent guy," Harford states. "And he was really the scapegoat. A lot of Russians blamed him for losing the moon race. But there was no way you could blame him. Korolev himself would not have been able to do this."

Unrepentant and disillusioned to the end of his eighty-four years, Vasily Mishin—perhaps unfairly regarded as the man who lost the moon race for the Soviet Union—passed away in Moscow on 10 October 2001.

3. Rendezvous on the Ocean of Storms

John Youskauskas

As the moon's fair image quaketh
In the raging waves of ocean,
Whilst she, in the vault of heaven,
Moves with silent peaceful motion.

Heinrich Heine (1797–1856)

Dick Gordon floated over to a window, peered through and marveled at what he saw. Hanging there in the brilliant sunlight against the slowly moving backdrop of the lunar surface was a spindly, gold Mylar-covered Lunar Module named *Intrepid*, inside of which were two of the closest buddies he had ever known. Pete Conrad and Al Bean were bound for a precisely targeted landing on the Ocean of Storms, defined by a cluster of craters known as "Snowman," and a small robotic probe named *Surveyor 3*, which had arrived there over two and a half years earlier.

For the first time since their launch four days earlier, Gordon was alone in *Yankee Clipper*, the command ship he would be responsible for until his crewmates returned from their lunar exploration. He was genuinely happy that the fateful rotation of crew assignments had brought them to this point in the program. All three were certainly fortunate to have come this far, considering that for each astronaut selected by NASA, scores of their piloting peers never got the chance. They may have been too short, too tall, too young, too old, or did not have sufficient experience. Others, sadly, had been killed in tragic accidents.

Gordon was far too occupied to be concerned about it at the time, but as he would later relate in an interview for NASA's Oral History Project, he knew that if he adhered to the space agency's selection policy, his own chance would come. "The name of the game as far as I was concerned was to walk on the moon," he ventured, "[but] at that time I was relegated not to do that. . . . I had a job and a function to perform."

With two crew members gone, the once-cramped interior of the CM now seemed quite roomy, and Gordon quickly got ahead of the timeline, removing his pressure suit in order to move around more freely. After performing a quick inspection of *Intrepid* and taking some photographs, he readied the ship for a small burn that would separate the CM from the lunar lander. On board *Intrepid* Conrad transmitted a brief farewell to his most trusted companion in the spaceflight business.

"Okay, Dick," he said. "Do your usual good job."

"My best," Gordon responded.

"I know that," came the chirpy response.

A little over an hour later Conrad and Bean fired up their descent engine to lower *Intrepid*'s orbit, beginning a slow descent to the moon's surface. Gordon watched through the eyepiece of his navigational optics, calling out to his commander, "Hey, I have you in the sextant; looking right up your descent engine. Fantastic!" And he meant it. As *Intrepid* disappeared in the bright blue plume of exhaust, Gordon felt little envy that his two best friends were headed off to land on and explore the moon. He certainly never expressed any to them or anyone else.

Long before the flight he had dismissed all of the nightmarish thoughts of returning home without his crewmates, never once doubting Pete Conrad's ability to fly the dauntingly complex descent to the lunar surface and return to rendezvous with the CM. Nevertheless, Gordon had practiced countless times the techniques needed to lower his orbit in the unlikely event he would have to rescue a crippled Lunar Module that had limped off the moon.

As the now-solo CMP, Gordon had many busy hours ahead of him working a flight plan crammed with maneuvers, photography, and observations. But at this moment, burdened by all of the mission responsibilities ahead of him, the hard-charging, perfectionist naval aviator kept one dream alive in the back of his mind—to pull this thing off, get home in one piece, and get back in line for his own chance to travel that last sixty miles onto the moon's surface.

Culled from the ranks of the military test flying community, many of the astronauts had known each other by reputation, if little else, before joining NASA. Lt. Cdr. Richard "Dick" Gordon had enjoyed a close friendship with the wise-cracking Lt. Charles "Pete" Conrad from their days flying for the navy and rooming together on board the USS *Ranger* during carrier qualifications. Lt. Al Bean had been a student in Conrad's aircraft performance class at the Patuxent River Test Pilot School in Maryland before being selected to the astronaut corps. The men each brought unique skills and personality to the *Apollo 12* crew, yet they shared a common bond, having been tested by years of hazardous flying duties both at sea and on land, wringing out new, unproven aircraft.

For his part Conrad took the art of having fun on the job to an almost mythical level. He had a way of finding humor even in the tensest of situations, and it was so infectious that it drew people to him. Slightly built, five foot six inches tall and balding, he would be described by Gordon as "very unassuming," yet he could fill any room with his larger-than-life personality, gap-toothed grin, and unbridled cackle of a laugh. Conrad also freely displayed a salty, colorful command of the English language, which he had to force himself to suppress when speaking in public or broadcasting from space.

Apollo 12's backup CMP, Al Worden, remembered Conrad as the kind of person who had "a million friends," in fact so many friends that he was a hard person to get to know. "He was a unique individual," Worden recalled, "a good commander. He took care of his crew. Their crew was very cohesive. If one of them did something, they all did it."

As a youth, Gordon had seriously considered becoming a dentist, but once he began flying, he loved it too much to contemplate doing anything else. Graduating from the University of Washington with a bachelor of science degree at the height of the Korean War, he had an obligation to enter military service. After joining the navy and earning his wings as a naval aviator in 1953, he entered transitional jet training and four years later was accepted into the Navy's Test Pilot School at Patuxent River. While there, he conducted test flights on the F8U Crusader, F11F Tiger, FJ Fury, and the A4D Skyhawk. While serving as the first project test pilot for the F4H Phantom, he even set a new transcontinental supersonic speed record of two hours and forty-seven minutes. He had applied with his friend Pete

11. The crew of *Apollo 12* in a jovial mood beside one of NASA's T-38 jet airplanes. *From left*: CDR Pete Conrad, CMP Dick Gordon, and LMP Alan Bean. Courtesy NASA.

Conrad for the second group of astronauts, but while Conrad was accepted, Gordon was turned down.

"I was pissed off," he admitted feeling at the time of his non-selection. "Guys like [Pete] were selected and I wasn't. . . . I was so much more experienced than he was. . . . I was mad [and] very disappointed [because] we were in the squadron together and he left."

With Conrad off being introduced to the nation as one of the New Nine astronauts, Gordon eventually swallowed his disappointment and set out to gain more operational experience and education: "I calmed down after a little bit of time and then deployed with my squadron to the Far East on the [USS] *Ranger* in '62 and '63; and then was assigned to post-graduate school in Monterey, California, in mid '63." After learning that NASA was seeking applications for a third group of astronauts, he reapplied and this time was accepted. In October 1963 the newest group of astronauts was announced; one of them was Al Bean.

Gordon was immediately put to work on cockpit layout and design for both the Apollo Command and Lunar modules. That work was interrupted in 1965 by what he recalls as "the best thing that ever happened to me," when he was first of all assigned with Conrad to the backup crew for

Gemini 8 then flying in space for the first time as a prime crew member on *Gemini 11*. The three-day mission solidified Gordon's friendship with Conrad, and the two men would later become one of only three Gemini crews to fly an Apollo mission together.

Conrad and Gordon quickly built up a reputation as being two of the cockiest pilots ever to team up for a flight. They were good, and they knew it. *Really* good. Everyone else in the Astronaut Office knew it too, but the thing that drove all of them crazy was that while they worked their tails off in the simulators and the classroom, these two navy guys made it all look so easy and had fun doing it. "Pete and I had been together so long, trained together for so long [that] we got to the point where we knew what each other was gonna say before we said it," Gordon recalled.

Al Bean was not so easily integrated into the ranks of the astronauts. Unlike his days at Pax River, being a new astronaut meant a restricted amount of flying, with the majority of his time spent attending endless meetings and engineering reviews. He had a difficult time expressing his opinions and interacting with the managers, engineers, and other astronauts in attendance. It seemed as if no one would ever see things his way, even when he went to extraordinary lengths to prepare his arguments. He felt intimidated walking into meetings, knowing full well that most people in the room knew more about a given subject than he did. "I can't say that I was a good contributor," he would recall years later. "I wanted to be every day, but I never figured it out."

Pete Conrad knew something was up. Stopping Bean in the hallway one day, he asked in his forthright way, "Why is Al Shepard so mad at you?" Shepard, then the head of the Astronaut Office, and Deke Slayton, in charge of Flight Crew Operations, held virtually all of the cards when it came to crew assignments. Bean confessed he did not really know why. Conrad then offered advice on how to get along better with those who determined his fate, but it would take years before Bean fully appreciated what his fellow astronaut was trying to do for him.

Things appeared to improve when he was assigned the coveted capsule communicator (CapCom) role for *Gemini 6*. On the day of the scheduled liftoff Bean went about his job methodically, even though he says he felt a little unsure of himself sitting among some of the legends of mission control. As the clock reached zero, the Titan rocket with Wally Schirra and Tom Stafford perched atop roared to life and promptly shut down. With just a split second to decide whether to risk an explosive ejection or not, Schirra sat coolly in his rocket-powered seat and exhaled quietly. Bean never saw an indication of liftoff, so he reacted exactly as he should have. There was silence. As Stafford would relate to William Vantine in a 1997 interview, there were two heroes that day, Wally Schirra and Al Bean: "The only one who can call liftoff is a fellow astronaut looking at the base of a booster, you know, on a TV screen, and when he sees first motion, he's the only one that can call liftoff. It was Al Bean, and he didn't say a thing, and that's exactly what he should have done, is not say a thing."

Bean was excited when Deke Slayton later assigned him as the backup commander for *Gemini 10*, even though it was too late in that program for him to rotate onto a mission. Bean and fellow rookie C.C. Williams took their backup position seriously and set out to prove themselves worthy of a future flight assignment on Apollo. "That was a very good assignment," Bean would remember fondly. "I worked with C.C. Williams and we immediately started training, and were pretty good at it." Yet everything in the space business was so new that as Bean later explained to the author, there was some concern about going into orbit with so little operational experience:

It wasn't like flight training in the navy where the people that were training you had done it all before. Nobody knew that much about all this stuff. We had a lot of confidence and we thought we knew a lot, but in retrospect—it wasn't the blind leading the blind—but we were both doing the best we could. Looking back, I'm glad we weren't called on to fly that mission. I think if we were we could have done it by the book, by the numbers you might say, but we didn't have the feel for it in my opinion.

Unfortunately for Bean, his *Gemini 10* experience did not garner for him the attention he thought it might. Not long after John Young and Mike Collins splashed down, Slayton sent Bean on a course that led farther away from an actual spaceflight than he had ever been. The Apollo Applications Program (AAP, later renamed Skylab) was in its infancy, consisting of a small contingent of engineers and astronauts doing very preliminary studies on possible follow-on uses of Apollo hardware following the moon landing program.

Now here was Al Bean, who came into the program with visions of walking on the moon, trying desperately to get a spaceflight, any spaceflight, and he was forced to start thinking about programs *beyond* Apollo. All the while his astronaut classmates were being assigned "real" missions. According to Bean, it was almost too much to bear. "I began to forget about Apollo as much as I possibly could," he stated. In a JSC Oral History interview he expanded on his disappointment:

I'm sure they were thinking things like, "Let him go over and work on this new project. He's got a lot of ideas. Put him over there in that corner. We need somebody because we're going to eventually fly this thing. He knows what he's doing, although he can't seem to connect completely, particularly with the bosses." So I got shuffled over there, and I then didn't learn that much either, other than I was out in left field and I just had to accept it and had to make the best of it, so I began to do that.

While admitting that the late Gemini backup deal was a dead-end job, Slayton would claim he had sound reasons for sending Bean in the direction he had. In his 1994 memoir, *Deke!*, the Mercury veteran expressed confidence in Bean's abilities and his trust in the fact that he could handle the challenge of getting the Applications branch squared away.

The project office was overseen by Slayton, and this allowed more frequent contact between the Mercury veteran and his new expert on the space station concept. Bean found that working alongside Slayton was helpful, "because even though he doesn't say too much, just having him around seems to make things go better." Looking back, Bean says, "Now that I think about it, I probably worked more closely with Deke then than any other time in my career."

Pete Conrad had an uncanny knack for getting almost whatever he wanted from Slayton. It was well understood when he came off *Gemini 11* that he would be assigned to one of the early Apollo flights—perhaps even a lunar landing attempt. Slayton required that the CMPs, with responsibility for a possible rescue of a disabled Lunar Module, have previous spaceflight experience, and there was no chance Conrad was going to pick anyone to fly with him other than Dick Gordon. Conrad had also kept an eye on his old student from Pax River, Al Bean. He knew that Bean was talented, but the guy just had not been given the opportunity really to show it.

Gordon would approve of Conrad's decision, although he did not really know Bean all that well. "Pete was Alan's instructor at Test Pilot School," he recalled, "[but] I did not know him. We were together in the second selection, so I got to know Al a little bit at that time. But, hey—he was a naval aviator . . . test pilot. To me that meant everything. He was acceptable!" It therefore came as no real surprise to Gordon when, in the fall of 1967, Conrad quietly submitted his choice for an all-navy team to back up Jim McDivitt's crew of Dave Scott and Rusty Schweickart.

But then Deke Slayton said no. "Al Bean's not available," he told Conrad, handing him a piece of paper with names on it. "Here are these five guys you can take somebody from." He was intent on keeping Bean firmly planted in Apollo Applications and was not inclined to give any reason why. Conrad reluctantly looked over the list of potential LMPs and quickly settled on Bean's *Gemini 10* partner, C.C. Williams.

Williams was a very popular guy among the astronauts, a personable, fun-loving Marine fighter test pilot who almost did not make the cut as an applicant because he bordered on being too tall. He and Bean had become great friends while training as the *Gemini 10* backup crew, and he was also a good match for Conrad's more extroverted personality.

The three had barely begun working together when, on 5 October 1967, Williams was killed. Flying a brand-new T-38 out of the Cape, he experienced a flight control malfunction as he passed twenty-two thousand feet over Tallahassee. The aircraft began to roll uncontrollably and entered a steep dive. Unfortunately, as the jet accelerated to near supersonic speed, shock waves of compressed air on the wings slowed the roll rate, probably giving C.C. the false impression that he was regaining control.

C.C. was running out of altitude and rapidly gaining speed. A spinning panorama of Apalachicola forest and Florida coastline engulfed his field of view, reaching up to grab him. He had tried long enough—it was time to leave. Williams blasted the canopy off and pulled the ejection handle, lighting the rocket motor under his seat and propelling him violently away from the crippled jet, but it was far too late. Hurtling toward the ground alongside the corkscrewing T-38, there was not enough time for his parachute to deploy fully, and he slammed into the ground at high speed.

Four days later at Arlington National Cemetery, Conrad, Gordon, and Bean joined several other astronauts as pallbearers for their fallen comrade. "We'd all been in squadrons where we'd lost squadron mates," Bean reflected. "That was part of the business . . . just part of it."

Conrad now had a decision to make. The prime and backup crews for the mission that would become *Apollo 8* had not been publicly announced, so he approached Slayton once again, and this time he was not going to take no for an answer. "I want Al Bean," he said as firmly as he could muster. The gruff Mercury veteran glared at him, chomping on the stump of a cigar. Maybe he had seen enough of Bean as a result of his involvement with the AAP office to know that the teaming might just work. "Okay, go ahead," he finally growled in resignation.

In the hangar at Ellington Field, Alan Bean, just back from an afternoon proficiency flight, stood in utter silence, mouth agape. Pete Conrad was standing before him, in his trademark ball cap and with a parachute casually slung over his shoulder. Conrad had just taken Bean aside and given him some news, but he was too stunned to grasp the meaning of it. *What's he talking about?* Bean was vaguely aware that he had babbled some mundane response, trying to sound nonchalant, but whatever he had said was immediately lost to him. Conrad simply gave him a gap-toothed grin, slapped him on the shoulder, and spun around to head out to another T-38 for a quick hop.

Even then, a full realization of what Bean had been told failed to sink in. He was going to join Conrad and Gordon on the backup crew for Jim McDivitt's flight, and given Slayton's normal rotation of crews from backup to prime, it was not too difficult to figure out what this might lead to. At the time, however, it did not all add up for Bean, the former outcast. He was also worried that Conrad might have thought him flippant or even ungrateful. "I didn't think of myself as a potential crew member on anybody's crew," he told the author, "so I was kind of puzzled by what he meant. He probably thought I wasn't very grateful, [maybe] I wasn't. I wasn't excited at all. I mean, I didn't catch on so well. It was so far away from what I thought was possible. It wasn't like I was saying, 'Boy, now I'm going to get a chance to train, and fly . . . be a real astronaut . . . go and maybe walk on the moon.' Good news was a stranger to me, and that news was strange."

In fact, it was several days before Al fully appreciated the incredible chance he had been given, and he would always state that had it not been for Pete Conrad, he would never have been part of the lunar program, let alone one of its handful of moon walkers.

The announcement of McDivitt's crew with Conrad's as backup was made officially on 20 November 1967. One thing was certain—Apollo Applications would have to get along without Al Bean for a while. Working side by side with Jim McDivitt's crew would be a big step in the right direction for the three navy aviators who hoped to mount a lunar expedition of their own someday. Bean felt the excitement every day he went to work. "We're backing it up," he later recalled, "so we're training for this mission, but we know we'd better get ready, because as soon as this mission goes, we're going to get one of the early shots at going to the moon. We knew this from day one, so we were always happy about it."

It became apparent as development progressed slowly through 1968 that the LM would never be ready in time for *Apollo 8*. When the decision was made to slip the test flight to *Apollo 9* and send *8* around the moon instead, McDivitt opted to keep his crew with their original mission. This move in turn bumped Conrad, Gordon, and Bean from being the projected crew of *Apollo 11* to *Apollo 12*. At the time all this was being sorted out, no one could know for certain which flight would ultimately be the first to land on the moon. It is only in hindsight that the shake-up reveals its true effect on history—a fact that Gordon still ribs McDivitt about today. Of course, Gordon holds no real grudge against the decision or the man but adds, "If I had known that *11* was going to be the first one on the moon and McDivitt had the chance to fly *8*, and we would've rotated to *11*, there would've been hell to pay if he turned it down!"

Apollo 9 went off without a hitch, clearing the way for NASA to assign the first all-navy prime crew to a spaceflight since Conrad and Gordon first flew together in September 1966.

"See you in town!" Dick Gordon yelled as he popped the clutch and mashed the accelerator in his '69 Corvette, desperately trying to muster all 390 horsepower in a fresh attempt to avoid being shown up yet again by Pete Conrad.

The massive Turbojet v-8 engine howled, and his tires screamed against the wet asphalt as a rooster tail of water sprayed up from behind the car. Thrust back deep in his seat, Gordon worked up through the four-speed gearbox, heading for the sharp right-hand turn just before the south gate of Patrick Air Force Base. They both knew from previous sprints to the gate that the front-heavy Corvette would want to keep going straight off the rain-soaked road, but he was not about to let up. As he entered the turn, Gordon felt the tires lose their grip on the road ever so slightly, forcing him to drop down to third gear. He looked up ahead . . . and there was Pete, smoothly negotiating the hairpin corner. *How the hell does he do that?*

The whole thing started with Alan Shepard. He had always loved the power of this streamlined American sports car, having owned two of them before joining NASA. The Mercury astronaut soon became friends with Melbourne Chevy dealer and Indianapolis 500 winner Jim Rathmann. Of course, General Motors president Ed Cole was more than happy to see such prominent national heroes driving around town in one of his products, so he worked out a deal with Rathmann to lease Corvettes to the astronauts for one dollar a year. When the *Apollo 12* trio began spending long weeks training as a prime crew in early 1969, it was their turn to enjoy one of the coolest perks of being an astronaut.

Almost all of the astronauts at one time or another took advantage of the lease deal from Rathmann and Cole, and the parking lot in front of Building 3 at the Manned Spacecraft Center in Houston was usually filled with a lineup of various colored Corvettes. Every time the locals heard one of the hotrods come rumbling down the street, heads would turn to see which one of the rocket jockeys was driving. Pete Conrad wanted his crew to stand out, to be distinctive, so when they approached Rathmann to sign the leases, they worked out a paint scheme that would set their cars apart from the rest of the Astrovettes. Although not intended, the matching Riverside Gold bodies with black "wings" wrapping around the aft fenders hinted at the colors of the ship they would fly to the moon. There was also a rectangular red, white, and blue label applied just above each of the side exhausts with CDR, CMP, or LMP painted in script, as a subtle way of denoting which crew member piloted the machine.

Al Bean appreciated the speedy coupe not so much for its power but for the sleek, distinctive lines of the Stingray body shape. Although he was trained as a technical pilot and engineer, he had an artistic eye for cars just as he did for high performance aircraft. Looks counted to Bean, and the personalized, black and gold Corvettes driven by *Apollo 12*'s crew were indeed beautiful designs.

In return for such an attractive benefit, Ed Cole hoped to get as much publicity mileage as possible out of the astronaut-Corvette connection. While NASA restricted its pilots from officially endorsing any products, one did not have to appear in a television commercial in order to increase a product's appeal. One September afternoon in Houston during their training for the moon landing, the crew decided to stage a little public relations coup. They wanted to get a photo of themselves with their personalized street racers, and Conrad hatched another plan that would become legendary.

The trio of astronauts had their traditional crew photo taken on what was known as the "Rock Pile" at the Manned Spacecraft Center, upon which was perched a mock-up of the LM. Once the official photos were taken, the three men, still in their space suits, drove their Stingrays onto the simulated moonscape. Amid this unlikely collection of sports cars and spacecraft, Conrad, Gordon, and Bean lined up grinning beside their respective coupes and had the ultimate "boys and their toys" shot taken.

Life was there as well, but NASA restricted the popular magazine from taking any photos on the taxpayer-funded government facility. So Ralph Morse, a respected *Life* photographer who shadowed the astronauts almost everywhere they went, came up with an alternate plan. Once the crew got out of their pressure suits, he rendezvoused with them in the parking lot and took a few pictures of them sitting atop their Corvettes, one of which appeared across two full pages in a December 1969 issue of the magazine. Ed Cole had his national advertisement—and it had not cost him a dime.

The backup crew of Dave Scott, Al Worden, and Jim Irwin—all U.S. Air Force officers—decided they were not about to be outdone by the U.S. Navy. They were competitive in almost everything they did in the space business, although attempting to keep up with Conrad's antics and racing prowess proved somewhat pointless. Worden admitted to having been a little envious of the fraternal nature of the prime crew, and his team tried to emulate them as best as their own personalities would allow. "We were all Air Force. So we were the perfect backup crew," he explained.

12. Conrad, Gordon, and Bean pose for Ralph Morse's camera with their matching Corvettes shortly after their formal crew photo was taken. The crew routinely had their General Motors–leased cars driven back and forth between Houston and the Cape by the secretaries in the Astronaut Office and at the Kennedy Space Center. Courtesy Getty Images.

They had their cars painted gold and black, and we had ours painted red, white, and blue. So [whatever] they did we did the same. They were a wonderful crew. Our crew . . . we were not close-knit at all. In fact, we didn't always see eye to eye on a lot of things, and so we weren't so tight that we all had to do the same thing. But Pete and his crew were, and it was a kick watching them drive down through Cocoa Beach; these three cars all gold and black, one right after the other, and they'd pull up there in the parking lot somewhere, one, two, three . . . like the Three Musketeers.

As soon as Conrad's crew became candidates for a moon landing, they spent more and more time on geology field trips, honing their skills in identifying and collecting the precious lunar treasure they would find at the Ocean of Storms. Of course, they were not professional geologists; they were highly skilled pilots, being trained to think and observe like geologists. Pete enjoyed ribbing the instructors in the field and most often referred to some innocuous pile of rocks as "stuff." Bean remembers that Pete "liked to do it and he would tease the geologists and they knew he was teasing them . . . he knew as much geology as any of us."

The crew spent much of their time at Cape Canaveral making use of the Command and Lunar Module simulators and were provided with their own living quarters and offices for the long stays. Hazel Sekac Banks, one of a handful of secretaries assigned to the training building at the Cape, remembers Pete Conrad stopping in one day and asking for her assistance in replying to a particular piece of mail. The writer claimed to be of some relevance to Conrad's family, but Pete did not have a clue who the man was. "Can you write this guy and tell him I'm not a twig off of his tree?" he asked. Hazel tried to think of a way to soften the reply, but in the end simply typed out, "Dear Sir, Thanks for your letter, but I'm not a twig off your tree." Conrad loved it. He laughed as he signed the note and dropped it in the mail for dispatch to his "long-lost" relative.

As the long summer months of 1969 passed, it looked like all of the pieces might fall into place to make *Apollo 11* the first lunar landing attempt. On the afternoon of 20 July the crew next in line took a break from training and watched with the world as Armstrong and Aldrin guided *Eagle* onto the Sea of Tranquility. Gordon remembers: "Until *11* landed we had always thought that we had a shot. As a matter of fact the three of us were in the viewing room when they landed . . . we all kind of looked at each other and said 'Aw shucks!'"

One night in August the *Apollo 12* crew was having dinner with Scott, Irwin, and Worden at Alma's, just down the road from the Holiday Inn in Cocoa Beach. A local Italian favorite, Alma's was one of *the* places to go for the throngs of space workers and astronauts who passed through town.

As Al Bean twisted some spaghetti around his fork, the conversation took a turn from the usual talk of the day's technical issues. "So Pete," one of them asked, "what are you going to say when you step off onto the surface?" Conrad had not really given it much thought, but he knew everyone would be expecting something memorable to follow Neil Armstrong's immortal words. Pete, as he often did, cracked one corner of his mouth up into a wide half-smile and said, "Well, you know, I'll probably say something like, 'It was a small step for Armstrong, but it's a big one for a little fella like me!'" The rest of the patrons at Alma's probably wondered what the joke was when the table full of astronauts suddenly roared with laughter.

The grueling months of simulations, field trips, and public relations events during that summer seemed to fly by. All attention was now focused on them, and as the next crew to launch, they got first crack at the simulators and aircraft. But they were also burdened with seemingly endless changes to the flight plan. As summer faded into fall, *12*'s astronauts could see the conveyor belt of activity moving steadily along as final preparations were being made for humankind's second voyage to another world. Bean once described the days leading up to launch as being "like Christmas and your birthday rolled into one. I mean, can you think of anything better?"

Pete Conrad knew he had a great crew and that they were as ready as anyone could be for such a complex undertaking. As he would do throughout the mission, he offered the guidance and encouragement of a seasoned commander to his two shipmates. Bean recalls, "Funny thing is, Pete had told us before flight we'll have some things go wrong and they'll be things we'd never done or trained for." None of them had any idea how prophetic that statement would turn out to be.

John Aaron had a natural curiosity for almost anything technical. A brilliant engineer from southwestern Oklahoma, he had joined NASA straight out of college. Working his way up through the ranks of Mission Control, he would land himself a plum assignment as an EECOM, in charge of Electrical, Environmental, and Communications systems for the Command and Service modules. His curiosity compelled him to learn his systems to a level of detail unmatched by most of his peers. He would play "What if?" games over and over with other engineers, not so much in planning for far-fetched anomalies but just because he wanted to know.

One night Aaron decided to stay and work with the crew of the overnight third shift at the Cape while they conducted a test of an Apollo CM. The test dragged on, and the late hour did nothing to help keep the team sharp. Almost inevitably, the technicians got out of sequence and dropped power to the ship, which immediately switched to its internal batteries. Aaron looked at a monitor screen, where, given the loss of electricity, he expected to see a display of all zeros but did not. In a January 2000 interview with historian Kevin Rusnak he described his reaction: "This pattern of numbers came up and I was intrigued by them, because they didn't go to zero. They were at 6.7, 12.3. I mean some squirrelly kind of numbers. I drove home that night, thinking where did those squirrelly numbers come from? And the next morning I came in the office and I sat down with Dick

Brown . . . a North American engineer that worked in our office. We sat down and went through all the circuitry to find out just how does this thing work. Why would those patterns of numbers have come up?"

Aaron would spend many more nights like this, working near the spacecraft or hunched over his console in Mission Control. Staring groggily at reams of data through thick black-framed glasses with his headset clipped to one rim, he was the personification of the flight controller image that existed within this isolated chamber of spaceflight brainpower.

For three naval aviators, flying in rough weather seemed almost routine. But as they walked across the swing-arm on 14 November to board *Yankee Clipper*, with beads of raindrops drizzling down their bubble helmets, Dick Gordon took a few moments to give consideration to Mother Nature and the monster rocket he was about to board: "I got to sit outside while those two guys were getting strapped in. I got to stand up there and look at the weather and look at the coast of Florida and watch the black clouds come in and say, 'Hey, this is real.' That beast below us, that Saturn V, is a living, breathing object. It's venting vapors and ice is falling off of it. And it's a creature that's just about to come alive."

Earlier that morning Tom Stafford, who had taken over as head of the Astronaut Office while Alan Shepard trained for *Apollo 14*, awakened the *12* crew for the Big Day. Entering the breakfast area, Conrad burst out laughing when he saw a life-sized stuffed gorilla dubbed Irving, adorned in a standard NASA white smock and construction helmet, seated in the corner. Once again, Conrad's love of a good laugh had inspired someone, probably Rathmann, to go out of his way to impress him. It was not the last of the pranks they would encounter as they made their way toward their waiting moonship.

As Gordon and Bean meticulously went about donning their bulky pressure suits, a gregarious Pete Conrad appeared relaxed, reclined in his lounger, breaking the tense silence of the room with his typical wisecracks and laughter. The tedious process of suiting up was complete only after one of the technicians had stuffed a plastic-wrapped ham sandwich in a lower leg pocket of each astronaut's space suit. In the years since John Young's corned beef sandwich incident on *Gemini 3*, which resulted in the two crew members being severely reprimanded for lightheartedly smuggling an unlisted snack item aboard their flight, caution had given way to necessity. The handy meals would serve the crew well in the busy hours after launch, before they had a chance to unpack any of their on-board supplies.

Dick Gordon lay in the center seat of *Yankee Clipper*, crammed shoulder to shoulder between his two crewmates, surveying the vast array of switches, gauges, and warning lights before him. The gentle hiss of oxygen in his helmet was occasionally interrupted by bursts of chatter through his "Snoopy" communication cap, but he was constantly aware of the steady beat of rain on the outside of the hull. Conrad concerned himself with the intrusion of windblown rainwater streaming down the outside of the windows, still shrouded by a supposedly airtight boost protective cover. He also had visions of the upward-facing cups of the reaction control jet nozzles brimming with water and began considering the adverse effects if the nozzles froze up during ascent.

President Richard Nixon decided to take the opportunity to witness his first Apollo launch, and as *Air Force One* approached the Cape, it passed through a cold front that was producing the driving rain. The pilot radioed ahead and gave the launch team his assessment of the weather conditions aloft. While some rumored that the presence of the chief executive placed undue pressure on first-time launch director Walt Kapryan to proceed with the launch, Gordon disagrees, arguing, "That's *not* the way we did business."

With the carefully calculated liftoff time approaching, the weather continued to worsen as the front approached the Kennedy Space Center (KSC), but remained within the rules governing the launch of the Saturn V. Meanwhile, Pete Conrad cheerfully went about his work high atop the rocket, reporting on gauge readings and setting his switches according to the lengthy checklist. In the final moments of the countdown the commander reached across the cockpit and had his crew clasp hands in a brief gesture of camaraderie. *Good luck. Do a good job. Don't mess anything up that I've got to correct!*

In his post-mission report Conrad stated that "the Navy prides itself on its all-weather operations, and when Air Force Colonel Stafford told us the weather was suitable to launch, we went." It was a call that Tom Stafford would later refer to as "the worst operational decision I ever made."

The five thundering F-1 engines at the base of the 363-foot Saturn V spewed a torrent of flame into the trench beneath the launchpad, and as the hold-down clamps released the massive rocket, servicing arms on the gantry swung aside to clear the way for its ascent. The four outboard motors of the first stage gimbaled to one side, guiding nearly seven million pounds of booster and spacecraft away from the launch tower.

Once the rocket climbed above the looming rust-colored structure, control of the mission switched to Houston, and a very confident Pete Conrad reported, "I got a pitch and a roll program, and this baby's *really* goin'." Barely eight seconds later, thoroughly pleased with the Saturn's impressive ride, he exclaimed, "That's a lovely liftoff, that's not bad at all!" The blinding fire of the mighty first-stage engines quickly disappeared into the gray overcast, trailing a long, ionized column of superheated air and smoke all the way down to the steel launchpad.

Thirty-six seconds after leaving Pad 39-A, with the powerful rocket pushing them ever faster, Conrad saw an abrupt flash of white through his uncovered window. He also *felt* something. A burst of static on the radio was immediately followed by a blaring master alarm and a litany of warning lights in front of Gordon's face. "What the hell was that?" Conrad exclaimed. Moments later he added, "I lost a whole bunch of stuff; I don't know . . ."

Bean had no idea what had happened, reporting postflight, "My first thought was that we might have aborted, but I didn't feel any g's, so I didn't think that was what had happened. My second thought was that somehow the electrical connection between the Command Module and the Service Module had separated, because all three fuel cells had plopped off."

Twenty seconds after the first flash and over a half-mile high, Conrad saw another, less intense flicker of bright white. His eyes went straight to the "eight ball" attitude indicator, and he watched it roll over and start tumbling. He glanced over at the caution and warning panel and found it "a sight to behold" but steadily read off the list of malfunctions to CapCom Jerry Carr. "Okay," he reported, "we just lost the platform, gang. I don't know what happened here; we had everything in the world drop out . . . I got three fuel cell lights, an AC bus light, a fuel cell disconnect, AC bus overload 1 and 2, Main Bus A and B out." His voice was strained but had an even cadence to it. This was where operational test flight experience paid off.

Bean was busily working the problem from his side of the spacecraft, trying to ascertain why the electricity-producing fuel cells had dropped offline. To his surprise he found that the main buses, bridges of electrical connections throughout the ship, still had some current flowing through them, but at twenty-four volts it was way too low. Gordon kept a close eye on what Bean was doing. They had trained endlessly to handle almost any type of emergency and had a routine worked out wherein Gordon called out failures and Bean took the action to move the switches but only after telling him what he was doing. What they were confronted with now, however, defied all logic. Pete Conrad had been right—they had never seen anything like this before. Gordon remembers thinking, "I don't know what the hell to do."

John Aaron needed data, anything that would give him some idea of what was going on with the spacecraft. What he had in front of him now at the EECOM console was a jumble of electrical readouts that made virtually no sense at all. The Saturn V appeared to be continuing on its trajectory skyward, unaffected by whatever had jolted every system of the Command Module. Yet something about the pattern of numbers he was looking at seemed familiar. He *had* seen this somewhere, months earlier, during that botched test at the Cape. *Oh, I got it!*

As Flight Director Gerry Griffin pressed him for a solution, Aaron delivered a terse reply in the typical understated jargon of Mission Control. "Flight, EECOM. Try SCE [Signal Conditioning Equipment] to Aux." Griffin had no idea what that was, and neither did Carr, who relayed the command up to Conrad. Even the commander took several tries to get the acronym straight, replying, "Try NCE to auxiliary." Conrad unkeyed the mike and then wondered aloud to his crewmates, "What the hell is *that*?"

Several long seconds passed, and Aaron still had no telemetry from *Yankee Clipper*. The flight director grew impatient. The rocket was now almost twenty miles high, and no one wanted to end up with an electrically dead spacecraft in orbit around the Earth, with no way to return. "What panel EECOM?" Griffin asked, intending to point the crew to the exact location of this obscure switch. No one, including the astronauts, seemed to know anything about its function, except for his electrical systems expert seated a few feet away.

Meanwhile, Gordon was pushing Bean to get power back to the spacecraft. "Try the buses," he urged. "Get the buses back on line!" Conrad looked across at him, catching the glow of bright orange warning lights reflecting in Gordon's fishbowl helmet, and finally managed to get out, "*S-C-E* to Aux." Bean strained against the mounting g-forces as he reached for the switch in the lower left corner of his panel. "I got to it . . . I knew where it was," he would later explain. "It was a switch in front of me where I sat. Dick didn't fool with those switches that much, and I didn't fool with his or Pete's. That wasn't his switch; he would never have reached over and fooled with it."

John Aaron's screen instantly flowed with the electronic vital signs of the CM. The SCE was now doing its job in the backup mode, and he saw that the fuel cells were alive and well but had somehow been isolated from the buses. The twenty-four volts of electrical power Bean had noticed were being fed from the CM's reentry batteries, but they needed to be saved for the end of the mission. Aaron knew he needed the power of those fuel cells back on line and soon. The center engine of the cluster at the base of the rocket shut down right on time in order to reduce the loads on the vehicle just prior to separation of the first stage.

Jerry Carr radioed up to the crew, "*Apollo 12*, Houston, try to reset your fuel cells now." Bean repeated the words and was almost ready to do it when Gordon stopped him, saying, "Wait for staging." Conrad rapidly echoed his words. They had both ridden rockets before, but previous Apollo crews had told them that the Saturn was much more spectacular at staging than the Titan had been. They knew it was no time to be reaching for switches trying to repair critical systems.

As the spent first stage fell away, the three men were thrown forward toward the instrument panel by a teeth-jarring ripple up the booster then slammed back into their couches as the five J-2 engines of the second stage came up to thrust. "Got a good S-II, gang," Conrad reported. "Now we'll straighten out our problems here. I don't know what happened; I'm not sure we didn't get hit by lightning." It would turn out he was right on the money. Bean got to work reconnecting the fuel cells to the electrical system, but his own experience suggested that he take it a little slower than Mission Control might like, as he explained in a 1998 interview:

Well, I didn't much want to reset the fuel cells, because I didn't think three fuel cells had failed, and so you don't want to put power on a short. That's a dumb idea, get a fire going. At the same time, you need power, and the primary rule of flight that a lot of people don't know is, if you're going the right direction, don't do too much, because at least you're going the right direction. [It's] only a myth: test pilots make snap judgments. Not the ones still around. So I'm saying to myself, "That's a good idea for them, but I think I'll think about this a minute." Okay, I'll reset one of these, and if it trips off again, then I'll reset another one on the other bus to see if I can get something going." But I wasn't in any big hurry, because we were headed up to orbit, and I didn't want to screw that up by messing around over here. So I tried one and it stayed on. So I said, "Wow." I checked the amps and volts. It worked good. I put on another one. It did the same thing; it worked great. I put on the third one. Each time I was waiting for something to go "Beep!" you know, and everything go off again. It never did.

As Bean performed the delicate procedure, Pete Conrad radioed down to Houston, "Think we need to do a little more all-weather testing!" They were beginning to get a handle on things, and Carr could almost see the grin on Conrad's face just by the relaxed tone of his voice.

With the power producing fuel cells back online, the crew began reflecting on their brush with disaster. Even though they were still racing into orbit aboard a powerful rocket, there were brief periods with nothing to do but enjoy the ride. Conrad started to laugh, and soon the sound rose to his familiar high-pitched cackle as he and his crew shared their disbelief at what had just happened.

Now that the ship had power and he could see that it was functioning well, John Aaron breathed a sigh of relief. The inertial platform that would guide the ship through space would have to be realigned, but that was out of his hands. *Apollo 12* slipped precisely into orbit around the Earth, having survived the massive discharge of static electrical energy as it climbed through the rain-laden clouds over the Cape. Gordon and Bean immediately went to work on the platform, as *Yankee Clipper*'s systems were scrutinized for damage by Aaron and the teams of controllers on the ground. Incredibly, none was found, and after only one circuit of the home planet, *Apollo 12* and its three navy crewmen were headed for the moon.

Pete Conrad often joked that his CMP was "the driver of the rig." Now gliding on a trajectory away from Earth, Gordon's major task was to dock his craft to the Lunar Module *Intrepid*, a responsibility that gave him his first experience in the commander's seat: "Once we got into orbit, the Command Module was mine to fly. So I did the transposition and docking with the Lunar Module. I was the active participant in the docking. And I occupied the left hand seat when I wasn't down in the lower equipment bay doing star sightings or whatever. So I was the bus driver, yeah." Gordon separated smoothly from the spent third stage of the Saturn V. By employing a display intended for use during reentry, he attempted to gauge their velocity as they slipped away from the booster. In his mind it was an exercise in overcomplicating a simple maneuver. The display had such a lag in it that it caused him to use more fuel than planned, and he felt that he could have been much more precise by using his wristwatch to time his distance from the stack.

Nevertheless, Al Bean complemented his crewmate's performance as he delicately pitched around and guided the nose of *Yankee Clipper* into the docking drogue of *Intrepid*. "This Dick Gordon's smooth as silk!" he radioed to Houston. With a volley of snaps from the docking latches, the two craft became firmly mated together, but the test pilot in Gordon was immediately concerned with how precise he had really been. "How much fuel did I waste during that docking?" he asked CapCom Jerry Carr. When Carr replied that he had used seventy pounds, Gordon replied dejectedly, "That's too much . . . too much."

Neil Armstrong and Buzz Aldrin had monitored the launch from the viewing area of Mission Control. They now hovered around the CapCom console, offering advice and relating experiences from their July flight for Carr to pass up to the crew. Glancing up at the big screen at the front of the room, they suddenly found it filled with the smiling face of Dick Gordon bobbing weightlessly in the commander's couch. In his space suit, "Snoopy" cap, and sunglasses, he looked somewhat like a space age barnstormer, missing only one critical piece of equipment.

"Hey, Red Baron," Carr jokingly inquired. "Where's your scarf?"

"I think I forgot it during that boost phase," a smiling Gordon responded, adding, "That's a terrible way to break Al Bean into space flight, I'll tell you!"

Watching the Earth recede in the window, the astronauts noticed its curvature become increasingly distinct, until each end of the horizon framed in the window wrapped completely around and met the other, forming a brilliant blue and white ball. Gordon could even make out a sweeping panorama of clouds associated with the weather system that had wreaked so much havoc during their launch.

As the day finally wound down and the crew got ready for some sleep, Bean wished that he could go back and experience another liftoff, one he could enjoy instead of spending it trying to piece the electrical system back together. But now they were safely on their way, and they laughed about it some more as they prepared to bed down and get some much-needed rest, their spacecraft closing in on the moon. Just as Gordon drifted into sleep, he winced at the flashes of orange and red behind his closed eyes. There they were again—*all those damned lights*!

Oceanus Procellarum—the Ocean of Storms—is a vast lunar mare, situated on the western side of the moon as seen from the Earth. There, on the eastern slope of a shallow crater, *Surveyor 3* had rested lifelessly since May 1967, having successfully completed two weeks of lunar photography and sampling. Named for its robotic visitor, Surveyor Crater was part of a cluster of depressions that, in previous orbital photographs, vaguely resembled a snowman in the gray, pockmarked landscape. As the moon traveled around the blue planet, the baking sunlight of the day and the intense cold of the darkness took a toll on the fragile little spacecraft. Brittle, discolored, and cracked, it waited out its days in this hostile environment for the inevitable moment when it would crumble into a pile of rubble.

But this lunar day would be different, as a brilliant spear of blue flame appeared in the eastern sky, cutting through the inky blackness with the intensity of a welder's torch. It slowly descended toward the Surveyor, revealing an insect-like form silhouetted against the rising sun. As it neared the north rim of the crater, the sparkling gold and silver machine swept around in a left-hand circuit of the bowl then hauled back at an ungainly angle in an effort to stop its forward motion. Hovering there for a moment in the vacuum-induced silence, the spacecraft's translucent flame touched the ground and began spraying the loose regolith in every direction, pelting *Surveyor 3* with a fine shower of pulverized dust.

Four dangling legs reached out for a level plateau among the field of craters, and small rockets sporadically flickered in short bursts of balancing thrust. As the long probes of the footpads settled into the veil of flying rocket-blown debris, the blue flame suddenly vanished, and the radiating sheets of rock and moondust fell slowly to the surface of this airless world.

"Good landing, Pete! Outstanding, man!" exclaimed Bean as he and Conrad ran through the post-landing checklist. Pete radioed to Houston his impressions of the blinding dust that had kicked up as he maneuvered for touchdown. "I think we're in a place that's a lot dustier than Neil's. It's a good thing we had a simulator, because that was an IFR landing," he said, referring to the instrument-only method of flying he had used countless times to land airplanes in poor weather back home. Minutes later they heard from Dick Gordon as he passed high overhead the landing site.

"*Intrepid*. Congratulations from *Yankee Clipper*," said an elated CMP.

"Thank you, sir," responded Conrad. "We'll see you in thirty-two hours."

"Okay," said Gordon. "Have a ball."

The two moon landers took a few moments to survey their surroundings, amazed by the desolate sunlit field of craters, but were also immediately taken by the comic effects of the low lunar gravity upon seemingly simple tasks. Conrad attempted to pass a checklist book to Bean, but he missed the handoff completely, and the book went sailing across the cabin, bouncing off the adjacent circuit breaker panel. Accidentally knocking a pencil off the small work desk sent Bean on a search of the floor, only to find it had arced slowly to the back of the cockpit, rather than straight down. "Man, this is going to be fun!" he thought.

As his crewmates prepared for the first of their lunar EVAs, Gordon continued his own observations from orbit and even managed to spot the tiny Lunar Module among the craters of Snowman.

I could see the crater in my optics. I could identify the crater and I saw a source of light, reflected light on the rim of that crater. And I didn't see the shape of the LM or anything like that. But I knew that they were going to land on that crater and I knew the only thing that was going to look like that, my own interpretation was that had to be the Lunar Module. Well I stuck my neck out to about here and said, "I see the Lunar Module." And it was but it was just a source of reflected light off of the Lunar Module that I saw.

As he told Mission Control, one other feature pointed to his two buddies and the flying machine they had just landed on the moon—the shadow of the LM. "It looks like it is about . . . a third of a crater in diameter."

Pete Conrad bounded down the ladder without wasting a moment. As he took a long, slow leap from the last rung, he blurted out an unrestrained "Whoopee! Man, that may have been a small one for Neil, but that's a long one for me!" No one who knew in advance should have ever doubted that he would, indeed, say it. It has since been told that the diminutive Conrad had a five hundred–dollar wager riding on his willingness to be so publicly self-deprecating, but it was a bet that he never did collect on.

In fact Conrad had been so excited to get out on the surface that he missed a critical step on the checklist. Thinking his life support system fan was malfunctioning, he had Bean check his connections only to find that he had forgotten to attach the umbilical hoses from his backpack to his suit. Now at the footpad of *Intrepid*, Conrad peered around the side of the LM into the blinding sunlight, and the gleeful astronaut found his target.

"Boy, you'll never believe it," he exclaimed. "Guess what I see sitting on the side of the crater!"

"The old Surveyor, right?" replied Bean.

"The old Surveyor. Yes, sir," came the excited confirmation. "Does that look neat! It can't be any further than 600 feet from here. How *about* that?"

As Bean prepared to join his commander on the surface, Conrad bounced around the strange terrain, spotting beads of heat-fused glass all over the surrounding craters. He happily hummed a tune as he went about collecting samples. *Dum dee dum dum dum.* Bean then made his way down the ladder and immediately got to work, trying to stay on the tightly scheduled timeline.

Bang! Bang! Bang! Bean rapped on the top of the color TV camera with a tool intended for breaking rocks. "I hit it on the top with my hammer. I figured we didn't have a thing to lose," he later radioed to Houston. The camera had suddenly failed as he carried it out away from the LM and accidentally pointed it directly at the sun, irreparably frying its video tube. That would be the end of live television from the Ocean of Storms. He was extremely disappointed, but time was an issue. They had to get moving, and

13. With LM *Intrepid* perched behind him, Pete Conrad makes contact with the *Surveyor 3* probe. Courtesy NASA.

he tiptoed lightly over to Conrad, who was having troubles of his own setting up the flag.

Bean snapped a few photos of his mission commander holding up the telescoping pole designed to hold the flag outstretched. The pole refused to latch in the extended position, but again, there just was not time to deal with such a relatively insignificant problem. Bean was so engrossed in keeping to his cuff-mounted checklist and wristwatch that he forgot to have Conrad take a photo of him with Old Glory.

The pair busily set about getting their Apollo Lunar Surface Experiment Package (ALSEP) deployed, taking free moments to amuse themselves by throwing packing material and sheets of protective Mylar for incredibly long distances in the absence of air and reduced gravity. Conrad chuckled and sang, but when he attempted to whistle, he found it impossible in the low pressure of the suit, blowing helplessly into his microphone.

As Conrad flipped a page of his cuff checklist over, he suddenly giggled out loud at the sight of a voluptuous Playboy Playmate within the spiral binding, with a caption reading, "Seen any interesting hills and valleys?" Bean found that his checklist was similarly endowed with the photos. The backup crew had covertly carried on the astronaut tradition of the "gotcha" to the delight of the two moon walkers.

Before they knew it, it was all over. Bean had to hustle to get a core tube sample and took one last try at repairing the TV camera, leaving the rock-laden Conrad waiting at the base of *Intrepid*'s ladder.

"Kinda feel like the guy in the shopping center waiting for his wife," chided Conrad. "I'm standing here holding two bags, buddy."

"I'm coming! Coming!" Bean responded. Pete started to laugh, and Al could not help but join in.

"*Clipper*, you were sort of a forgotten man for a little while . . . all eyes are on you now. We're with you," CapCom Ed Gibson radioed to the CM. Gordon found the view of the lunar surface from sixty miles up as "just an awesome sight. . . . It was the first look at the moon from up close and personal." With his extensive training in lunar geography, it seemed like an almost familiar scene—"other than the fact that we hadn't seen it before," he remembers.

Gordon had been awake for nearly nineteen hours but still had a major engine burn to perform. As *Intrepid* rested on the surface, the moon rotated underneath *Clipper*'s orbit, in effect carrying it farther and farther away with each pass. The burn would change the plane of *Clipper*'s orbit with respect to the moon's equator, shifting it back to a path that would intercept *Intrepid*'s as it blasted from the surface the following day. Gordon ran through the now familiar checklist in preparation for the maneuver and executed it with his typical precision. Once he got the cabin cleaned up and had a bite to eat, the exhausted pilot turned down all the lights, covering the bright red master alarm light with a cue card, and strapped himself into the commander's couch for the night.

Meanwhile, on board *Intrepid* Al Bean lay awake listening to the distracting sounds of the ship. The steady hum of machinery was interrupted by creaks and pops of the thin hull and the occasional change in pitch of one pump or another. The tiny LM was not a comfortable place to sleep, even in one-sixth gravity. Lightly hanging in one of two hammocks strung across the cabin, he rubbed some of the greasy, graphite-like lunar soil from his space suit between his fingers and breathed in the gunpowder-like scent. He thought about where he was and all the things that had to go right in order for them to get home as he finally dozed off. After about four hours of drifting dreamlessly in and out of sleep, Pete finally woke Al to help loosen one leg of his space suit, which had been adjusted too short before the flight. Then it was back to work.

"Right now, this stuff—," Pete stopped mid-sentence, then exaggerated his correction. "This *ma-ter-i-al* around the spacecraft reminds me . . . looking into the Sun, [it's] a very rich brown color . . . kind of reminds me of a good plowed field." The irrepressible Conrad was ribbing the geologists once again, only this time from 240,000 miles away. As Bean gathered up the TV camera for engineers to evaluate back home, Conrad took a moment to roll a grapefruit-sized rock down into Head Crater for the seismologists to monitor on their instruments, delighting in watching it slowly bounce down the crater wall. Bean then joined him as they grabbed rocks, shoveled, trenched, and photographed as much as they could in the limited time they had at each stop.

Pete kept pushing Al to get moving, as they still had to lift off and rendezvous with *Yankee Clipper* before the day was out. Only during the brief traverses between craters did Bean have a chance to look up and admire the Earth, hanging in the shiny blackness of space. Finally arriving at Surveyor Crater, they stepped gingerly around its sloping walls and down to *Surveyor 3*, to grab some parts for study back on Earth. And the duo had one other plan in mind . . .

On a page of their cuff checklists was the cryptic notation "Perform D.P." D.P. was an abbreviation for "dual photo." Bean had repeatedly practiced with a small Hasselblad camera timer in hopes of getting a photo of him and Conrad together in front of *Surveyor 3*. He had stuffed the timer in the leg pocket of his suit before launch, but now, just below the rim of the crater and filthy with moondust, he could not find it anywhere. After several frustrating moments he had to quit looking and move on.

Hauling their load of rocks, soil, and *Surveyor* parts back to *Intrepid*, Houston told them that the time had come to leave. Packing the rock boxes up at the base of the ladder, Conrad radioed Bean: "Get the rocks over here. Come on. We can't baloney all day. We got to get out of here." Emptying out one of the sample bags, Bean found the timer sitting neatly atop a pile of rocks. With every minute now precious, he tossed it over his shoulder without a thought of the great photo they could have then taken, shaking hands together in front of *Intrepid*.

Brushing off Conrad's blackened space suit, Bean's gloved hand swept across the crew patch printed on his chest. It was a beautiful emblem, picturing a great clipper ship sailing from the Earth to the moon on a wake of stars. In the patch's black background, above the moon, were four individual stars—one for each of them and one for C.C. Williams, who gave his life in the program, opening the door for Al to be there on the moon with Pete. Before they left Earth, Conrad and Bean had decided that they could honor Williams's memory in another subtle way. Al pulled the metallic gold navy wings of their fellow astronaut from a pocket on his suit and let them fall slowly into the lunar dust.

The single hypergolic-fueled engine of the ascent stage had performed perfectly. Now orbiting in a chase with Dick Gordon and *Yankee Clipper*, Conrad and Bean, filthy as two coal-blackened miners, slipped once again into the shadow of the moon. A great cloud of debris surrounded *Intrepid*, sporadically illuminated by the flashing rendezvous strobe light, and the cabin air was full of weightless moondust. The cockpit was stuffed with rock boxes, Surveyor parts, the broken TV camera, and several of their lunar tools, including a hammer and rock tongs, which had accidentally made their way aboard in the pair's haste to climb the ladder. Al was busily working his backup rendezvous charts and the computer when Pete interrupted him with an unforgettable offer, as Bean related to the author:

He said to me, "Bean, you're working too hard . . . why don't you just quit doing the backup charts and look out the window?" All of the sudden he says, "How would you like to fly this thing?" And I said, "Well [chuckle] *okay." He says "You've got it," and I said "I've got it." I say to him, "I don't think the people at Mission Control are gonna like this" because every time we did something that they weren't aware [of] they'd call us up and ask us about it. He says, "We're on the back side . . . they'll never know!" That was the difference between Pete and other astronauts . . . he went out of his way to have me enjoy the flight.*

Bean was immediately impressed by the crisp feel of *Intrepid*'s ascent stage and was amused to find that when he maneuvered the tiny ship, it rotated *around* him until a bulkhead or panel bumped his arm, allowing his weightless body to catch up. Conrad used the computer to measure each change in velocity that Bean put in, and when he was finished, Pete canceled them out completely so no one would be the wiser. "No other commander ever let his Lunar Module pilot do that," Bean remembers today.

The reunion of the three navy men was a joyous one, following a flawless docking by Dick Gordon. Now heading back to Earth, the astronauts were in a relaxed and reflective mood. Conrad had presented his crewmates with white beta cloth baseball caps early in the mission, adorned with navy wings, "scrambled egg" visors, and topped with a single spinning propeller, which they proudly displayed on television during an in-flight press conference.

Bean found that filling the time on the way home was tough and wished he had brought a book along. "The moon—you'd been looking at it for two days and you didn't want to look at it again," he reported after the flight. Conrad's only apparent regret was that they "should have come home in two days instead of three."

One more stunning event awaited them. It was calculated before the flight that *Clipper* would pass through the Earth's shadow on the return trip, offering the crew the first opportunity to witness an eclipse of the sun by the home planet. Unfortunately, no one had expended a lot of thought on how to photograph the phenomena. Moreover, they had used up virtually all of their color Hasselblad film magazines. Gordon dimmed the lights to afford a better view, and looking through a dark eyepiece placed against the window, he was amazed by the scene. "We're getting a spectacular view at eclipse," he reported. "We're using the Sun filter for the G&N optics, looking through, and it's unbelievable."

The sun took on a diamond ring effect as it set behind the planet, illuminating the atmosphere from behind in an increasingly larger arc until the bright band of air encircled an empty black disc in the center. Then something totally unexpected occurred as the luminous ring of light burst into a kaleidoscope of colors, only rather than being banded like terrestrial rainbows, it was segmented around its circumference in shades of blues, greens, pinks, and reds. The planet Venus hung like a single bright jewel in the blackness of space just below the huge fiery ring of color that filled the hatch porthole. With the sun now hidden, their eyes slowly adapted to the darkness, revealing an ominous sphere cast in bluish-white moonlight within the thin circle of dazzling hues. Tiny white sparkles of lightning hung around the equator of the Earth like a necklace, and a diffuse white glow of the full moon behind them reflected off the ocean far below. Bean recalls seeing "these little flashes down on the Earth," adding, "I think it was Dick Gordon who said those look like the flashes you see of thunderstorms when you fly over them on a T-38."

Conrad tried to describe the scene to those in Mission Control who did not have the privilege of seeing it firsthand. "Houston, *12* . . . we're better night-adapted now, and by golly, we can see India, and we can see the Red Sea, and we can see the Indian Ocean quite clearly," he reported. "We can see Burma and the clouds going around the coastline of Burma, and we can see Africa and the Gulf of Aqaba; it looks like the same photograph Dick and I took on [*Gemini*] *11*."

Just three and a half hours before reentry, and still struggling to get some 16-mm film of the event, they had to turn their attention to one last small correction burn as the blinding sun reappeared and washed away the incredible view of Earth. Dick Gordon was now left with one final responsibility as the Command Module pilot—to bring *Yankee Clipper* home.

Traditionally, reentry was one of the phases of an Apollo mission in which the commander of the flight gave up his left-hand seat for the CMP to pilot the spacecraft. Pete Conrad had such a bond of trust with Gordon that he allowed him to call virtually all of the shots, preferring to give him the experience of command while he monitored and observed. Gordon remembers that his commander and friend "expected me to do that and I expected him to let me do that. It went without saying . . . it was not spoken. That typified our relationship."

As *Yankee Clipper* plunged toward Earth, Gordon was focused on guiding the ship to his own pinpoint landing, only this one was on a patch of

South Pacific Ocean near the waiting aircraft carrier *Hornet*. Gordon recalls the critical nature of his task: "The consequences of being more than a half degree off in either direction are not very wholesome. Talk about structural failures and skipping back out." His crewmates, however, had the luxury of a few brief seconds to look out the windows and marvel at the incredible speed at which they were traveling as they streaked across the upper reaches of the atmosphere.

"It's coming pretty fast," reported Conrad. "We is flat smoking the biscuit . . . Whooee!"

Bean was equally descriptive. "Okay. Gee, we look low. Boy, we are hauling ass!"

"You're going to slow down in just a minute," Gordon reminded his crewmates.

It was not a gentle buildup of heat and deceleration forces. At 400,000 feet over the ocean, *Clipper* slammed into the atmosphere and was immediately swallowed by fire. The friction of the spacecraft hitting the wall of air at over 25,000 miles per hour generated a blazing hot trail of ionized plasma in the process of slowing the ship down. Dick Gordon describes what a distractingly awesome sight it was:

The ablative material comes off as the heat builds . . . and because the spacecraft is maneuvering for entry, it's kind of a corkscrew out there behind you. And the material is burning off at different temperatures and there [are] yellows and reds and greens and purples and they're all mixed up. So it's like, "Wow, look at that!" But then . . . your head's inside monitoring the performance because you're trying to navigate back to where the ship is. You have a specific point in the Pacific Ocean that you're supposed to land and you want to land there as close as you can just to get the hell out of it.

As the g load increased to over six times the force of Earth's gravity, drops of water that had condensed on the walls of the tunnel above the crew began raining down on Pete. "Here comes the water," he laughed with surprise. "I'm getting the water all over me . . . Man, am I getting the water . . . I'm getting soaking wet!"

Gordon kept an intent eye on the Entry Monitoring System (EMS) guiding *Clipper* home; with his hand gently wrapped around the control stick, he quietly talked to his ship, urging it to keep them on course for the aim point near the carrier. *Come on, mama. Hold the lift vector down.* As they slowed to a speed at which they would not skip off the atmosphere, the CM rotated 180 degrees to lift itself slightly, shallowing its descent rate and reducing some of the crushing g load on the crew.

Flying up over the top of the huge arc, the ship turned its lift vector earthward again and plowed back into the thickening air. Conrad called out the required bank angles to Gordon as he continued to watch the EMS. With *Clipper* now firmly in the grasp of the home planet, Gordon concentrated his efforts on navigating through the air in an accurate path toward the splashdown zone. The g-forces decreased as they slowed, and the sheath of flame around them diminished until the ship was essentially a smoking meteor, free-falling to the deep blue water below.

As the three orange-and-white main parachutes blossomed above them, Gordon was ecstatic, calling out to Mission Control and the recovery forces: "There they go! Hello, Houston; *Apollo 12*. Three gorgeous, beautiful chutes . . . and we're at 8,000 feet on the way down in great shape!"

At some point during the descent, Al Bean was supposed to remove a 16-mm movie camera from a bracket in the window that had been used to photograph the reentry. But everything was going so well, and they were so happy to be arriving home safely, that somehow Bean missed this task, as he related to the author: "We're coming down in the chutes and everything is going perfectly, and I'm supposed to take the camera, remove it and sort of turn around and put it in the bag behind my head. Well by then we're starting to pull g's, so I'm starting to get heavy and for some reason, I decide . . . don't ask me why . . . that it's okay to leave it there. Didn't seem like a big deal to me."

Bean's most immediate job following splashdown was to throw the circuit breaker switches that inflated three airbags after Conrad had initiated cutting the three main parachutes loose. This procedure would prevent the spacecraft from flipping over with the pointed nose of the docking ring underwater, leaving the crew hanging in their couches, a position referred to as "Stable Two." Things did not go exactly according to plan, as Conrad reported postflight:

We really hit flatter than a pancake, and it was a tremendous impact, much greater than anything I'd experienced in Gemini. The 16-mm camera, which was on the bracket . . . whistled off and clanked Al on the head to the tune of six stitches. It cold-cocked him, which is why we were in Stable 2. Although he doesn't realize it, he was out to lunch for about five seconds. Dick was hollering for him to punch in the breakers, and in the meantime, I'd seen this thing whistle off out of the corner of my eye and he was blankly staring at the instrument panel. I was convinced he was dead over there in the right seat, but he wasn't, and finally got the breakers in.

Al Bean's recollection of the splashdown is understandably different:

I remember hitting the water and the impact . . . when I wake up, Pete and Dick are looking over at me, leaning forward, telling me to punch in the circuit breakers. So I'm saying to them, "What's the hurry? We just hit the water!" So I lean up, I punch the circuit breakers; they throw the thing but we go over into Stable Two. Well I don't know why we went to Stable Two—I just guess we did 'cause I don't know anything [was] wrong. And then, either Pete or Dick look over and they see that I'm bleeding. They then begin to realize that I was hit in the head . . . that's why I didn't punch in the circuit breakers because I was unconscious at that time.

From almost 240,000 miles out in space, Dick Gordon had parked *Yankee Clipper* just two and a half miles from the carrier *Hornet.*

Safely aboard the recovery ship, Conrad and Gordon relaxed with a few martinis and the first cigarettes they had smoked in over ten days. Confined in the Mobile Quarantine Facility—a converted Airstream trailer—the pair snickered as they watched over Al Bean lying on a cold steel table in his fresh blue flight suit, his head being tended to by a physician assigned to care for them while they were locked up awaiting quarantine clearance from any invasive "moon bugs." Bean remembers the scene fondly: "Of course now the doctor's happy 'cause he's got something to do. So they're laughing it up . . . I'm lying there and they're looking at me, and the doc said, 'I've got to sterilize this cut,' and Pete and Dick say, 'Well let's use some of this alcohol!' So these guys are pouring their martinis on my head . . . it's just fun . . . one of the fun memories I've got of the mission. The doctor probably got upset."

Hazel Sekac Banks recalls that "there was always a party after the flight, and everybody was always trying to outdo the last party." One afternoon at the Cape, as she was busy handling all of the arrangements for the astronauts who were in town, *Apollo 12*'s backup crew stopped by her office. They were putting together a gag film for the prime crew's postflight bash, and they needed a pretty young lady for a supporting role in their epic movie. Hazel recalls, "As far as the storyline . . . they dreamed up all of that." The idea, she said, was that she would be "the stowage equipment." When she asked what she should wear for her role, one of the astronauts replied, "Well, wear your bikini!" She had to laugh at the thought of that. "Oh, okay!" she responded.

The melodramatically narrated film was a big hit in Houston when the Astronaut Office later got together to roast the crew of *Apollo 12*. It portrayed the pair of bumbling, giggling navy pilots tripping over rocks, beating on the television camera, and clumsily breaking several of the lunar experiments while earthbound long-haired scientists shook their heads at the incredible waste of resources on the moon. When the scene switched to the CM, Jim Irwin played the role of a lonely Dick Gordon in a space suit and gorilla mask. Suddenly, the scantily clad secretary from the Cape appeared from the lower equipment bay in her polka-dot bikini and joined Irwin on the center couch. The room rolled with laughter as the movie ended, and a huge round of applause greeted Al Bean as he was awarded the gold pin of an astronaut who had now flown in space. A few weeks earlier he had respectfully thrown his old silver astronaut pin as far as he could into some unnamed crater on the Ocean of Storms.

Al Bean never lost the timer for the Hasselblad. He pulled it out of the leg pocket of his space suit just like he had planned, bounced over to the hand tool carrier, and attached it to the camera that Pete had just finished setting up in the soft lunar soil. They both turned back around toward the Lunar Module *Intrepid*, where another astronaut waited impatiently. The nametag on the chest pack of his dirty gray space suit read, "R. Gordon."

"Let's get this over with . . . we've got to get to work!" Dick said in the clipped tone of his communication cap. Pete stood alongside his dust-covered crewmate, as Al hurriedly set the timer and joined them. Gordon, ever the showman, spread his arms out wide as if to say, "Look at me! I made it!" Bean could not help himself. Just before the timer wound down, he held two fingers up behind Dick's helmet in a bunny ear salute. *Click.*

One of the great things about being an artist, Al Bean reasoned, is that you can make anything you want become a reality. In the final days of his astronaut career he was already thinking of how much fun it would be someday

14. In the summer of 1999 the crew of *Apollo 12* visited Penwal Industries of Rancho Cucamonga, California. Here seen with owner Chris Pennington and his wife, Debra, next to a replica of the F-1 engine built by Penwal as part of a full-scale Saturn V rocket constructed for the U.S. Space and Rocket Center in Huntsville, Alabama. It would be the last time the three men were together. Pete Conrad died from injuries sustained in a motorcycle accident on 8 July 1999. Courtesy Chris and Debra Pennington.

to produce a painting that depicted Dick Gordon making it to the moon's surface. Many years and many paintings later, as he was putting the finishing touches on *Conrad, Gordon and Bean: The Fantasy*, Al invited Pete and Dick to his home in Houston. More than two decades had passed since they sailed to the moon together, but the three retired navy captains still remained the closest of all the Apollo crews, never going more than a few weeks without talking to each other about their latest adventures.

Bean had worked for months on this painting, scarring the plaster medium with the old geology hammer he had once used on (and inadvertently

15. Al Bean, astronaut, moon walker, and artist. Holding the hammer he used on *Apollo 12*, he was just finishing his fine work *Conrad, Gordon and Bean: The Fantasy* at his Houston home in the summer of 1992. Courtesy John Youskauskas.

rescued from) the Ocean of Storms, mixing moondust-stained pieces of his suit patches and sprinkling bits of *Clipper*'s heat shield into the wet acrylic paint, perfecting every detail of the three space suits and *Intrepid*'s wrinkled black and gold descent stage. It would become his signature work of art.

The flight of *Apollo 12* might have been considered by some to be the ultimate anticlimax. The lack of television coverage due to the failed camera likely contributed to this perception, but for the crew it was undoubtedly the adventure of a lifetime. With the goal of a moon landing already accomplished by *Apollo 11*, they had the mixed fortune of flying just after the mission every astronaut dreamed of and just before the one no astronaut would ever have wanted to fly.

Dick Gordon had held out for a command slot in Apollo, while his *12* crewmates left for the Skylab program. He was assigned as backup to Dave Scott for *Apollo 15*, which had him pointed at command of *18*. Unfortunately, dwindling public interest, congressional budget cuts, and NASA management's trepidation over pushing their luck with Apollo hardware too far all conspired to end the program with *Apollo 17*. He would not get his shot in real life, but through Al Bean's painting he could, in his mind's eye, fulfill his dream of crossing that last sixty miles to the surface of the moon and walking among its dusty craters. And he did it with the two best friends he ever had in *either* world.

The first time Al Bean presented *The Fantasy* to his crewmates from that fantastic voyage of 1969, his old CMP had looked at the painting with unconcealed awe, reflecting on what great times they had had and what could have been. As he quietly ran his fingers across the heavily textured image of his space-suited figure amid the lunar craters, he suddenly paused. Dick Gordon then turned to the two "other" moon walkers, smiled, and said softly, "Wow, that's neat . . . *really* neat." It was all he could say. It was all he *needed* to say.

4. *Apollo 13,* We Have a Solution

Stephen Cass

> Nothing is improbable until it moves into past tense.
>
> George Ade (1866–1944)

"Houston, we've had a problem." On 13 April 1970 these words marked the start of a crisis that nearly killed three astronauts in outer space. In the four days that followed, the world was transfixed as the crew of *Apollo 13*—Jim Lovell, Jack Swigert, and Fred Haise—fought cold, fatigue, and uncertainty to bring their crippled spacecraft home.

But the crew had an angel on their shoulders—in fact thousands of them—in the form of the flight controllers of NASA's Mission Control and supporting engineers scattered across the United States. To the outsider it looked like a stream of engineering miracles was being pulled out of some magician's hat as Mission Control identified, diagnosed, and somehow remedied life-threatening problem after life-threatening problem on the long road back to Earth.

From the navigation of a badly damaged spacecraft to impending carbon dioxide poisoning, NASA's ground team worked around the clock to give the *Apollo 13* astronauts a fighting chance. But what was going on behind the doors of the Manned Spacecraft Center in Houston—now the Lyndon B. Johnson Space Center—was not a trick or even a case of engineers on an incredible lucky streak. It was the manifestation of years of training, teamwork, discipline, and foresight that to this day serves as a perfect example of how to do high-risk endeavors right.

Many people are familiar with *Apollo 13*, thanks to the 1995 Ron Howard movie of the same name. But as Howard himself was quick to point out

16. *From left*, the crew of *Apollo 13*: LMP Fred Haise, CMP Jack Swigert, and CDR Jim Lovell. Haise and Lovell lost their chance to walk on the moon following a violent explosion within the Service Module. Courtesy NASA.

when the movie was released, it is a dramatization, not a documentary, and many of the elements that mark the difference between Hollywood and real life are omitted or altered. Therefore, to get the real story of how they saved the day, several key figures in Mission Control were interviewed to establish the way in which these dramatic events unfolded on the ground.

First, a little refresher on moon shot hardware: a powerful, 110.6-meter tall, three-stage Saturn V booster launched in each Apollo mission from Cape Canaveral in Florida. Atop the Saturn V rode the Apollo stack, which was composed of two spacecraft: a three-person mother ship to go to the moon and back, called the Command and Service Module, or CSM; and a two-person landing craft, called the Lunar Module, or LM, to travel between the CSM and the surface of the moon.

The two spacecraft in turn each had two parts. The CSM divided into a cylindrical Service Module (SM) and a conical Command Module (CM). The SM housed the main engine and supplied all the oxygen, electricity, and drinking water that the crew needed for the long voyage. It took about six days for a round trip between the Earth and the moon. The crew lived in the cramped CM, which also housed the flight computer and navigation equipment. The CM was the only part of the Apollo stack that was intended to come back safely to Earth. It would plummet through the atmosphere, the blunt end of its cone designed to withstand the immense heat generated by the descent, and then deploy parachutes and splash down in the ocean.

The LM consisted of an ascent stage and a descent stage. The descent stage had a powerful engine used to land the LM on the moon. After the expedition to the surface was complete, it served as a launchpad for the ascent stage. The ascent stage housed the astronauts and had a smaller engine that allowed them to blast off and rendezvous with the CSM in lunar orbit.

For most of the way to the moon the CSM and the LM—dubbed *Odyssey* and *Aquarius*, respectively, on the *Apollo 13* mission—were docked nose to nose. The astronauts could pass from one spaceship to the other, but they generally remained in the CM because the LM was turned off to preserve power.

Most of that power came from a cluster of three fuel cells in the SM. The fuel cells were fed hydrogen and oxygen from cryogenic tanks. Inside the fuel cells, the hydrogen and oxygen was combined, producing electricity and water. There were three small batteries on board the CM, but they were intended for only a few hours' use during reentry, after the SM was jettisoned close to Earth.

It was one of the cryogenic tanks that would reveal itself as the *Odyssey*'s Achilles' heel. On 13 April 1970, around 9:00 p.m. Houston time and almost fifty-six hours into *Apollo 13*'s flight, Mission Control asked the crew to turn on fans in all the cryogenic tanks. The fans stirred up the tanks' contents, allowing Mission Control to get accurate quantity readings. Due to a series of prelaunch mishaps, turning on the fan sparked a short circuit between exposed wires within oxygen tank 2, causing it to explode. But why did this happen? The crisis on board *Apollo 13* can be traced back to events that occurred months and even years before the mission.

In 1962 North American Aviation (now part of Boeing) was awarded the contract to build the Apollo CSM, which would ferry astronauts to the moon and back. North American Aviation subcontracted the design and construction of the SM's cryogenic tanks to Beech Aircraft in Boulder, Colorado, specifying that the wiring inside the tanks be rated to function at 28 volts—this is the voltage at which the tanks would be operated in space.

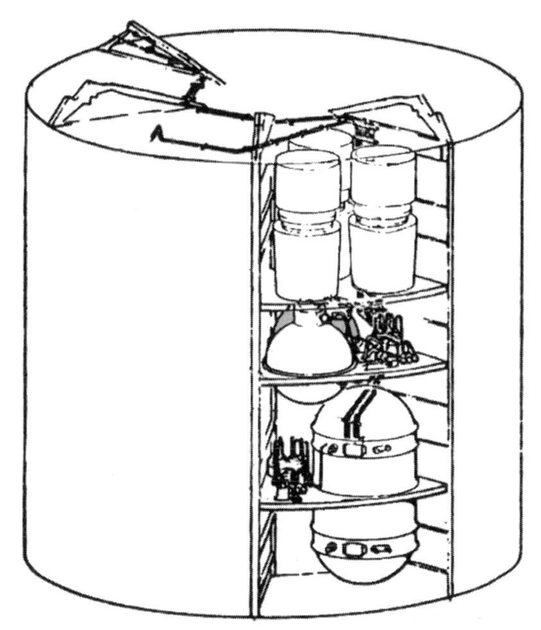

17. A cutaway diagram of the Service Module. The three cylindrical fuel cells at top provided water and electricity by combining oxygen and hydrogen stored in cryogenic tanks (*center and bottom*). The spherical Oxygen Tank 2 (*center, foreground*) exploded during the *Apollo 13* mission, almost killing the crew. Courtesy NASA.

Beech in turn subcontracted for the thermostatic safety switches used inside the cryogenic tanks, specifying a 28-volt operating environment.

Three years later NASA decided that it wanted the electrical systems of the Apollo spacecraft to be compatible with the 65 volts used in the ground support and test equipment at Cape Canaveral in Florida, in addition to the lower levels used in space. North American informed Beech Aircraft of the change, but Beech failed to inform its subcontractor, so the thermostatic safety switches were never upgraded to operate above 28 volts.

Still, the *Apollo 13* Service Module probably could have gotten away with using the underrated switches, if not for an accident that occurred in 1968. Along with its twin, oxygen tank 1, oxygen tank 2 was installed in the SM that would fly as part of the *Apollo 10* mission. Because of an unrelated problem, oxygen tanks 1 and 2 in *Apollo 10* were swapped out for another pair, and the originals went into *Apollo 13*. But during the swap oxygen tank 2 was dropped about two inches. The tank was inspected and appeared to be fine, and so it was installed in *Apollo 13*'s Service Module. During a test about a month before *Apollo 13*'s launch in 1970, the tank was filled with liquid oxygen. All went well until gas was pumped into the tank to empty it by forcing the liquid oxygen out under pressure. Oxygen tank 2 failed to empty. The official investigation into the *Apollo 13* crisis later surmised that the two-inch drop in 1968 had knocked a fitting loose inside the tank, preventing normal emptying.

Stuck with a tank full of liquid oxygen, the test engineers eventually decided to turn on the heaters built into the tank to boil off the oxygen. The heaters were connected to the 65-volt ground power supply, which—unknown to anyone—irreversibly fused the thermostatic switches shut. Instead of tripping open when temperatures in the tank reached 27°C, the switches allowed the heaters to keep running as temperatures inside the tank soared to more than 500°C. This high temperature damaged the Teflon insulation coating the wires near the heaters and left exposed wiring within the tank. Later, when the spacecraft was already on its way to the moon, it was this wiring that sparked when the tank fans were turned on. In turn the short-heated Teflon wiring insulation ignited and burned in the pure oxygen environment. The pressure and temperature rose rapidly and ruptured the tank.

The resulting explosion blew off a good portion of the SM's aluminum outer skin and severely damaged the plumbing that supplied the fuel cells. One oxygen tank was out of commission, the other was leaking, and soon the fuel cells would be starved and stop generating power.

Odyssey was dying, but no one knew it yet. Even the crew was unaware of the gravity of the situation. In the Ron Howard movie the oxygen tank 2 explosion is accompanied by a whole series of bangs and creaks, while the astronauts are suddenly jostled around. But in real life, according to *Apollo 13*'s commander, Lovell, "there was a dull but definite bang—not much of a vibration though . . . just a noise." Then *Odyssey*'s caution and warning lights lit up like a Christmas tree.

On the ground Mission Control was initially unperturbed. During a cryogenic tank stir the flight controller in charge of the fuel cells and the tanks, Sy Liebergot, had his attention focused on the oxygen and hydrogen tank readings. Liebergot was an EECOM, a job title that dated back to the Mercury program days of the early 1960s. It originally meant the person was responsible for all electrical, environmental, and communications systems on board the CSM. The communications responsibilities had recently been stripped out of the EECOM's job, but the name remained.

In an unfortunate coincidence oxygen tank 2's quantity sensor had failed earlier, but it was interconnected with its twin, tank 1, the reason Liebergot requested an extra stir. He was watching the hydrogen tank readings and instructed his backroom specialists to watch the oxygen tank readings—all to verify that the stir had taken place.

As he sat in Mission Control at his console, with its mosaic of push buttons and black-and-white computer displays, Liebergot was not alone in tending to *Odyssey*'s electronic and life-support systems. He was in voice contact with three other controllers in a staff support room across the hall. Each flight controller in Mission Control was connected via so-called voice loops—preestablished audio-conferencing channels—to a number of supporting specialists in back rooms who watched over one subsystem or another and who sat at similar consoles to those in Mission Control.

Liebergot's wingmen that day were Dick Brown, a power systems specialist, and George Bliss and Larry Sheaks, both life-support specialists. As the pressure rapidly rose in oxygen tank 2 and then abruptly fell within seconds, their eyes were also fixed on the hydrogen cryogenic tank readouts, and they all missed the signs that tank 2 had just exploded.

Suddenly the radio link from the crew crackled to life. "Okay Houston, we've had a problem here," reported CM pilot Swigert as he surveyed *Odyssey*'s instruments. "Houston, we've had a problem," repeated Lovell a few seconds later, adding that the voltage of one of the two main power distribution circuits, or buses, that powered the spacecraft's systems was too low. But a few seconds later the voltage righted itself, so the crew began chasing down what seemed to be the big problems: the jolt of the explosion had caused their computer to reset and had knocked a number of valves closed in the attitude control system that kept *Odyssey* pointed in the right direction.

In Mission Control, however, things were not adding up. The spacecraft's high-gain directional antenna had stopped transmitting, and *Odyssey* had automatically fallen back to its low-gain omnidirectional antennas. Liebergot and his team were seeing a lot of screwy data, dozens of measurements out of whack. Fuel cells 1 and 3 had lost pressure and were no longer supplying current, leaving only fuel cell 2 to pick up the load; oxygen tank 2's pressure was reading zero; the pressure in oxygen tank 1 was rapidly failing; and *Odyssey* had completely lost one of its electrical distribution buses along with all the equipment powered by it. The crew connected one of their reentry batteries to the remaining bus in a bid to keep the CM's systems up and running.

Liebergot's training kicked in. Simulation after simulation had taught controllers not to make rash decisions based on a few seconds of oddball data—measurements were made by imperfect sensors and had to pass through a lot of space, with a lot of opportunities to get mangled, before they turned up on a controller's screen. "Engineers that work in this business are well schooled to think first in terms of instrumentation," explained Arnold Aldrich, chief of the CSM systems branch during *Apollo 13*. He was in Mission Control at the time of the explosion and recalls that "it wasn't immediately clear how one particular thing could have caused so many things to start looking peculiar."

So when Gene Kranz, the flight director in charge of the mission (referred to as "Flight" on the voice loops), pointedly asked Liebergot what was happening on board *Odyssey*, the EECOM responded, "We may have had an instrumentation problem, Flight." Forty years on Liebergot still ruefully remembers his initial assessment. "It was the understatement of the manned space program. I never did live that down," he chuckled.

To Kranz the answer sounded reasonable, as he had already had some electrical problems with *Odyssey* on his shift, including one involving the high-gain antenna. "I thought we had another electrical glitch and we were going to solve the problem rapidly and get back on track. That phase lasted for three to five minutes," said Kranz. Then "we realized we'd got some problem here we didn't fully understand, and we ought to proceed pretty damn carefully."

Kranz's word was law. "The flight director probably has the simplest mission job description in all America," he pointed out to me. "It's only one sentence long: 'The flight director may take any action necessary for crew safety and mission success.'" The only way for NASA to overrule a flight director during a mission was to fire him on the spot.

The rule vesting ultimate authority in the flight director during a mission was on the books thanks to Christopher Columbus Kraft Jr., who founded Mission Control as NASA's first flight director and who was deputy director of the Manned Spacecraft Center during *Apollo 13*. He had written the rule following an incident during the Mercury program, when Kraft, as flight director, had been second-guessed by management. This time, as the crisis unfolded, no one had any doubts about who was in charge. While other flight directors took shifts during *Apollo 13*, as the lead flight director Kranz bore most of the responsibility for getting the crew home.

Mission Control and the astronauts tried various fuel cell and power bus configurations to restore *Odyssey* to health, but anyone's remaining hope that the problem was something that could be shrugged off was dashed when Lovell radioed down, "It looks to me, looking out of the hatch, that we are venting something out into space." It was liquid oxygen spewing out from the wounded SM.

The problems were piling up at Liebergot's door. Although his voice is impressively calm throughout the recordings of the voice loops from Mission Control, Liebergot admits that he was almost overwhelmed when he realized "it was not an instrumentation problem but some kind of a monster systems failure that I couldn't sort out. . . . It was probably the most stressful time in my life. There was a point where panic almost overcame me."

Liebergot gives credit to the endless emergency simulation training for getting him through the moment—as well as to the big handles that flanked each Mission Control console, intended to make servicing easier and jokingly dubbed "security handles" by the controllers. "I shoved the panic down and grabbed the security handles with both hands and hung on. I decided to settle down and work the problem with my backroom guys. Not to say that the thought of getting up and going home didn't pass my mind," he remembers.

Kraft told me that the emergency simulations, or "sims," had also taught controllers "to be very careful how you made decisions, because if you jumped to the end, the sims taught you how devastating that could be. You could do wrong things and not be able to undo them."

As controllers scrambled to track down the source of the venting, flight director Kranz echoed this thinking to all his controllers. "Okay, let's everybody keep cool. . . . Let's solve the problem, but let's not make it any worse by guessin'," he broadcast over the voice loops, practically spitting the word *guessin'*, and he reminded them that, just in case, they had an undamaged LM attached to *Odyssey* that could be used to sustain the crew.

For now Liebergot and his back room concentrated on ways to ease the ailing CM's power problem until they could figure out what was wrong, and the crew started powering down nonessential equipment to reduce the load temporarily. The goal was to stabilize the situation pending a solution that would get *Odyssey* back on track. But Liebergot, who was starting to realize the full depth of the problem, unhappily told Kranz, "Flight, I got a feeling we've lost two fuel cells. I hate to put it that way, but I don't know why we've lost them."

Liebergot began to suspect that the venting Lovell had reported was coming from the cryogenic oxygen system, an idea bolstered when Bliss, one of Liebergot's backroom life-support specialists, asked Liebergot worriedly, "Are you going to isolate that surge tank?" The surge tank was the small reserve tank of oxygen that the crew would breathe during reentry, but the massive leak in the SM's cryogenic system meant that the remaining fuel cell was starting to draw on the surge tank's small supply of oxygen to keep power flowing.

Drawing on the CM's limited reserves, such as its battery power or oxygen, was usually a reasonable thing to do in sticky situations—assuming the problem was relatively short-lived and the reserves could be replenished from the SM later. But Liebergot was now worried that the SM was permanently running out of power and oxygen. Once he confirmed that the surge tank was being tapped, he revised his priorities, from stabilizing *Odyssey* to preserving the CM's reentry reserves. The decision caught Kranz momentarily off-guard.

"Let's isolate the surge tank in the CM," Liebergot told Kranz.

"Why that? I don't understand that, Sy," Kranz replied, noting that isolating that tank was the very opposite of what was needed to do to keep the last fuel cell running.

In effect Liebergot's request was a vote of no confidence in the SM, and if the SM could not be relied on, the mission was in deep trouble. "We want to save the surge tank, which we will need for entry," Liebergot prompted. The implication immediately sank in. "Okay, I'm with you. I'm with you," said Kranz resignedly, and he ordered the crew to isolate the surge tank.

For a few more minutes Liebergot and his backroom guys fought the good fight to keep the remaining fuel cell online, but it was looking grim. Without the fuel cell he was going to have to power down even more CM systems in order to keep the most essential system running: the guidance system. The guidance system mainly comprised the onboard computer and a gyroscope-based inertial measurement system that kept track of which way the spacecraft was pointing. Without it the crew would not be able to navigate in space. But turning off nearly everything else in the CM was going to make it a pretty inhospitable place for the astronauts.

"You'd better think about getting into the LM," Liebergot told Kranz. It was now about forty-five minutes since the explosion, and Liebergot's backroom team estimated that at the oxygen supply's current rate of decay, they would lose the last fuel cell in less than two hours. "That's the end right there," said Liebergot.

Kranz called Bob Heselmeyer on his loop. Heselmeyer sat two consoles over from Liebergot, and his job title was TELMU, which stood for Telemetery, Environmental, Electrical, and Extravehicular Mobility Unit. What that mouthful boils down to is that the TELMU was the equivalent of the EECOM for the LM, with the added responsibility of monitoring the astronauts' space suits. Like Liebergot, Heselmeyer had a posse of backroom guys—Fred Frere, Bob Legler, Hershel Perkins, and Bill Reeves—and Kranz was about to hand them all a job. "I want you to get some guys figuring out minimum power in the LM to sustain life," Kranz ordered Heselmeyer.

It does not sound like a tall order—the LM had big, charged batteries and full oxygen tanks all designed to last the duration of *Apollo 13*'s lunar excursion, some thirty-three hours on the surface—so one would assume it should have been a simple matter of hopping into *Aquarius*, flipping a few switches to turn on the power, and getting the life-support system running. Unfortunately, spaceships do not work like that. They have complicated, interdependent systems that have to be turned on in just the right sequence as dictated by lengthy checklists. Miss a step, and you can do irreparable damage.

What follows is a little known story, even to many involved in the *Apollo 13* mission. While they have been complimented on rapidly getting the LM into lifeboat mode, stretching its resources to keep the crew alive for the journey back to Earth, few realize the LM controllers first had to overcome a basic problem: how to get the LM to turn on at all. Over the last four decades the incredible efforts of the LM flight controllers have been somewhat overlooked, ironically because *Aquarius* performed so well. It did everything asked of it, whether designed to or not. So the attention has focused on the titanic struggle over the crippled *Odyssey*. But without the LM controllers' dedication, foresight, and years of work, Lovell, Haise, and Swigert would not have had a chance.

A fundamental problem stood in the way of getting the LM online. Call it the "step-zero problem." They could not even turn on the first piece of equipment in the lifeboat checklist because of the way *Aquarius* had been designed to handle the coast between the Earth and the moon. Remember that for most of this coast the LM and the CSM were docked, connected by a narrow transfer tunnel, with almost everything on the LM turned off to save power. A number of critical systems in the LM were protected from freezing by thermostatically controlled heaters. During the coast these heaters were powered via two umbilicals from the CM, which in turn got its power from the SM.

Within *Odyssey* the umbilicals were connected to a power distribution switch that shifted the LM between drawing power from *Odyssey* and drawing power from its own batteries, the bulk of which were located in the descent stage. Here was the hitch: the distribution switch itself needed electricity to operate, which *Odyssey* could no longer supply. With the last fuel cell running out of oxygen, the astronauts needed another way to get the LM batteries online, fast.

The LM controllers were already on the case when Kranz's order came through. Back in the staff support room, the LM consoles were right beside the EECOM's support controllers' consoles, separated by a paper strip chart that recorded the activity of the LM heaters. From the start of the crisis they had front-row seats as Bliss, Brown, and Sheaks tried to save the CSM with Liebergot. It had not been long before Brown turned to the LM controllers and said, "I'll bet anything that oxygen tank blew up," recalled LM controller Legler. "Bill Reeves and I put a lot of stock in what Dick Brown said, and if that was true, the CSM was going to be out of power before long and we were going to have to use the LM as a lifeboat."

Looking at their strip chart, Legler and Reeves could see the LM heater activity had flatlined, meaning the electrical bus in *Odyssey* that was connected to the umbilicals was no longer supplying power to *Aquarius*. "We had lost power to the switch that was used to transfer power from the LM descent batteries. So they would have been unable to turn on the LM," said Legler.

The large batteries in the descent stage were essential to powering up most of the LM's systems. They were connected to the LM's power distribution system via relays, which required power to operate—power that was no longer available from the CM. Fortunately, smaller batteries in the LM's ascent stage could be tapped independently of the switch in *Odyssey*, but they could power only some systems for a limited amount of time. In order to get systems such as life support and the computer running, the ascent batteries had to be connected to the power distribution system, energizing the relays and so allowing the descent batteries to be brought online.

Nobody had ever planned for this situation. Legler and Reeves began working out a set of ad hoc procedures—step-by-step, switch-by-switch instructions for the astronauts—that would coax some power through the maze of circuits in *Aquarius* from the ascent batteries to the relays. Working from wiring and equipment diagrams of the LM, it took them about thirty minutes from the time of Brown's warning about the state of the CM to finish the list of instructions. The final list involved about "ten to fifteen" switch throws and circuit breaker pulls for the crew, as Legler recalls. Once the relays had electricity, the crew could switch over from *Odyssey*'s now-dead umbilicals and start powering up the LM's life-support systems in lifeboat mode, an even more complicated process.

Fortunately, somebody had already been working on *that* problem for months. A year earlier, in the run-up to the *Apollo 10* mission, the flight controllers and astronauts had been thrown a curveball during a simulation. "The simulation guys failed those fuel cells at almost the same spot [as when *Apollo 13*'s oxygen tank exploded in real life]," according to James "Jim" Hannigan, the LM branch chief. "It was uncanny."

Legler had been present for the *Apollo 10* simulation when the LM was suddenly in demand as a lifeboat. While such procedures had already been worked out for earlier missions, none addressed having to use the LM as a lifeboat with a damaged CM attached. Although Legler called in reinforcements from among the other LM flight controllers, they were unable to get the spacecraft powered up in time, and the *Apollo 10* simulation had finished with a dead crew.

"Many people had discussed the use of the LM as lifeboat, but we found out in this sim [that exactly how to do it could not be worked out in real time]," Legler recalls. At the time, the simulation was rejected as unrealistic, and it was soon forgotten by most of the people involved. Hannigan explained that NASA "didn't consider that an authentic failure case" because it involved the simultaneous failure of so many systems.

The simulation nevertheless nagged at the LM controllers. They had been caught unprepared, and an Apollo crew had died, albeit only virtually. "You lose a crew, even in a simulation, and it's doom," Hannigan emphasized. He therefore tasked his deputy, Donald Puddy, with forming a team to come up with a set of lifeboat procedures that would work, even with a crippled CM in the mix.

"Bob Legler was one of the key guys [on that team]," Hannigan opined. He also said that as part of his work, Legler had "figured out how to reverse the power flow, so it could go from the LM back to the CSM" through the umbilicals. "That had never been done," said Hannigan. "Nothing had been designed to do that." Reversing the power flow was a trick that would be critical to the final stages of *Apollo 13*'s return to Earth.

For the next few months after the *Apollo 10* simulation, even as *Apollo 11* made the first lunar landing and *Apollo 12* returned to the moon, Puddy's team worked on the procedures, looking at many different failure scenarios and coming up with solutions. Although the results had not yet been formally certified and incorporated into NASA's official procedures, the LM controllers quickly pulled these new techniques off the shelf after the *Apollo 13* explosion. The crew had a copy of the official emergency LM activation checklist on board, but the controllers needed to cut the thirty-minute procedure to the bare minimum.

The LM team's head start stood them in good stead. Although Liebergot and his team had initially estimated two hours of life left in the last fuel cell when Kranz asked Heselmeyer and his team to start working on how to get life support running in the LM, the situation was rapidly worsening. By the time the crew actually got into *Aquarius* and started turning it on, the backroom controllers estimated there were just fifteen minutes of life left in the last fuel cell on board *Odyssey*. With the LM's life-support systems coming online, the immediate threat of death to the crew had been suspended, and it was time to start thinking about how to get the astronauts home.

Jerry Bostick was the chief of the flight dynamics branch, the part of Mission Control that looks after a spacecraft's trajectory—where it is, where it is going, where it *should* be, and how to get it there. The controllers of the flight dynamics branch sat in the front row of Mission Control, which they had proudly dubbed "the Trench." As they listened to the crew in space and the systems controllers in the row behind them struggle with the explosion's aftermath, Bostick recalls, "We went into the mode of okay, well, can we come back home immediately?" The Trench soon calculated that if the crew used *Odyssey*'s main engine and burned every last drop of fuel, they could turn around and come straight back to Earth, in a procedure known as a direct abort.

But the main engine was in the SM, and who knew what damage had been done to it? The engine might malfunction: in the worst case firing it up could result in another explosion and kill the crew instantly. The other option was to let *Apollo 13*, carried forward by its momentum and the moon's gravity, go around the moon. There gravity would pull *Apollo 13* around the back side of the moon, accelerate it, and sling the spacecraft back toward Earth. This journey would take several days, however, and the LM was intended to support only two men for two days—not three men for four. If the crew did not get home fast, they would run out of power and die.

A candid Kranz says this was his toughest call on *Apollo 13*. "My team was pretty much split down the middle. Many of my systems controllers wanted to get home in the fastest fashion possible. The trajectory team did not want to execute a direct abort because it had to be executed perfectly. If we didn't get the full maneuver, more than likely we would crash into the moon," he explained. "I was of the frame of mind that said, 'Hey, we don't understand what happened here . . . and if we execute a direct abort, we're not going to have much time to think about it.' . . . We needed to buy some time so that when we did make a move, it would be the proper move."

Weighing the concern that *Aquarius* would not cut it on a longer return journey, Kranz stressed that he had "a lot of confidence in my Lunar Module team." He had previously been involved in a number of missions involving an LM. "I knew it was a very substantial spacecraft. . . . I was pretty much betting that this control team could pull me out of the woods once we decided to go around the moon." Kranz made his decision: the main engine was out. *Apollo 13* was going around the moon.

There was, of course, a proverbial fly in the ointment. During earlier Apollo missions the outgoing trajectory of the spacecraft had been selected so that if the SM's main engine failed for any reason, going around the moon would produce a slingshot effect that should aim the CSM perfectly at Earth, on a so-called free-return trajectory. But this trajectory put very tight constraints on the mission timeline, and for *Apollo 13* it had been abandoned. "We were on a non-free-return trajectory," the Trench's Bostick explained. "If we did nothing, we'd whip back towards the Earth but miss it by several thousand miles."

As the question of trajectory was being decided, a shift change was taking place at Mission Control. When the explosion occurred, Kranz and his controllers, collectively known as the White Team, had been about an hour away from the end of their shift. As was common, most of the next shift—the Black Team, led by Glynn Lunney—had already shown up, so as to be able to take over running the mission seamlessly from their predecessors, and they had been on hand throughout the crisis.

As Kranz's team gathered up to leave Mission Control, Bostick went to speak to the incoming flight director, Lunney. By good fortune Kranz and Lunney were perfectly matched to the different phases of the crisis they would be faced with. Kranz was a systems guy—he knew the internals of the spacecraft better than any other flight director, the ideal person to cope with the second-by-second equipment failures and reconfigurations triggered by the explosion. Lunney had come up through the flight dynamics branch, making him ideally suited to get the spacecraft headed in the right direction. According to their boss at the time, Chris Kraft: "Kranz was there at the right time to make the decisions that had to be made rapidly, and then when Lunney took over he brought a calmness to the control center to do the right things once they had gotten stabilized. . . . They turned out to be a wonderful pair."

So Bostick was speaking to the perfect audience when he voiced his concerns. "We need to get this thing back to a free-return trajectory," he told Lunney. Lunney instantly agreed, but the decision left Bostick with a problem. Getting *Apollo 13* onto a free-return trajectory required a solid push from a big engine. With *Odyssey* and *Aquarius* docked together and the main SM engine out, that left only the engine attached to the LM's descent stage, designed to be used only for the relatively short period of time needed to land

Aquarius on the moon, and a smaller ascent engine that would have been used for liftoff from the lunar surface. "It was a problem, because we didn't have capability in the control center to calculate the result of a docked maneuver [using the descent engine]," said Bostick.

During a mission controllers called on a bank of mainframe computers in a Manned Spacecraft Center facility set up and maintained by IBM, known as the Real Time Computer Complex (RTCC), to calculate the length and direction of engine burns needed to produce a given trajectory. To do these calculations, the mainframes were programmed with information about the spacecraft, such as their mass, center of gravity, how much thrust the engine produced, and so on. Unfortunately for *Apollo 13*, the program to calculate how the conjoined Command and Lunar Module could be maneuvered using just the descent engine simply did not exist.

"So the first thing we did was call our computer guys and say 'Hey, call all the IBM guys in and start writing some software!'" recalled Bostick with a laugh. As a backup, the mission planners who originally put together the *Apollo 13* mission were called in to double-check the RTCC's results. "In two or three hours we were able to come up with a free-return maneuver."

Kranz's team had not gone home after its shift. The White Team now formed the nucleus of a new Tiger Team, dedicated to figuring out the fastest way possible to get the crew home, given that the spacecraft was going around the moon. They also had to work out how to stretch the LM's consumables to last the entire trip and how to get the CM reactivated and configured to survive a reentry—the astronauts' only way to get home alive.

Arnie Aldrich, the CSM branch chief, had joined the Tiger Team, along with another EECOM, John Aaron. An hour earlier Aaron had been at home, standing in front of the mirror shaving as he prepared to come in for his shift, when his wife brought him the phone, saying his boss was on the line. "John, I need to ask you some questions," Aaron recalls Aldrich telling him. "There's something significant that's happened out here and these guys can't quite figure it out. It's not going well."

Aldrich had called Aaron for a couple of reasons. One was that Aaron was an expert on the CSM's instrumentation system. The other was that Aaron was one of the best mission controllers in NASA, and four months earlier he had saved the *Apollo 12* mission. The incident cemented Aaron's reputation as a "steely eyed missile man." So, when *Apollo 13* ran into trouble, Aaron was Aldrich's go-to guy. "I had a very good group of people working for me at the time of the explosion, but we were scratching our heads, and the very best person I had was John Aaron," said Aldrich.

After the explosion Aldrich had moved into the spacecraft analysis (SPAN) room, located across from Mission Control. The SPAN room was fitted out with more consoles and acted as a bridge between the flight controllers and the army of engineers who had actually designed and built the spacecraft. "In there were supervisors like me and executives from the engineering organizations in NASA and the manufacturers, and this group would sit together and monitor the flights," said Aldrich. The SPAN room had come into being, according to Kranz, because "we learned during Mercury that we wanted immediate access to the manufacturers, that we needed clear and unfiltered data very rapidly."

Over the phone Aaron asked Aldrich to walk around behind the consoles in the SPAN room and describe what he saw. "I started asking him: tell me what this measurement says, tell me what that measurement says," Aaron remembered. "And that went on for about ten minutes." He was looking for a pattern in the data that would map out any failures in the instrumentation system on board *Odyssey*, but he was coming up empty. "I told Arnie, 'Well, I'll be right there. In the meantime tell those guys they've got a real problem on their hands,'" recalled Aaron.

As the LM controllers raced to power up *Aquarius*, Aaron had made it in to Mission Control. "When I walked in the room, I intentionally did not put a headset on because I could see each of the flight controllers had zoomed in and were trying to sort the problem out from the perspective of their individual subsystem," he says. He walked behind the controllers, looked at their data, and listened to what they were saying to the back rooms. Finally, he sat down beside the embattled CSM controller Liebergot, plugged his headset in, and said, "Sy, we've got to power the Command Module down."

Aaron did not just want the CM powered down to minimal systems only. He meant powered down as in *off.* No guidance system, no heaters to keep back the cold of space, no telemetry to help controllers diagnose the problem. Nothing. He was concerned that even a minimal power draw from the batteries would leave them without anything for reentry.

Gary Coen, one of the controllers with responsibility for *Odyssey*'s guidance system, disagreed with Aaron. "He was pleading with me to leave the heater circuit on in the inertial platform in the CM," Aaron recalls. The inertial platform, which gave the computer raw data about which way the spacecraft was pointing, was never designed to handle extreme cold. "He said 'John, [the heater] only takes 0.4 amps . . . if we turn it off, the platform may never work again.' And I said, 'Well, Gary, just do the math. 0.4 amps times forty-eight hours—we gotta turn it off. If it doesn't work again, we'll just have to figure out how to get home without it.'"

But without *Odyssey*'s guidance system telling the crew precisely which way they were pointing in space, how would they be able to align the spacecraft correctly to perform the free-return trajectory maneuver? The answer was to rely on the LM's guidance system, which had at its heart an identical computer to the one in *Odyssey*'s guidance system. The LM's guidance system had been powered off, however, for most of the way to the moon—it had no clue about which way it was pointing. The crew would have to transfer the alignment information manually from the CM's computer to the LM's computer before pulling the plug in *Odyssey*. Doing so would require some good old-fashioned arithmetic.

"You could read the angles out of one computer and type them into the other, but you had to invert them," because *Odyssey* and *Aquarius* were docked head to head and therefore pointed in opposite directions, explains Aaron. The job fell to Lovell on board *Aquarius*. "Because I had made mistakes in the arithmetic several times during sims . . . I asked the ground to confirm my math," said the commander afterward. The Trench broke out pencil and paper and confirmed the angles.

As soon as possible after the crew aligned the LM's guidance system for the free-return trajectory maneuver, they shut down the CM completely. In the end the inertial platform heater circuit breaker was the last one they pulled. Aaron and the other members of the Tiger Team were gathered in a room near Mission Control. Kranz soon arrived and looked around the crowded space. The controllers were subdued and shaken—they had failed to contain the crisis, and the crew was still in extreme danger. But the last thing the astronauts needed was for controllers to begin second-guessing themselves.

Confidence was part of the bedrock upon which Mission Control was built. When prospective controllers joined NASA, often fresh out of college, they started out by being sent to contractors to collect blueprints and documents, which they then digested into information that mission controllers could use during a mission, such as the wiring diagrams the LM controllers had used to figure out how to power up *Aquarius*. After that the proto–flight controllers started participating in simulations. The principal problem NASA had with these neophytes was "one of self-confidence," explains Kranz. "We really worked to develop the confidence of the controllers so they could stand up and make these real-time decisions. Some people, no matter how hard we worked, never developed the confidence necessary for the job." Those not suited for Mission Control were generally washed out within a year.

Now Kranz feared his controllers, battered by the events of the last hour, would lose their nerve. What happened next was a spectacular moment of leadership. "It was a question of convincing the people that we were smart enough, sharp enough, fast enough. That as a team we could take an impossible situation and recover from it," says Kranz. He went to the front of the room and started speaking. His message was simple. "I said this crew is coming home. You have to believe it. Your people have to believe it. And we must make it happen," recalls Kranz.

In the Ron Howard movie this speech was "simplified into 'Failure is not an option,'" chuckles Kranz, who never actually uttered the now famous phrase during the *Apollo 13* mission. Still, Kranz liked it so much, because it so perfectly reflected the attitude of Mission Control, that he used it as the title of his 2000 autobiography. Kranz's speech electrified the room. "Everybody started talking and throwing ideas around," remembers Aaron.

Kranz appointed three flight controllers as his key lieutenants. Aldrich was put in charge of assembling the master checklist for powering the CM and other reentry procedures. LM controller William Peters was ordered to make sure *Aquarius* lasted long enough to get the crew close to Earth. And Aaron was put in charge of devising how electrical and other life-support systems would be used so that as the crew turned on the CM again prior to reentry, they would be able to get it up and running and complete the descent through the Earth's atmosphere before the batteries were exhausted.

The main problem for Aaron was that, as with *Aquarius*, powering up the CM was a complex procedure, made even more difficult by the fact that, unlike the LM, *Odyssey* was never supposed to be powered down at any point during the mission. "The only power-up sequence we knew was the one that started two days before launch," Aaron remembers. But judging by what was left in *Odyssey*'s batteries, "we had just a couple of hours at full power," he says.

Aaron listened to the hubbub of ideas on how to get the CM going and decided it was time to step in. "I started throwing some ideas out as to how the power-up sequence could be altered," he remembers. Controllers immediately started to object, explaining why it was vital that one aspect or another of the sequence remain untouched. Aaron decided to "chuck them all out of the room"—with the exception of Jim Kelly, a backroom CSM controller who specialized in the electrical power system—to give himself a chance to think. "I said 'Go get some coffee and come back here in forty-five minutes, and Jim and I will have a timeline of what we can turn on and when for a rudimentary re-entry sequence.'"

The two men took some paper and started sketching out a timeline, blocking out how much power each system in the CM would use as it was brought online. "We didn't have any computer programs to do this," says Aaron. But, thanks to the simulations, the pair had been trained in "all kinds of situations where power failures happened." "Mostly," says Aaron, "we were just sketching the timeline out from memory and what we had learned from training."

The other controllers returned to find a big block diagram drawn on the blackboard. "They came back in, and I started describing [the timeline]," recalls Aaron. "That started the brokering process, because every controller still wanted their favorite piece of equipment on, and the earlier the better." The brokering process, with Aaron acting as the final arbiter, would continue for another two or three days, refining the timeline and fleshing it out until the sequence was finally ready. Aaron's work would raise his stock among his colleagues even higher. "[He] just had a knack for the job," states Kraft. "He was always thinking ahead, always capable of making the best of a tough situation and getting us out of it."

Integrating the power-up sequence with other tasks that would have to be done before entry into a set of procedures that could be read up to the crew was Aldrich's job. The result, for a time the most precious document in the U.S. space program, started out as a typewritten document but was repeatedly "updated in pen and pencil." "It was five pages long," says Aldrich, who still has the final checklist in his possession. "I haven't looked at it in quite a long time. I know where it is, but it's buried!" protested Aldrich when pressed for more details. He revealed that he has assiduously kept a scrapbook from his decades in the space business.

Kranz's Tiger Team worked closely with the inhabitants of the Mission Evaluation Room (MER), who were located in the building next to Mission Control. While the SPAN room was designed to act as a communications conduit between Mission Control and the engineers who had actually built and designed the spacecraft, the MER was where the problems posed by Mission Control actually began to be solved.

The MER was established during the Mercury program. In the early days of the program the same people who built the spacecraft would staff the consoles in Mission Control. "[But] people didn't have time to be responsible for the engineering and also put all the time in learning how to operate missions," says Aldrich, "so there was a split." In Mission Control was a team that really understood "how the flight was to be executed, what the possible trouble spots might be, and was prepared to deal with things that came up," Aldrich explains, but questions that cropped up concerning one piece of equipment or the other required the help of engineers. In the MER would be "an engineering team that was pretty well informed, but which wasn't directly engaged with the flight on a first-hand basis," says Aldrich.

The MER was big enough to house dozens of engineers and if a problem could not be solved by those present, they could call on engineers throughout NASA's nationwide network of research and development centers as well as the engineers of the contractors that built the spacecraft. North American Rockwell, based in Downey, California, had built Apollo's CSM, while Grumman Aerospace, based in Bethpage, New York (now part of the Northrop Grumman Corporation), had built the LM.

As procedures for powering up *Odyssey* or stretching *Aquarius*'s life-support system were developed in Mission Control, hundreds of engineers in California and New York would test them out in the same factories where the spacecraft had been built. The *Apollo 13* movie shows Grumman Aerospace hedging its support for some of the risky tactics being employed by Mission Control, but in fact, remembers Kranz—who is nevertheless "very pleased" with the film—"that did not happen." "The contractor support was absolutely superb," he says determinedly. "The contractors knew what was at risk for every mission. If we had a problem and we turned to them, they gave us everything we needed."

Apollo 13 performed the free-return trajectory maneuver using the LM's descent engine without a hitch. "I think it made everybody feel a lot better—including the astronauts," says Bostick. He remembers talking to the crew after the mission. "When we executed the free-return burn it made them feel that they might get out of this thing alive," he says. Yet the debate continued about the fastest way to get the crew home. If no changes were made to the trajectory, the crew would splash down in the Indian Ocean in about four days. But there were no recovery forces to pick up the CM if it ended up in that part of the globe.

Bostick and his flight dynamics controllers immediately began working on how to shave some time off the return journey and have the splashdown take place in the Pacific, where all the recovery forces had already been deployed. "We concluded we could do that fairly easily and speed up [the splashdown] by about twelve hours, but we had also worked up an option that would get back to the Pacific and speed it up by thirty-six hours," says Bostick. But the thirty-six-hour option would have involved jettisoning the SM immediately, exposing the all-important reentry heat shield to space for a long time, and required nearly every drop of fuel left in *Aquarius*'s descent stage. Neither of these actions sounded appealing.

"By then," recalls Bostick, "the systems guys had really done a bang-up job of squeezing the consumables [in the LM]." They had done this principally by turning off nearly every system in *Aquarius* except for guidance, communications, and a water/glycol cooling system that was needed to stop certain systems from overheating. "Most of the water [on board] was used for cooling; it was our most critical resource," explains Legler, who was the lead controller responsible for managing *Aquarius*'s power and water usage. The two consumables were interrelated; the fewer systems that were turned on and drawing power, the less water would be needed for cooling. Normally, fully powered up, the *Aquarius*'s systems drew fifty to seventy-five amperes, and by dint of hard work the team "powered it down to about twelve amps," says Legler. Twelve amps is about as much power as a vacuum cleaner uses. Unfortunately for the crew, there was no power in the budget to run heaters to keep them warm, and temperatures inside the spacecraft began to drop sharply.

With *Aquarius* now expected to go the distance, the risky thirty-six-hour option was no longer needed, and the twelve-hour maneuver was chosen. This required another burn from the LM's descent engine, one that would take place two hours after *Apollo 13*'s closest approach to the moon. The point of closest approach was known as pericynthion, or PC, and so the trajectory adjustment was called the "PC + 2 burn."

The PC + 2 burn needed to happen exactly right, and the Trench insisted the LM's computer be used to control it. But the LM's guidance system used a lot of power, and the Trench agreed that if they could use it for the PC + 2 burn, they would not ask for it again. Almost exactly twenty-four hours after the oxygen tank explosion, the crew completed the burn and shut down the navigation system. From here on out the astronauts would be flying by the seat of their pants.

As *Apollo 13* sped toward Earth, Mission Control was beginning to worry about a new problem. While the LM had enough spare oxygen to accommodate Swigert as well as the intended LM crew of Lovell and Haise, carbon dioxide levels were beginning to build up. Normally, lithium hydroxide (LiOH) canisters absorbed the gas from the air and prevented it from reaching dangerous levels, but the canisters aboard *Aquarius* were being overwhelmed. *Odyssey* had more than enough spare LiOH canisters on board, but these canisters were square and could not fit into the round holes intended for the LM's round canisters. Mission Control needed a way to put a square peg into a round hole. Fortunately, as with the LM activation sequence, somebody was ahead of the game.

As reported in *Lost Moon*, Lovell's book about the *Apollo 13* mission (cowritten by Jeffrey Kluger and later republished as *Apollo 13*), Ed Smylie, one of the engineers who developed and tested life-support systems for NASA, had recognized that carbon dioxide was going to be a problem as soon as he heard the LM was being pressed into service after the explosion. For two days straight since then, his team had worked on how to jury-rig *Odyssey*'s canisters to *Aquarius*'s life-support system. Using materials known to be available on board the spacecraft—a sock, a plastic bag, the cover of a flight manual, lots of duct tape, and so on—the crew assembled Smylie's strange contraption and taped it into place. Carbon dioxide levels immediately began to fall into the safe range. Mission Control had served up another miracle.

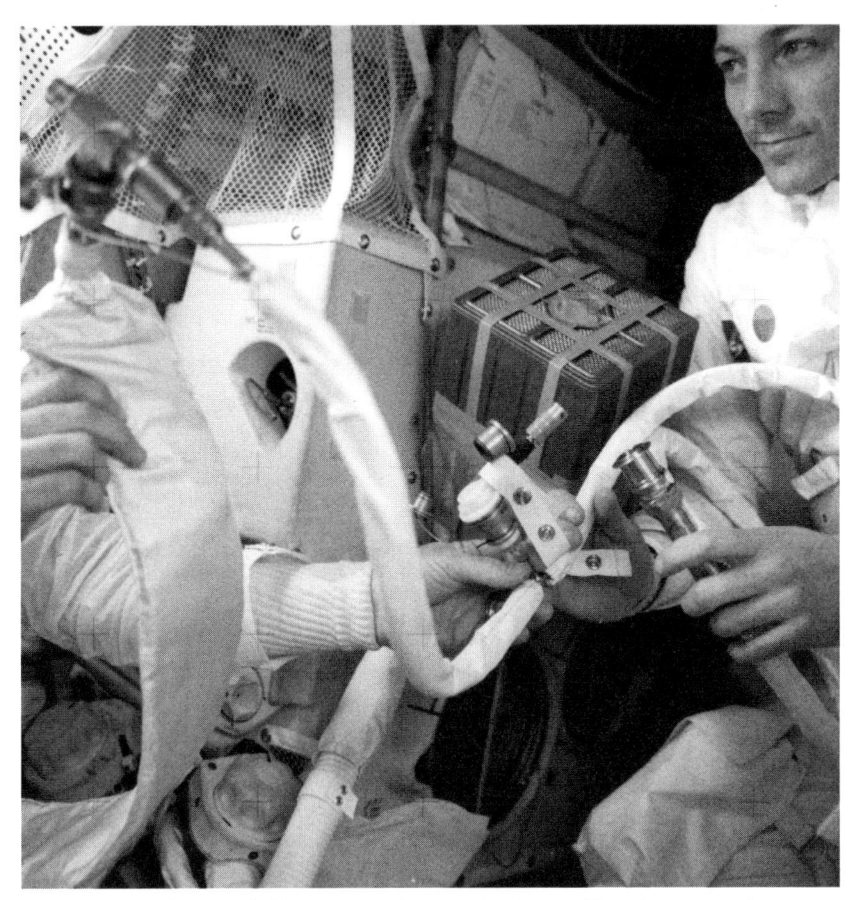

18. CMP Jack Swigert holds a jury-rigged contraption that would scrub the atmosphere inside LM *Aquarius* of potentially lethal carbon dioxide. It was pieced together by the crew following instructions relayed to them by Mission Control. Courtesy NASA.

Although the PC + 2 burn had been right on the money, the Trench was increasingly unhappy about *Odyssey* and *Aquarius*'s trajectory. Something was pushing the spacecraft off course (afterward it would be determined that a cooling water vent on *Aquarius* had been acting like a little rocket jet, gently sending *Apollo 13* in the wrong direction), and they needed another burn to correct the trajectory. But the Trench had given up the navigation system after the PC + 2 burn. "We had to come up with some way to align the spacecraft [properly for the corrective burn]," Bostick explained.

One of Bostick's controllers, Charles "Chuck" Dietrich, remembered an alignment technique that had been developed for earlier NASA missions in

Earth orbit. A spacecraft could be pointed in the right direction by using a portion of the surface of the Earth as a reference marker—in this case the terminator between night and day.

"All during the Mercury, Gemini, and Apollo Earth-orbit programs, that was a technique we had used, but never on a return from the moon. It was a little more dicey there," says Bostick. In Earth orbit a small alignment inaccuracy prior to a reentry would result in the spacecraft landing miles off target, but it usually was not life threatening. But if *Apollo 13* missed the trajectory it needed to take to reenter safely—known as the entry corridor—the results would be disastrous, the CM skipping into space or burning up in the atmosphere. "If you missed the entry corridor by a degree," observes Bostick, "that's a real bad day."

The crew was cold and exhausted by this point—temperatures on board had dropped almost to freezing point. The astronauts had gotten very little sleep since the explosion, and yet they pulled off the course-correction maneuver—and a second one a day later—perfectly.

This second small burn was at the behest of scientists at the Atomic Energy Commission (AEC), who were worried about the potential release of plutonium from a piece of equipment known as a radiothermal generator, or RTG. Despite their use of plutonium, RTGs are not nuclear reactors—they are actually a battery. A particular isotope, Pu, is used that does not permit chain reactions. Instead, electricity is generated from the heat produced by the slow radioactive decay of the plutonium atoms. RTGs were used on all the Apollo missions to the lunar surface to provide a long-lived source of electricity for instruments left behind by the astronauts. Because *Apollo 13* never landed on the moon, its RTG was still stored inside *Aquarius*, which was now irrevocably going to reenter the Earth's atmosphere. Unlike earlier RTGs, the SNAP-27 model RTG carried on board Apollo missions was designed to withstand reentering the Earth's atmosphere without releasing any radioactive material. But given the unusual circumstances of *Apollo 13*, the AEC wanted to make sure that whatever was left of *Aquarius* and its RTG after reentry wound up as far away from anywhere as possible.

Jerry Bostick recalls the period about eight to ten hours before *Apollo 13* was due to land: "I was down in the Trench; Lunney was on as flight director. He motioned for me to [come] up to the console. He said, 'Just got a call from AEC; they're worried about the RTG.' . . . I had been eyewitness to all these tests done on the RTG; it was indestructible. . . . I reminded Glynn of that, and he said, 'I know, I know, but we've got to put it in a safe place.' I said, 'Glynn, I will do the best I can do but the number one thing is getting the guys back home.' So we did move the landing point a little bit . . . to put the RTG in the deepest part of the Pacific we could find. We honored [the request] to the best of our ability. I still think it was a waste of time, but it placated them." The controllers' immediate concern was, as Lunney stressed, to "get the guys home." (The remains of *Aquarius* now lie at the bottom of the Tonga Trench, over six and a half miles down in the Pacific Ocean. A helicopter was sent out after the mission to check for radioactivity that might have been released, and none was found.)

It was now over three days since the explosion in oxygen tank 2, and it was time to get ready for reentry. The first step was to recharge the batteries in the CM, which had been significantly depleted before the LM came online. Remember how, while figuring out LM lifeboat procedures after the *Apollo 10* simulation, Legler had worked out a way to run power from the LM to the CM back along the electrical umbilicals that connected the spacecraft? That was about to come in handy now because that power could be used to recharge *Odyssey*'s batteries.

Aquarius had to serve as a lifeboat. "The biggest problem," remembers Aaron, "was that initially the Lunar Module guys didn't know how much power they were going to need." For the first thirty hours Aaron's power-up team did not think the LM guys were going to have any power to spare for *Odyssey* about twelve hours after the explosion. "We talked to them about getting some power," says Aaron. "They threw us out of the room."

But the PC + 2 burn had shortened *Apollo 13*'s return flight sufficiently that *Aquarius* would be able to supply the power needed to charge the batteries. Working with North American Aviation and Grumman to refine the procedure, through LM gurus Hannigan and Mel Brooks in the SPAN room, Legler and Bill Peters wrote up the needed instructions. The charging process was "only 20 to 25 percent efficient," remembers Legler, but it was enough.

Even with fully charged batteries, *Odyssey* risked running out of electricity before it splashed down. Batteries are rated using a term called "ampere-hours," or "amp-hours." If you start with a forty amp-hour reentry battery and then turn on a piece of equipment that uses one amp-hour and it takes eight hours to finish the reentry and splashdown, you have only thirty-two amp-hours left to power everything else. But if you can delay turning on that piece of equipment until two hours before splashdown, now you have *thirty-eight* amp-hours to go around. "It's not only a matter of how large a load is, but how long that load is on for," Aaron explains. Once a system had been turned on in *Odyssey*, it had to stay on, so "the only variables were how few systems could we turn on and how late could we wait?"

Aaron had an inspiration. Normally, in a spaceship power-up sequence, one of the first things turned on is the instrumentation system so everyone can be sure that the rest of the sequence is progressing normally. But for *Apollo 13* the instrumentation would be turned on last for a final check of *Odyssey* just before reentry began.

It was a gutsy move. It required the crew—in particular the CM pilot, Swigert—to perform the entire power-up procedure in the blind. If a crew member made a mistake, by the time the instrumentation was turned on and the error was detected, it could be too late to fix. But, as a good flight controller should, Aaron was confident his sequence was the right thing to do.

"I still wake up at nights in a cold sweat and wonder about that," an older and wiser Aaron says, "because the one thing I wasn't conscious of, and I prided myself on being conscious of everything, was the condition of the crew." Despite the cold, and the fatigue, and the stress, the crew had voiced few complaints. "You couldn't tell from listening to their voices how bad conditions had got. When they got back I realized, 'Oh my goodness, I built this incredible procedure that had to be executed perfectly, and I handed it off to a crew that hadn't had any sleep for three days,'" shudders Aaron. "I've thought about that a lot, ever since."

But Swigert and the rest of the crew powered up *Odyssey*, seemingly effortlessly. "Therein lies the reason we chose test pilots [to be astronauts]," says Kraft. "They were used to putting their lives on the line, used to making decisions, used to putting themselves in critical situations. You wanted people who would not panic under those circumstances. These three guys, having been test pilots, were the personification of that theory," explains Kraft.

As part of the reentry procedure, the crew jettisoned the damaged SM, quickly snapping pictures and taking footage of the huge gash in the side of

19. After jettisoning the SM, the full extent of the damage could briefly be observed and recorded by the *Apollo 13* crew as it drifted away. The Sector 4 panel was completely blown out, exposing the module's internal systems. Courtesy NASA.

the module as it tumbled into the distance. "There's one whole side of the spacecraft missing," radioed Lovell. "It looks like it got to the [main engine] bell, too," added Haise, validating Kranz's gut decision four days earlier to rule out using the main engine and go around the moon.

Then it was time to strap into the CM and abandon *Aquarius*. For the LM controllers it was a bittersweet moment. "We were proud of the *Aquarius* and very thankful—it had really performed, did everything we asked it to do," remembers Legler. "It's hard to describe that feeling," says Hannigan. "Thank God that we made it, but— "

"Farewell, *Aquarius*, and we thank you," radioed Lovell as the astronauts jettisoned the LM and watched it slowly drift away. Hannigan remembers hearing Lovell's unbidden requiem for the spacecraft. "He did a good job," he says of the mission commander.

It was about another hour before the CM, headed for the Pacific, met the first tenuous wisps of Earth's atmosphere. Soon, as *Odyssey* plunged into the atmosphere, those wisps would become a tremendous fireball of ionized air.

The ionization would block radio communications for several minutes. In the meantime the heat shield would be subjected to incredible temperatures and pressures, and if it had been cracked during the explosion four days earlier, the crew would burn up without ever being heard from again. Assuming the heat shield was okay, then the parachutes would deploy, slowing *Odyssey* to a gentle splashdown—if the parachutes had not frozen solid and the pyrotechnic charges intended to release them still worked. In a few more minutes Lovell, Haise, and Swigert would either be home free or dead.

But the astronaut's last words before reentry were not for themselves. They were for Mission Control. "I know all of us here want to thank all of you guys down there for the very fine job you did," Swigert transmitted. "That's affirm," chimed in Lovell. A few seconds later *Odyssey* disappeared into a sea of radio static.

By *Apollo 13* NASA had a pretty good handle on radio blackouts during reentry, and for a given trajectory it could work out how long—almost to the second—a spacecraft would be out of touch. In the case of *Odyssey* it should have been about three minutes.

The appointed time came and went, and as the seconds turned into minutes without any sign of *Odyssey*, the tension dragged out like a rusty blade through Mission Control. "It was the worst time of the whole mission," agrees Kranz. "The blackout was a very difficult time for every controller. You ask yourself 'did I give the crew everything I needed to and was my data right?' . . . It was just a difficult time."

Bostick, the trajectory specialist, was in hell. "It was probably the worst I ever felt in my life," he says. "My feeling was 'Oh my God, we have done the impossible: we got them all the way home . . . and now something goes wrong in *entry*?' . . . It was one of the most depressing [times] of my life." As he recalls the entry, Bostick's voice wavers for a moment, the memory still emotionally charged after thirty-five years. Then his voice strengthens into triumph. "But then, when we heard from them, it was the happiest moment of my life," he declares.

An antenna-laden plane, circling in the air as part of the recovery effort, had picked up the CM's signal: the crew had survived blackout. But even after radio contact was reestablished, the astronaut's lives were still in danger. The main parachutes still had to be deployed. Kranz and the controllers stood rooted to their consoles, watching the main display on the front

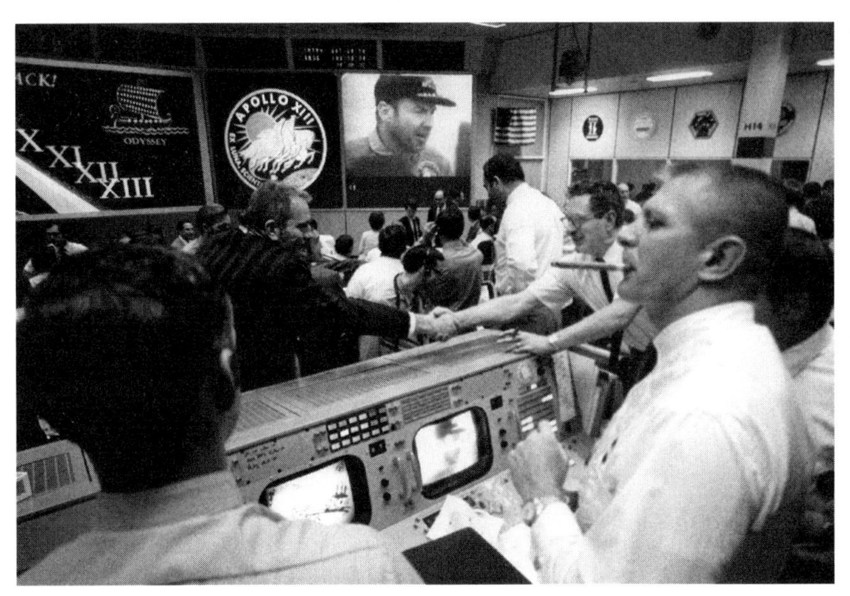

20. Jubilation erupts in the Mission Operations Control Room following the safe arrival of the *Apollo 13* crew aboard USS *Iwo Jima*. In the foreground Glynn Lunney (back to camera) sits at his console alongside Gene Kranz, who is about to light a celebratory cigar. In the background the director of Flight Crew Operations, Deke Slayton (in suit), shakes hands with Chester Lee of the Apollo Program Directorate. Standing behind Lee (also in a white shirt) is Apollo program director Rocco Petrone. Courtesy NASA.

wall of Mission Control. *Odyssey* was going to splash down, for good or ill, within sight of the live TV camera on board the aircraft carrier leading the recovery effort, the USS *Iwo Jima*. Suddenly, the parachutes—three red and white canopies—blossomed into view on the screen.

Pandemonium broke out in Mission Control. "I cried," says Kranz simply. "I think many of the controllers did. The emotional release at that instant was so intense many of us were unable to control our emotions. There were an awful lot of wet eyes that day."

Kraft was one of the few not swept away by the sight of *Odyssey* gently descending into the Pacific beneath its parachutes, suspending his celebration until the crew was safely aboard the *Iwo Jima*. On seeing the deployed parachutes, "I felt fine," he remembers. "But I felt a lot better when I saw them walking on the deck of the carrier. That's the way I always was. Too many things could happen between the parachutes and the deck." As I interviewed him thirty-five years later, Kraft pondered the memory of the crew walking in the open air on the *Iwo Jima*. "That was one of the most excellent things I've ever seen," he finally declared.

When the crew and the flight controllers were eventually reunited in Houston, there was, naturally, a raucous celebration, the highlight of which was the playing of an audiotape made by splicing together various Mission Control voice loop recordings. The creator was merciless, lampooning almost everyone involved, and got a great deal of mileage from Liebergot's dramatic sound bite "We may have had an instrumentation problem, Flight," and Kranz's later line "I don't understand that."

Despite President Nixon's award of the Presidential Medal of Freedom to those involved in saving *Apollo 13*, few dwelt on its significance at the time. They were busy building on the lessons learned from *Apollo 13*, and nine months later *Apollo 14* would blaze into the skies above Florida as it left for the moon. Indeed, for years, although many in Mission Control viewed it as the highlight of their careers, they detected a sense of embarrassment in NASA about the mission—they had failed to go the moon after all—a taint that would not fully be dispelled until the release of Ron Howard's movie in 1995.

The movie, starring Oscar-winning actor Tom Hanks as commander Jim Lovell, became a smash hit, earning $350 million in theaters worldwide, putting it in the top one hundred U.S. movies of all time. The filmmaker actually shot scenes in zero gravity on a set built inside a KC-135 airplane. Used to train astronauts, this plane flies parabolic arcs to produce weightlessness for brief periods. Howard's attention to recreating the look and feel of *Odyssey*, *Aquarius*, and Mission Control was impeccable.

Although largely based on *Lost Moon*, Lovell and Kluger's detailed book about the mission, the *Apollo 13* screenplay necessarily dramatized the crisis, condensing events and using composite characters to stand in for dozens of flight controllers. What do some of the real NASA controllers think of the film? According to Gene Kranz, the lead flight director of *Apollo 13*, it was "really well done": "I was very pleased at how it portrayed my flight controllers and flight directors. It provided the public with a glimpse into this business called mission control." Although Chris Kraft, deputy director of the Manned Spacecraft Center at the time and Kranz's boss, feels some parts of the film are "overdramatized," he believes the movie captures the right tone of the mission. "The movie was tremendous," he says.

LM branch chief Jim Hannigan watched the film with regular moviegoers. At one point he remembers calling out loudly to his wife, "We didn't do that!" Everyone in the theater "turned around and looked at me," he laughs. "The producers hammed it up with the astronauts, and Hollywoodized it," but Hannigan says he understands why: "It would be a boring movie if you just made it technical."

John Aaron, CSM flight controller, says, "Some guys I worked with, they worked a particular niche, and that niche wasn't depicted in the detail they would have liked to have seen it. But of all the movies I've seen about space, *Apollo 13* was the only one even close to being factual enough that I wasn't ashamed of it. . . . They did an excellent job." Aaron believes that although the filmmakers did not depict everything, they were very effective in conveying "the mood, the emotion . . . the sense of what went on."

Jerry Bostick, chief of the flight dynamics branch, who was working at Grumman in the 1990s, was a technical advisor for the movie. He become involved after his son, Michael Bostick, an associate producer for the film, phoned to ask him his opinion about buying the movie rights to Lovell and Kluger's book. "I said, 'Make a movie of that? What are you going to do, a documentary?' Then about two hours later I got another phone call from my son and Ron Howard," remembers Bostick, who reminded the moviemaking pair of the 1969 lemon *Marooned*, which featured three Apollo astronauts stranded in orbit—"I thought [it] was the worst thing ever made in my life," he says. But Howard was convinced he could do justice to *Apollo 13* and still make a mainstream film, and Jerry Bostick was convinced to come on board as an advisor. Bostick is pleased with the movie, although he acknowledges that the absence of some elements "made some people unhappy . . . but I learned a lot about making movies. If you're going to try to tell a four-day story in two hours, you have to condense a lot of things."

Bostick's main worry was how some of his coworkers at Grumman, the aerospace contractor that built the LM, would react In 1995 Bostick's immediate boss was *Apollo 13* astronaut Fred Haise, and his boss in turn was Tom Kelly, the chief designer and builder of the LMs. Bostick also worked with Ed Smylie, the man who led the effort to save the crew from carbon dioxide poisoning, and Ken Mattingly, an astronaut who was originally slated to be part of the *Apollo 13* crew. Mattingly, who is prominently portrayed in the movie by actor Gary Sinise, was replaced by Jack Swigert at the last moment. "Every other day I'd go to Ron and say, 'I've got four guys back at Grumman who'll kill me if this doesn't come out right!'" Bostick recalls.

In particular, according to Bostick, Haise was initially "leery" about making *Apollo 13* into a movie. At the premiere of the movie in Houston, Bostick sat beside Haise, as the former astronaut watched himself being portrayed by a laconic, gum-chewing Bill Paxton. "I was just sweating. He's going to hate this and fire me!" Bostick remembers thinking. But Haise made no comment until halfway through the bus ride back from the premiere. Suddenly Haise reached over and grabbed Bostick by the shoulder and said, "'I never chewed gum.' And that's all he ever said about it," laughs a relieved Bostick.

Over the years that followed, the controllers left NASA one by one, leaving Mission Control to a new crop of flight controllers. But a chain of excellence had been forged—to this day every flight director in NASA has come up through the ranks. Each one studies the trade from flight directors who have also come up through the ranks, back to the prototype flight director, Chris Kraft, learning the values of discipline, competence, confidence, shouldering responsibility, toughness, and teamwork that form the foundation of Mission Control's culture, demonstrated so ably during the *Apollo 13* crisis.

Kraft still sees the culture he helped forge in evidence today at Mission Control. "I think that's the one place in the space program that still has it," he says bluntly. "The people who are running the control center today are just as good as we ever had, and I can't praise them too highly." While agreeing that today's flight controllers are top-notch, some of the other *Apollo 13* veterans worry that their authority is being slowly undermined. "Over the years I've seen that authority deteriorate badly," sighs Bostick, pointing to the management structure displayed during the 2003 *Columbia* tragedy as a worrying example. There was a "team of program managers who would meet every day and do flight planning: 'Here's what we're going to do today,' and they would pass that on to the flight directors, making the flight directors just executors [of other people's decisions]," Bostick says.

Aaron believes the problem stems from a lack of leadership, from Capitol Hill on down. Without an urgent and agreed upon goal—such as beating the Soviets to the surface of the moon—NASA started being subjected to the conflicting demands of different individuals and political camps in Congress, says Aaron, who worked at NASA Headquarters in Washington DC in the 1980s. NASA's marching orders have become "diffused and muddled. . . . That then affects NASA management, who instead of being technical gurus, have to become amateur politicians." But—in what can only be good news for an agency now planning to return to the moon after a forty-year pause—Aaron, who retired from NASA in 2000, is convinced that the space program's engineers are still the best in the business. At the grassroots level at least, they have "still got the Right Stuff," he says.

5. Altered Directions

Colin Burgess

I believe there is no philosophical high-road in science, with epistemological signposts. No, we are in a jungle and find our way by trial and error, building our road behind us as we proceed.

Max Born (1882–1970)

The earlier triumph of the orbital docking and crew transfer between *Soyuz 4* and *5* in 1969 had given considerable encouragement to Soviet space officials, still suffering from badly deflated spirits after three NASA astronauts had circled the moon the Christmas before in their *Apollo 8* spacecraft. Anticipating a need to become increasingly bolder in their plans and expectations, they stepped up plans for their Soyuz series of manned flights.

Also buoyed by the achievements and positive propaganda resulting from the joint Soyuz mission, Vasily Mishin boldly decided to dispense with plans for a thirty-day Soyuz endurance mission and conduct what would become known as the "troika" flight, involving three separate manned spacecraft in orbit around the Earth at the same time. After a period of evaluation he placed his proposal before a meeting of the Soviet State Commission on 25 April.

Mishin's concept was actually an extension of a much earlier proposition by his predecessor Sergei Korolev following the successful *Vostok 2* flight piloted by Gherman Titov. In a radical plan the former chief designer of the Soviet space program had suggested launching three separate manned Vostok spacecraft into orbit within days of each other. Considered far too ambitious by the state commission, Korolev's plan was subsequently abandoned in favor of a more conservative approach, and a tandem flight was subse-

21. The *Soyuz 6* crew of Valery Kubasov (*left*) and Georgi Shonin. Courtesy Colin Burgess Collection.

quently conducted by just two spacecraft, *Vostok 3* and *4*. What Mishin now envisaged was to launch a two-man *Soyuz 6* into orbit, a mission in which the crew would evaluate the use and practicalities of a welding experiment named Vulkan. *Soyuz 7* and *8* would later follow this craft into orbit before docking with each other—a procedure to be filmed by the crew of *Soyuz 6*, which was deliberately not fitted with a docking device. The two Soyuz craft would remain docked for three days, and then all three would return individually to Earth. Tentative plans were then set in place for *Soyuz 9* and *10* to carry out on-orbit tests of the Kontact docking equipment and life support systems that would be used in the L3 lunar landing program.

The innovative triple flight program was eventually approved by the state commission and began with the launch of *Soyuz 6* from Tyuratam on 11 October 1969, successfully carrying Georgi Shonin and Valery Kubasov into orbit. This launch was followed the next day by a second one, this time with a three-man crew of Anatoli Filipchenko, Vladislav Volkov, and Viktor Gorbatko.

All five cosmonauts were making their first flights into space. Georgi Shonin, Ukrainian by birth, had served as a fighter pilot with the Soviet and Baltic fleets before being selected in the first cosmonaut detachment in 1960. He had subsequently graduated with honors from the Zhukovsky Air Force Engineering Academy. Russian-born Valery Kubasov had earned a master's degree in technical sciences at the Bauman Aviation Institute in

22. Viktor Gorbatko (*left*), Anatoli Filipchenko, and Vladislav Volkov enjoying a few moments of levity. Courtesy Colin Burgess Collection.

Moscow and worked as a design engineer before becoming a cosmonaut. He was an expert in spacecraft systems, ballistics, astronomy, and planetary physics.

The commander of the second craft into orbit, forty-one-year-old Anatoli Filipchenko, had logged over 1,500 hours flying time in the Soviet air force when he joined the cosmonaut corps in 1963 and had previously served as a backup pilot for *Soyuz 4*. Moscow-born Vladislav Volkov had at first wanted to be a pilot, but his interest later turned to aeronautical engineering, and he had also received training at the Bauman Aviation Institute. He was working in Korolev's OKB-1 design bureau when he applied to be a civilian cosmonaut and was selected in May 1966. The third crew member, Viktor Gorbatko, had, like Shonin, been selected in the first cosmonaut air force group in 1960, graduating from the Zhukovsky Air Force Engineering Academy in 1968. He had previously served as a backup pilot for *Soyuz 2* and *4*.

With five cosmonauts now circling the planet, many Western experts had reasonably predicted that the Soviet Union might be planning to construct a large orbiting space laboratory or at least test procedures for building such a vehicle. This was partially dispelled in a television broadcast from *Soyuz 6* in which Valery Kubasov revealed that his spacecraft not only lacked the equipment necessary for a docking but was carrying a vast array of scientific equipment and additional fuel for maneuvering in space.

As *Soyuz 7* settled into orbit, there was already fresh speculation in the Western press of a third manned mission. It was confirmed on the next

23. The crew of *Soyuz 8*, Vladimir Shatalov (*left*) and Alexei Yeliseyev, failed to achieve docking with the *Soyuz 7* spacecraft. Courtesy Colin Burgess Collection.

day, 13 October, when *Soyuz 8* left the launchpad carrying Vladimir Shatalov and Alexei Yeliseyev. Unfortunately, plans for a docking between *Soyuz 7* and *8* would have to be abandoned when a manual control problem prevented the two craft from linking up. The closest they came to each other, as observed by the *Soyuz 6* crew, was around 1,600 feet. As nothing further could be achieved, the three craft gradually drifted apart to allow the crew members to carry out assigned experiments in navigation, orientation, science, and astronomical observation.

Georgi Shonin, a member of the first group of twenty Soviet cosmonauts, was asked about the main purpose of the flight of *Soyuz 6*. "It was our task to make a film of the docking," he responded.

It was very interesting for me and Kubasov to do it. Two spacecraft were approaching each other. For me as a professional pilot and as a cosmonaut it was very interesting to make the calculations and to maneuver. They all concentrated on the docking of Soyuz 7 *and* 8. *Since Shatalov had a lot of trouble with the [*Igla*] docking system they were all concentrating on solving that problem. When they found out that our ship was closing in on those two others the ground said*

"Wow, how did they manage to do it?" But unfortunately we couldn't make a photo because we were in the Earth's shade.

Shonin was asked if this meant there were no photos or film available of the attempted docking. "No, nothing," he confirmed.

One of the real success stories of the troika mission was the series of Vulkan experiments carried out on *Soyuz 6*. In preparation for this, Shonin and Kubasov had sealed themselves in the descent module and depressurized the adjoining orbital module. After this they activated the Vulkan apparatus, and while they observed and remotely managed the process through a window, three automated welding processes were carried out in the orbital module—electron beam, fusible electrode, and compressed arc. The cosmonauts were then able to report that the latter two techniques did not work well in a vacuum, while the electron beam experiment had proven highly successful. Such a process might be crucial in the future construction or repair of huge orbiting space stations, and the results thrilled those on the ground who had prepared the Vulkan experiments.

All three spacecraft then returned to Earth, with *Soyuz 6* recording a total mission elapsed time of four days, twenty-two hours, and forty-two minutes. Somewhat remarkably, both *Soyuz 7* and *8* recorded precisely the same mission flight time, just one minute short of that attributed to *Soyuz 6*.

Originally, veteran *Vostok 3* cosmonaut Andrian Nikolayev and rookie flight engineer Vitaly Sevastyanov had been the designated prime crew for *Soyuz 8*, but these plans were abandoned when the two men—Nikolayev in particular—did poorly in docking simulation tests six months before the mission. Both men were subsequently dropped to the position of backup crew and their places taken by Vladimir Shatalov and Alexei Yeliseyev. Gen. Nikolai Kamanin advised Nikolayev that it was highly unlikely they would be restored to the flight, a stance that angered several cosmonauts and officials. Valentina Tereshkova, the world's first woman in space and at that time married to Nikolayev, is said to have stormed into Kamanin's office that evening, threatening to take the matter to higher officials in the space hierarchy, including the influential Dmitri Ustinov, secretary of the Central Committee. Kamanin would not be intimidated and briskly told Tereshkova that if she went above his head, it would prove to be a huge mistake.

The inference was there; he still had a major role in future crew selections. Sense must have prevailed, because the two men were subsequently reassigned to *Soyuz 9*, a long-duration mission and one not involving any rendezvous or docking with a second craft.

The experience gained in the multiple-flight missions, and in particular the docking and crew transfer carried out between *Soyuz 4* and *5*, amply demonstrated that the Soviet Union was achieving a grasp on the technology required to place a space station into orbit that could be visited by a series of crews. Along with the question of crew compatibility on lengthy missions, there were also serious concerns about the physiological effects of prolonged spaceflight on space travelers in a weightless environment. Monitoring this aspect would be one of the main objectives of the flight.

On 1 June 1970 a very special guest was welcomed as a visitor to the cosmonaut training center in Moscow's Star City by Gen. Aleksandr N. Yefimov, deputy commander in chief of the Soviet air force. That person was Neil Armstrong, who had returned from the moon just eleven months earlier. His was the second in a number of reciprocal visits being carried out by the astronauts and cosmonauts from both space-faring nations. Frank Borman had visited Moscow and Star City, and by way of reply Georgi Beregovoi and Konstantin Feoktistov had been invited guests of the United States government, meeting several astronauts, having a private audience with President Richard Nixon, and even paying a visit to Disneyland.

Armstrong's hostess in Star City was Valentina Nikolayeva, but curiously, her husband, Andrian, was not among the cosmonauts introduced to the moon walker. One of Armstrong's first official functions was to pay a respectful visit to Gagarin's study, preserved just as he had left it on the day he died, and sign the guestbook. He later gave a well-received talk to a large crowd—including a number of cosmonauts—crammed into a reception hall, describing his *Apollo 11* mission and showing a film of the flight. Following the talk, and by way of thanks, Beregovoi and Feoktistov presented Armstrong with a model of two docked Soyuz spacecraft, after which Valentina Tereshkova made her way onto the stage. Here amid rapturous applause she presented the moon walker with a huge bouquet of flowers and a badge commemorating his visit. In offering his thanks, Armstrong told the audience that he had placed medallions on the surface of the moon honoring the Soviet cosmonauts known to have lost their lives, including Yuri Gagarin and Vladimir Komarov. He then invited the two widows of these cosmonauts onto the stage, embraced both with tears in his eyes, and presented them with duplicates of the medallions. It was an incredibly poignant occasion, Armstrong later saying he had been "most emotionally moved." It was a gesture greatly appreciated by the large crowd. At first they had been respectfully silent during the presentation to the two space widows, but then they erupted in cheers and applause. Another set of identical medallions were later presented to the Star City museum.

It had been a long day full of emotion and officialdom, and Armstrong was looking forward to kicking back and unwinding at dinner with a number of the cosmonauts. It was an all-male affair, with numerous vodka toasts, at the end of which Armstrong was presented with a gift—a beautifully crafted double-barreled shotgun with his name carved into the stock.

Around midnight Beregovoi insisted that Armstrong accompany him to his Star City flat for coffee, and the astronaut could not refuse. When they got there around midnight, the cosmonaut excused himself for a few moments, saying he had to make an urgent phone call. Armstrong did not speak the language, so he waited patiently as the call was made, and then a couple of minutes later the phone rang with a return call. Beregovoi quickly acknowledged something and then smiled as he hung up. "Neil," he said. "Two of our boys are about to lift off. Would you like to watch it?" A surprised Armstrong then sat down as Beregovoi switched on his television set, and the two men watched a delayed telecast of the launch of *Soyuz 9* from two hours earlier. "This is especially in honor of your visit here," the cosmonaut informed him as the tape began, and later the two men celebrated the successful launch with more vodka. "That is one of the best gifts you can give me," said an overwhelmed but delighted Armstrong, who insisted on having his good wishes communicated to the crew.

Two men were on board the *Soyuz 9* spacecraft, Armstrong was told, space rookie Vitaly Sevastyanov and *Vostok 3* veteran Andrian Nikolayev, Valentina's husband. "So I had spent the whole day with Tereshkova and the whole evening with all the colleagues of the two cosmonauts," the astronaut later revealed in his biography, *First Man*, "and it was never mentioned once that they were having a launch that day. I concluded that Valentina was either awfully good at keeping a secret or she was dreadfully misinformed."

It is now known that Nikolayev and Sevastyanov had earlier come perilously close to losing this mission, when Nikolayev was caught smoking just a week before the flight in violation of direct orders from Nikolai Kamanin. Sevastyanov, also a heavy smoker, would later confess that he too had been smoking against orders. Kamanin was furious, and in his diary he noted his frustration at this rebellious attitude. "If I had learned of this a month ago," he wrote, "I would have been against allowing Nikolayev and Sevastyanov to fly, but now, when there are only a few days left until the launch, and Nikolayev's crew has already been confirmed in fact as the primary crew in the Party's Central Committee and the government, it is impossible to raise the matter of replacing the cosmonauts with their backups."

There was a surprising and unprecedented openness about the principal objectives of this new mission—the first in eight months by the Soviet Union. As the Tass News Agency reported, the *Soyuz 9* crew would undertake medical and biological research pertaining to long-duration space flight. In fact, Tass put "medico-biological research" first on the list of mission objectives. Other stated goals were to study geographical and geological objects as well as the snow and ice cover of the planet, useful in helping to predict flood dangers, and to carry out meteorological research on "a solitary orbital flight in near Earth space," a strong indication that no other spacecraft would be launched. Along with providing mission details, a large number of launch-pad photographs of the Soyuz rocket and spacecraft, code-named "Falcon," were also released to the public. The news agency later quoted Nikolayev as saying that the orbiting crew were "coping well with conditions of weightlessness and [were] carrying out the flight program."

Eight years after his historic mission aboard *Vostok 3*, forty-year-old Nikolayev was back in orbit again. For flight engineer Sevastyanov, however, it was his first flight into the cosmos. At thirty-four years of age, Vitaly Ivanovich Sevastyanov was a highly accomplished young scientist, and a candidate of technical sciences (albeit the lower of two levels of doctorate in the Soviet Union). Six years younger than Nikolayev, he was born in the small Ural Mountains town of Krasnouralsk in the Sverdlovsk Region of Russia on 8 July 1935, the son of Ivan, a cab driver, and Tatyana Sevastyanov. Vitaly was eight years old when he began his formal education at the Krasnouralsk Secondary School No. 1, but two years later, in 1945, the family moved to Maykop, a thousand miles south of Moscow, where he continued his education. The family then relocated to the pleasant Black Sea resort town of Sochi, where he attended the Nikolai Ostrovsky Secondary School No. 9. During his later school vacation periods he took on temporary work as a deckhand aboard a pleasure launch. The launch, known only by the number *10*, would ply a daily route back and forth between Sochi and the nearby ancient city of Khosta. Sevastyanov handled the ropes, swabbed the decks, polished the brass railings, and eagerly performed other shipboard duties to earn a little spending money. It may have been hard and often repetitive work, but he loved being out in the fresh salt air and on the sparkling waters of the Black Sea.

Having completed school with the distinction of a gold medal for his work, Sevastyanov then enrolled at the Moscow Aviation Institute, where he began his doctorate studies in aviation and space engineering. After graduating in 1959, he joined Sergei Korolev's OKB-1 design bureau, where he became part of the design team for the Vostok and Zond spacecraft. In April 1961 he took great pride in being part of the design team responsible for sending the first spaceman, Yuri Gagarin, into orbit. "I had the good luck of listening to the broadcast of the preparations for the launch, and the blast-off itself, while at the Flight Control Center in Moscow as a representative of the design office," he later revealed.

Now married to the former Alevtina Ivanovna Butusova and with a daughter, Natasha, born in 1962, Sevastyanov undertook a postgraduate course at the Aviation Institute and was awarded his candidate of technical sciences degree in 1964. Meanwhile, he gave lectures and taught classes on the physics of spaceflight, rockets, and celestial mechanics at the cosmonaut training center—one of the first to do so—before formally applying to become a cosmonaut himself. After being accepted for training in January 1967, he was immediately assigned to work on a manned Zond mission (*Zond 9*) then scheduled for July 1969 as part of the Soviet Union's lunar program, which to Sevastyanov's regret was canceled the following year. Initially, he and Pavel Popovich had been assigned as the second backup crew for the first circumlunar mission, but he was replaced late in 1967 by Pavel Belyayev, before the official breakup of this ambitious program, so he could concentrate his efforts on the Soyuz program.

On 20 April 1969 Vitaly Sevastyanov was paired with veteran cosmonaut pilot Andrian Nikolayev to begin training for an early Soyuz mission. Three days into their flight, the two *Soyuz 9* cosmonauts had to counteract an unexpectedly high buildup of electric current, which was creating an excess of power and might cause problems in the spacecraft's systems. The problem was related to the very brief periods of "night time" in each orbit, due to the particular orbital inclination of the spacecraft, which was almost parallel with the Earth's terminator. The energy-absorbing solar panels were proving to be overly efficient, so ground controllers recommended to the crew that they initiate a slow roll of the spacecraft at a rate of a half-degree per second in order to give the solar arrays less constant exposure to the sun. This procedure seemed to rectify the problem.

By the following day things seemed to be progressing well for the crew of *Soyuz 9*. They had carried out manual orbital maneuvers, placing their spacecraft into a near-circular orbit 160 miles above the Earth, and photographed vital geographic and geologic areas of the Earth's surface. Sevastyanov had also been analyzing a puzzling smearing of his porthole but determined that it had occurred during the earlier course-correcting engine burn and was not a critical concern. Relatively simple but significant biomedical tests were also going on aboard the craft.

Each day the two cosmonauts were scheduled to spend two one-hour periods performing a number of exercises designed to prevent or minimize degeneration of their muscles, bones, and circulatory system. Elastic torso straps, harnesses, and simple chest expanders provided the means for conducting exercise regimes in their relatively cramped spacecraft. They would each monitor devices that kept check on their heart rate, blood pressure, respiration, and other vital physiological functions. Doctors on the ground, however, were harboring concerns that the cosmonauts might not be doing all their physical exercises. They had discovered the level of carbon dioxide exhalations were not matching the prescribed periods of physical exertion.

When quizzed by doctors, Nikolayev was ready to confess. "We have no time to do all the exercises," he responded, "because we are busy with all the experiments for the program." According to plan, the cosmonauts were working up to sixteen hours per Earth day. The reply, a mild rebuke, came quickly: "Physical exercises are a must at the expense of any experiments." Their admonitions, however, would have little effect, as the two cosmonauts never adhered to their fitness schedule, and their apathy would later result in dire physical consequences.

One hitherto unknown physiological effect had been noted and reported by both men and, in the absence of any sound reason, attributed to a lack of gravity. It involved a slight deterioration in their vision, and they mentioned poor eye muscle coordination and a certain difficulty in discerning colors. This phenomenon apparently had very little impact on their regular activities but was deemed of interest to medical controllers on the ground. New equipment that would be tested during their flight included a spacecraft orientation system based on star-lock navigation, which would use the prominent stars Vega and Canopus in the northern and southern celestial skies.

On 8 June, the sixth birthday of Nikolayev's daughter Elena, his wife Valentina was given permission to bring the little girl into the ground control center to allow the cosmonaut the opportunity to wish her a happy birthday by radio and television. They were able to observe him and Sevastyanov on a television monitor and enjoyed a moving few minutes in conversation.

On the tenth day of their mission, 10 June, Nikolayev and Sevastyanov enjoyed the first of their prescribed "days off." No experiments were scheduled on these days, and radio communications to the spacecraft would be kept to a minimum. Both keen chess players, they had issued, before leaving for space, a friendly challenge to a long-distance game to Gen. Nikolai Kamanin and CapCom Viktor Gorbatko. The cosmonauts would be Team Space, leading off using the white chess pieces, and those on the ground would be Team Earth, playing with the black. A game strategy known as Queen's Gambit Accepted applied, but after a lengthy game played over four communication periods and three orbits, the match ended in a draw on the thirty-sixth move.

Three days later those in touch with them, and others watching them on television, noticed that the crew seemed to be quite fatigued and making uncharacteristic and elementary errors, such as programming their television camera on the wrong setting. They were still carrying out a fairly rigorous work program and had decided between them to shorten their afternoon rest periods (two hours after lunch) in order to get everything done. The following day Sevastyanov accidentally engaged the craft's ASP automatic landing system. It was not a critical error and could easily be rectified—it had also been done by Anatoli Filipchenko on board *Soyuz 7*—but to the medical observers it was further indication of flagging attentiveness by the crew. Later that day Nikolayev formally requested a two-day extension to the flight, but Kamanin denied it.

On the seventeenth day of their mission Nikolayev and Sevastyanov set a new space endurance record on their 220th orbit of the Earth. The record they broke had belonged to U.S. astronauts James Lovell and Frank Borman, who in 1965 had flown for thirteen days, eighteen hours, and thirty-five minutes aboard *Gemini 7*. It was one of the flight objectives set before their launch. In recognition of this record-breaking feat the two astronauts graciously sent their counterparts a message: "We congratulate you and pass on our best wishes as you pass new milestones in space exploration."

Now nearing the end of their eighteen-day mission, Nikolayev initiated the landing process six thousand miles from their landing site. After a fiery reentry the *Soyuz 9* craft touched down in daylight, right on target, in a plowed field at Intumak, some forty-seven miles west of the city of Karaganda in Kazakhstan. Recovery and medical staff were quickly on the scene, ready to transport the cosmonauts to the medical facility at Star City for extensive checkups and postflight recuperation. They had fully expected the two men to display some signs of fatigue and weakness after their long-duration flight, but to their alarm they found both cosmonauts in much worse shape, physically incapable of any real exertion and complaining of feeling incredibly heavy. They had to be gently extracted from their couches and out of the descent module. An official cameraman was at the site, and in compliance with the man's request Nikolayev managed to smile broadly, but he then slumped back into the recovery couch with his energy completely sapped. Sevastyanov was in no better shape, and the two attending physicians held grave fears for the cosmonauts' well-being. Nevertheless, the Tass News Agency issued a bulletin shortly after the landing, announcing that "a quick medical check-up of the cosmonauts showed that they had withstood the prolonged space flight well." It was far from the truth. There would be none of the customary official postflight functions or receptions, as the crew was almost incapable of movement for several days.

Nikolai Kamanin met the cosmonauts aboard the recovery aircraft, and as he later recorded in his diary, he was shocked at their appearance. "When I entered the aircraft's cabin, Sevastyanov was sitting on the sofa, while Nikolayev was at a small table. I knew they were having a hard time enduring the return to the ground, but I had not counted on seeing them in such a sorry state. Pale, puffy, apathetic, without the spark of vitality in their eyes—they gave the impression of completely emaciated, sick people." The plan to fly the cosmonauts to Star City was immediately postponed until further tests could be carried out at Karaganda. Following these tests, they would be carefully driven by limousine to the cosmonaut training center in Moscow.

"For lack of muscle, we couldn't walk for seven days," Sevastyanov recalled with candor in 2001. "Our thigh girths had shrunk by five-and-a-half centimeters. Cardiograms were suggestive of something very close to a severe attack. Actual lesions, however, could not be detected. The hearts had simply adapted to weightlessness and took normal gravity as [an] abnormality and a serious challenge. We had returned to Mother Earth as seasoned space dwellers who had gone a long way to adapt to the conditions of outer space. Normal health did not start to return before our second week back home."

Nikolayev also recalled problems in readapting to gravity. "Orbital stations were unheard of, and our mission was supposed to test uncharted waters ahead of the first such station, the Soviet Salyut. Cramped conditions aboard our Soyuz craft ruled out appropriate exercise, spelling quick physical degeneration in every respect. My heart lost twelve percent of its initial volume over our eighteen days in flight.

"The first day back home was quite an ordeal," he continued.

A few days later, however, Sevastyanov offered me to share a cigarette with him, in a washroom where he had hidden it during medical check-ups before going up. We had to lean against the wall crawling to our not-so-far-away destination. I smoked first, as mission commander, and then passed the half-smoked cigarette to Vitaly. At this moment our doctor came in. And instead of scolding us, which we certainly well deserved, he burst into jubilation. He told us he felt assured of our quick recovery seeing us smoking.

Both men had lost weight on the flight, which took some days to regain. Their cardiovascular systems were within normal limits, although concerned physicians had observed what they termed "a certain abnormality" in both cosmonauts.

Professor Anatoli Grigoryev, a full member of the Russian Academy of Sciences and director of the Institute of Medical and Biological Problems

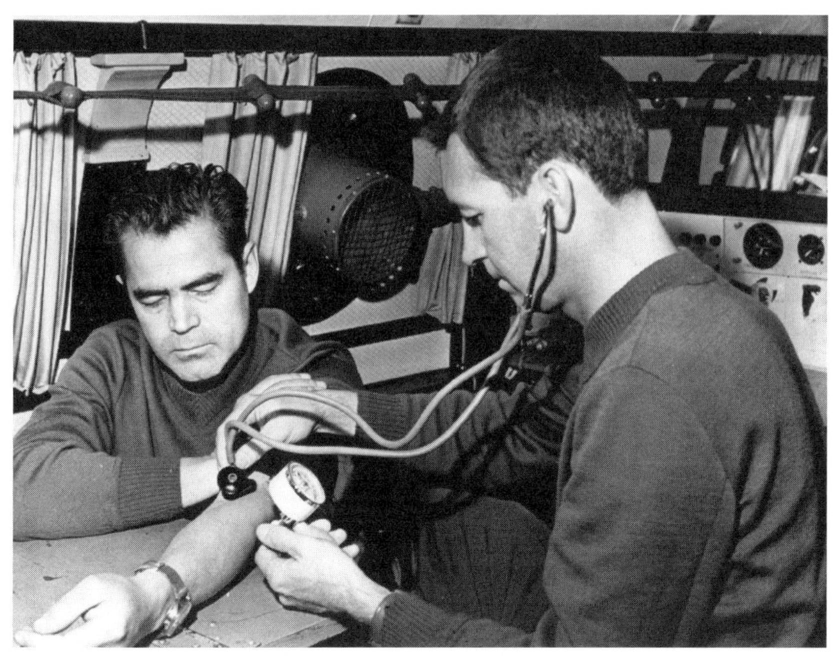

24. The *Soyuz 9* crew of Andrian Nikolayev and Vitaly Sevastyanov returned to Earth in very poor physical condition. Courtesy Colin Burgess Collection.

in Moscow, stated: "I saw both men shortly after their flight. That 1970 mission was a bold venture into the unknown. The Soyuz craft offered too little room inside, severely restricting exercise. For bigger spaceships that followed, my Institute developed in-cabin training machines. It also came up with techniques for cosmonauts to keep fit through exercise." The debilitated condition of the cosmonauts—which came to be known as the "Nikolayev Effect"—demonstrated the need to set up suitable exercise regimes in future space stations.

One researcher into early Soviet spaceflight, Asif Siddiqi, said that apart from a lack of exercise by the crew, there was evidence, discovered after their return to Star City, of other factors to explain the debilitated state of the two cosmonauts. Over the next few days it became increasingly clear that part of the reason for the very poor shape of the *Soyuz 9* crew was the slow spin of the spacecraft throughout the mission. The spinning also produced a weak field of artificial gravity, which affected the clarity of results of several experiments aboard the ship. Nikolayev and Sevastyanov spent several days in quarantine, not only to protect their weak bodies from infections but also, as it turns out, because a mutation of two microbes not occurring on Earth had been found in their metabolic systems. For five days after their return, the microbes spread very rapidly but then died from the effects of gravity.

The slow spinning of *Soyuz 9* had been introduced to keep the craft's nonsteerable solar panels perpendicular to the sun in order to maximize the generation of electricity. The slow rolling, however, also produced a mild form of motion sickness in the cosmonauts.

On 3 July, following a two-week period of isolation, recuperation, and medical tests, Nikolayev and Sevastyanov attended a formal reception at the Georgiyevsky Hall in the Great Kremlin Palace. Nikolayev was awarded his second gold medal as Hero of the Soviet Union and informed, to his delight, that he had been promoted to the rank of major general. Sevastyanov also received the gold star of Hero of the Soviet Union—his first—and the title Flier-Cosmonaut of the USSR.

Soyuz 9 would prove to be the final human space mission for 1970. Andrian Nikolayev never flew into space again, and on 3 July 2004 he died. But Vitaly Sevastyanov operated a second mission in 1975, paired with Lt. Colonel Pyotr Klimuk on *Soyuz 18*. Launched from Tyuratam on 24 May, they would safely dock with the *Salyut 4* space station and spend some fifty-nine days carrying out an extensive workload that included multi-spectral photography of the Earth, operating a solar telescope, medical experiments, and atmospheric measurement. Interspersed with all this activity, however, were ten days of relaxation for the two men. On one occasion they came within two hundred miles of the docked *Soyuz 19* and Apollo ASTP spacecraft. On 26 July, after a highly successful and productive mission, *Soyuz 18* undocked from *Salyut 4* and the two cosmonauts landed thirty-four miles northeast of the town of Arkalyk.

Soyuz 18 was the final spaceflight for Sevastyanov, who the following year was removed from active flight status. He then worked in ground control for the *Salyut 6* space station before returning to spacecraft design in the Buran Soviet shuttle project in the 1980s. In December 1993 he took his final leave of the space program, having spent a total of eighty days in orbit over two Soyuz flights, and was elected to the State Duma (the lower chamber of the Russian parliament) in 1994. By 2003 he was Duma deputy and an influential member of the Duma International Affairs Committee, positions of great political importance he retains to this day.

The Soviet Union was forging ahead on this new path of extended flight and science in orbit, but all too soon a misguided overconfidence would result in tragedy and concerns for the very future of human space exploration.

6. Science, and a Little Golf

Philip Baker

Whoever wishes to foresee the future must consult the past;
for human events ever resemble those of preceding times.
This arises from the fact that they are produced by men who
ever have been, and ever shall be, animated by the same passions,
and thus they necessarily have the same results.

Niccolò Machiavelli (1469–1527)

The eighteen men gathered in the small conference room on the third floor of Building 4 at Houston's Manned Spacecraft Center in April 1967 knew that this was an important moment in their lives. Most of them understood that an Apollo flight was in their immediate futures, although they did not know which one. Most important, they did not know which mission would prove to be the prize goal, the first lunar landing.

Deke Slayton got straight down to it. "The guys who are going to fly the first lunar missions are the guys in this room," he said, looking at each man in turn.

The G mission would prove to be the one that they all wanted. A few months later, in September, NASA proposed a progressive list of missions leading up to the first manned landing, setting them out in alphabetical order ranging from *A* (an unmanned Command and Service Module [CSM] test) through *F* (manned CSM and Lunar Module [LM] in lunar orbit). But the G mission was the one intended to *attempt* to place the first crew on the moon. The first attempt did not necessarily mean the first landing, and everyone understood that.

Much to their acute disappointment, two men in that room knew for sure that the first manned lunar landing mission would not be theirs. Nor, in fact, would any mission in the immediate future. Deke Slayton and Alan Shepard were both astronauts but in name only. They were both grounded from flying, even solo in airplanes, by different medical maladies. Shepard had at least managed to fly once as the first American in space on the *Mercury-Redstone* 3 (MA-3) mission back in May 1961, before being grounded by a debilitating inner ear disorder. Slayton had been suspended from active astronaut duties just weeks before making his scheduled *Mercury-Atlas 7* (MA-7) flight aboard a spacecraft he had already named *Delta 7*. A heart fibrillation problem had caused nervous NASA and Air Force flight surgeons to play it safe and replace Slayton rather than risk sending him on such a stressful mission. Both Shepard and Slayton still entertained high hopes that they would fly again, however, and both were determined to overcome the decisions that had nailed them to the ground.

Test pilots and astronauts share a common trait—that of complete faith in their own abilities. They can be no other way, when their life depends on their own actions or their ability to overcome the inefficient actions of others. Any lack of faith would have long ago meant that they could never climb into an aircraft cockpit again. Another similar breed of individuals are race car drivers and for much the same reasons. Some elite pilots such as Alan Shepard are chosen to make the first flight of a brand-new vehicle. In his case it was not just any vehicle but the very first manned craft planned to fly into space. Shepard himself probably described the achievement best when he was interviewed in 1991: "That was competition at its best, not because of the fame or the recognition that went with it, but because of the fact that America's best test pilots went through this selection process, down to seven guys, and of those seven, I was the first one to go."

He did indeed become the first man to ride a Redstone rocket in the Mercury capsule, but Soviet cosmonaut Yuri Gagarin had already become the first human in space three weeks earlier. Shepard had lost that particular place in history due to the caution of the planners, who chose to run one more test launch before committing him to history. The test flight was a success, but Gagarin flew shortly afterward, and a furious Shepard's chance to be first was over. Soon thereafter he became the first American in space, but that accomplishment was not enough for Alan Shepard. To make things even worse in his mind, John Glenn flew not the next mission but the one after that and became an instant American hero. Glenn was the first American to orbit the Earth, albeit the third person to do so, moving Shepard still further down the line of achievement as he saw it.

Shepard campaigned for some time for a further Mercury mission that would see him fly again, this time for three days, on a proposed mission to push the Mercury design to its limits and fill the time before the Gemini spacecraft was ready to fly. NASA administrator James Webb felt, however, that the Mercury program had run its course and that it would just be inviting disaster to push it any further. *Mercury 10* would not fly.

Despite the Mercury program ending, Shepard would be somewhat compensated by his appointment to command the very first flight of the next generation of American spacecraft, Gemini. But that was to be the last of the good things for him for a while.

As early as 1963, Shepard began to experience bouts of dizziness and nausea, initially not lasting very long but always returning after a few days. He hoped, as anyone would, that these were isolated incidents and certainly not an indication of a bigger problem. Denial was playing a heavy hand, but the facts became increasingly difficult to ignore. The original symptoms came and went repeatedly and were joined by a ringing in his left ear, vomiting, and complete loss of balance. Clearly, this problem was not going away by itself. Reluctantly, Shepard had to consult with a flight surgeon and submit to a diagnosis by people whose opinions he did not trust.

Like anyone going to see his or her doctor, Shepard knew that he was likely to hear things that he really did not want to hear. Chief among them was the very real likelihood that he would be permanently grounded. He went anyway, anticipating the worst but hoping against hope that his condition would turn out to be something simple—something that could be treated simply. But that was not to be. The famed astronaut was quickly diagnosed with Ménière's disease, a disorder caused by excess fluid in the inner ear. Shepard was grounded pending further tests and treatment.

Shepard's initial treatments went well, and he was returned to flight status long enough to be given command of the first Gemini mission, GT-3, along with copilot Tom Stafford. The symptoms returned, however, and in February 1964 Shepard was grounded. Much to his chagrin he was removed from the maiden Gemini flight, with Gus Grissom and rookie astronaut John Young selected to replace him and Stafford. He was not allowed to fly aircraft solo anymore either. "Can you imagine the world's greatest test pilot has to have some young guy in the back flying along with you?" he reflected some years later. "I mean, talk about embarrassing situations!"

At least Shepard had something to do within the astronaut corps. In November 1963 Deke Slayton approached him with the suggestion that Shepard replace him as chief of the Astronaut Office. Slayton was now the head of Flight Crew Operations and simply did not have the time to devote to the Astronaut Office exclusively. Shepard was the obvious choice as the most senior of astronauts then on flight status. At the time Shepard had not been sure about taking a desk role, even though he was still on flight status, perhaps because the role had originally been created for Deke when he became grounded by ill health. Shepard took the job out of respect for his friend and enjoyed the different perspective that it gave him. But now that it was his only job within NASA, the role had a hollow feel to it. Slayton knew exactly how Shepard felt; it had been his task to select the crews for missions and then walk with them to the launchpad after joining them at the breakfast table on launch day. Now Alan Shepard had that less-than-envious responsibility, but both felt strongly that one day they would be able to select themselves for a mission.

On 4 April 1966 NASA announced a fifth group of astronauts. It was a significant selection in many ways, particularly for the Apollo program. At the time the new astronaut group was nicknamed the "Original Nineteen," as "a takeoff on the Original Seven, a title given the first group of astronauts who had been selected in 1959," according to group member Charlie Duke in his 1990 autobiography, *Moonwalker*. "Our group was selected in particular for the Apollo moon program, but it wasn't long before I began counting noses and realized there were a lot more astronauts than there were seats to the moon."

The unsuspected strength of Group 5, however, came from just that fact. With all of the more senior astronauts already assigned to missions, this new group represented the Astronaut Office at major contractors' factories such as North American and Grumman. The knowledge that they gained from working so closely with the engineers that designed and built the Command and Lunar modules gave them a unique advantage in terms of knowledge and therefore power. They were ideally suited to fill positions on later Apollo flights. The hours upon hours of testing made them the best possible choice to accompany the more experienced astronauts as Lunar Module pilots (LMPs) or Command Module pilots (CMPs). Of all of the LMPs who walked on the moon, half came from the Original Nineteen. All of the surviving members of this group, apart from the medically disqualified John Bull, would fly in space during the Apollo, Skylab, or Space Shuttle programs. Among the members of that group were Stu Roosa and Ed Mitchell.

Born 16 August 1933, in Durango, Colorado, Stuart Allen Roosa began his career as a smoke jumper for the U.S. Forest Service, before joining the U.S. Air Force in 1953. During his psychology induction interview at the Air Force cadet academy, he was asked what he had previously been doing. According to fellow smoke jumper and friend Jimmie Dollard: "When he asked Stuart what he had been doing, Stuart said that he had been a smoke-jumper and was asked to explain. When Stuart told him what smokejumpers did, the shrink stared in disbelief and said, 'Okay, you shouldn't have any trouble with flight training,' and the planned two-hour interview was over in 10 minutes."

After earning his wings at Williams Air Force Base, Arizona, in 1955, Roosa was assigned to a fighter-bomber squadron based at Langley Air Force Base in Virginia, where he flew F-84F and F-100 aircraft and trained to deliver nuclear payloads into the heart of the Soviet Union. He later served as chief of service engineering (AFLC) at Tachikawa Air Base in Japan for two years following his graduation from the University of Colorado under the Air Force Institute of Technology Program. From July 1962 to August 1964 Roosa served as a maintenance flight test pilot at Olmstead Air Force Base in Pennsylvania, flying F-101 aircraft. Having graduated as a member of Class 64C with the Aerospace Research Pilot School (ARPS) at Edwards Air Force Base in California, he then flew as an experimental test pilot at Edwards from September 1965 to May 1966.

For Stu Roosa the reality of being an astronaut came into sharp focus nine months after his selection. On 27 January 1967 he was the capsule communicator (CapCom, also known by the transmission code "Stony") at the Cape Canaveral launch blockhouse on the day that his fellow astronauts Gus Grissom, Ed White, and Roger Chaffee lost their lives. He was one of the first to get to the charred Command Module after the hatch was opened, along with Deke Slayton, and also one of the first to view the tragic effects of the fire.

Edgar Dean Mitchell was born in Texas on 17 September 1930. In his youth he was active in the Boy Scouts of America, in which he achieved the organization's second-highest rank, Life Scout. Later, in 1952, he gained a bachelor of science degree in industrial management from Carnegie Mellon University and subsequently joined the U.S. Navy, undergoing basic training at the San Diego Recruit Depot. In May 1953, having completed instruction at the Officers' Candidate School at Newport, Rhode Island, he was commissioned as an ensign. He finished his flight training in July 1954 at Hutchinson, Kansas, and subsequently was assigned to Patrol Squadron 29, which would be deployed to Okinawa, Japan.

From 1957 to 1958 Mitchell flew A3 aircraft while assigned to Heavy Attack Squadron 2 aboard the USS *Bon Homme Richard* and then the USS *Ticonderoga*. He would then serve another year as a research project pilot with Air Development Squadron 5. As part of his development as a test pilot, he earned a bachelor of science degree in aeronautics from the U.S. Naval Postgraduate School in 1961 and a doctorate of science degree in aeronautics and astronautics from the Massachusetts Institute of Technology (MIT) in 1964. He gained his first experience of the spaceflight world while serving as chief of the Project Management Division of the Navy Field Office for the Manned Orbiting Laboratory (MOL) project until 1965.

For the next two years Mitchell trained at the U.S. Air Force's ARPS at Edwards under the command of Colonel Charles "Chuck" Yeager, in preparation both for air force astronaut duties and certification as a test pilot. He graduated from ARPS Class 65B along with another future Group 5 astronaut, Ken Mattingly. Mitchell later served as an instructor in advanced mathematics and navigation theory for astronaut candidates before his selection as a NASA astronaut.

Once Roosa and Mitchell had completed their initial astronaut training, both were given specific assignments within the Apollo program. Roosa was handed responsibility for the Apollo CM at the North American Aviation plant in California, while Ed Mitchell was assigned to travel to the Grumman plant in New York to work with the LM.

In 1968 Alan Shepard had to face the daunting fact that he might never be cured of his disease. It had been five years since his initial diagnosis, and no new treatment possibilities had presented themselves. In 1964 Tom Stafford came to him, after their removal from *Gemini 3*, and told him of a doctor named William House, based in Los Angeles, who was conducting a trial for a new method of surgery designed to cure Ménière's disease. At the time Shepard was not ready to consider surgery, hoping that another, noninvasive treatment would be found. In the years since his grounding, however, the disease had advanced, and he was almost deaf in one ear. With Project Apollo now moving into full swing, Shepard would come to regard House's radical surgery as the only real option left to him and arranged an appointment to see if the procedure could help him regain his flight status. House explained that there were no guarantees that his procedure would work, but a determined Shepard decided to proceed. "With NASA's permission I went out to California," Shepard told Commander Ted Wilbur during a 1970 interview for *Naval Aviator News*. "In order to keep the whole business quiet, Dr. House and I agreed that I should check into the hospital under an assumed name. It was the doctor's secretary who came up with it. So, as Victor Poulos, I had the operation."

The procedure involved inserting a small tube into Shepard's inner ear and extending it to his spinal column, where the excess fluid would harmlessly drain away. House warned Shepard that although the operation had gone well, it might be some time before he felt any benefit from it. Shepard went away satisfied that he had done all he could. Now it was just a waiting game.

Over the next weeks and months the ringing in Shepard's ears began to diminish, along with the dizzy spells. Eventually, filled with hope and trepidation, he went to see the NASA flight surgeon, looking to regain his flight status. Following a barrage of tests, much to his delight and relief, Shepard was given the all clear and was permitted to return to flight status. The long exile was over, and Shepard didn't waste any time. He immediately told Deke Slayton the good news and requested assignment to the first available Apollo flight. At that time, March 1969, the next available slot was *Apollo 13*. Shepard would accept nothing less than command of a moon mission, and he wanted a good crew.

As Shepard would later recount to Roy Neal in 1998: "I had picked two bright, young guys—one of them a PhD, and one of them a heck of a lot smarter than I was—and made up a team to go for an Apollo flight. I said [to Slayton], 'I would like to recommend that I get *Apollo 13*, with Stu Roosa as Command Module pilot and Ed Mitchell as Lunar Module pilot.'" Deke felt that Shepard had paid his dues, and only his medical problems had kept him from flying several more missions before now. As far as he was concerned, his fellow Mercury astronaut was still first in line to fly, and he promised to see what he could do.

Stu Roosa had been steadily climbing the ladder within the Astronaut Office. His intimate familiarity with the CM meant that he had been noticed among the senior astronauts, managers, and engineers with whom he had worked. He had been made a member of the support crew team for *Apollo 9* by Alan Shepard, who had been impressed with his work. When Shepard had invited Roosa to join the *Apollo 9* support crew, he also said, "Just be patient, I've got something in the works." At the time the comment barely registered with Roosa; he was just delighted to have a crew assignment, some recognition for his efforts so far. Officially, his role was as CapCom for this mission, as he would be working entirely in Mission Control, coordinating the various mission activities. When *Apollo 9* flew, and changes had to made to the flight plan because of Rusty Schweickart's space sickness, Roosa stayed on top of the situation and communicated the changes from the ground to the crew. Deke Slayton was impressed by Roosa's performance and made a mental note to include him in his plans.

Ed Mitchell had certainly been making the most of his opportunities since his selection in 1966. He too was appointed as a team crew member for *Apollo 9*, supporting the LMP. Just as Stu Roosa had become an expert on the North American CM, Ed Mitchell had gained considerable time and expertise on the Grumman LM, and his skills were in constant demand. He went straight from the *Apollo 9* assignment to that of backup LMP for *Apollo 10*, which would become the dress rehearsal for landing the first astronauts on the moon. He joined backup mission commander Gordon Cooper, along with Donn Eisele, who was fresh from flying *Apollo 7*, as CMP. Cooper hoped that this backup assignment would see him named commander of *Apollo 13* after a previous but dead-end backup command role on *Gemini 12*. He wanted to equal Wally Schirra's feat in flying Mercury, Gemini, and

Apollo missions, and he wanted to go even one better than Wally, by walking on the moon.

When Shepard had knocked on his door, Slayton was sure that he was the perfect man to lead *Apollo 13*, but selecting his suggested crew was not as straightforward. He agreed with Shepard's assessment of both Roosa and Mitchell, but there were some complications to appointing that particular crew to *Apollo 13*. An unofficial schedule of crew rotation was firmly entrenched in the crew selection process, one dictating that a crew would back up one mission, skip the next two, and then fly as prime crew on the following mission. Slayton had always tried to adhere to that rotation plan as much as he could. Under normal circumstances this system meant that the backup crew from *Apollo 10* should become the prime crew for *Apollo 13*. Gordon Cooper, Donn Eisele, and Ed Mitchell were that backup crew, but if Shepard was to command *Apollo 13*, he would be forgoing the proven rotation system and jumping the queue, ahead of Cooper's crew. There was no way that Cooper would serve under Shepard as LMP, and besides, Ed Mitchell had done a good job on his previous assignments and deserved promotion to a prime crew. So Cooper was the odd one out here.

Other backup commanders such as Jim Lovell deserved future command of a prime crew, and Slayton felt that Cooper had simply not done enough to warrant the command of *Apollo 13*. His lack of commitment and casual attitude toward simulator work and training in general were now conspiring against him. Likewise, Donn Eisele had become difficult to recommend for another prime crew assignment. While he had done an excellent job flying on *Apollo 7*, he had also fallen victim to the terse air-to-ground arguments that his commander, Wally Schirra, had maintained with ground controllers. It meant that he and fellow *Apollo 7* crewmate Walt Cunningham had incurred the wrath of Chris Kraft, then the influential head of Flight Crew Operations, who would later write in his memoirs: "I told Deke straight out that this crew shouldn't fly again. With Schirra, it was no problem; he was leaving. But Cunningham and Eisele had me worried. I didn't want to see either of them in a command position, and in my talks with Deke, he felt the same."

As a consequence, Donn Eisele would be hard to sell as the prime CMP for *Apollo 13*. Picking an entirely subordinate crew for Alan Shepard, rookie astronauts, seemed like the best way forward, so Slayton went along with

Shepard's choice of Ed Mitchell as LMP and Stu Roosa as CMP. He knew that meant breaking the generally hard-and-fast crew rotation cycle; Roosa had not yet served as a backup on any crew, but Slayton definitely liked the man's work and attitude.

When both Roosa and Mitchell were summoned to Shepard's office, they had no idea why. An Apollo flight for any of the Group 5 astronauts was thought to be some time away—particularly as all six Group 4 scientist-astronauts were still waiting for an assignment to a crew. Shepard, however, was happier than either of them could ever remember seeing him, grinning expansively. "If you guys don't mind flying with an old retread," he ventured, "we're the prime crew of *Apollo 13*."

Roosa in particular could not believe his ears. Prime crew! It was most unusual for an astronaut to be named on a crew without first having served as a backup, although it had been done before. Mitchell had previous backup crew experience, but Roosa would be the first—and as it turned out, the only—Group 5 astronaut to move directly onto a prime crew.

As expected, not everyone was happy. Gordon Cooper in particular was furious. He had understandably believed that *Apollo 13* was his and that he deserved it. It was very clear to him that the failure to get this assignment meant that he would be very unlikely to get another. His astronaut career was essentially over. Others within the Astronaut Office felt that Shepard simply reappearing and taking the first available moon-landing mission as commander after four years on the ground was unfair to everyone else and undermined the rotation process. Shepard was jumping the queue. Slayton simply said that Shepard had always been in line, and that was that.

Yet the final act in this drama of office politics had not yet been played out. Slayton submitted the crew of Shepard, Roosa, and Mitchell to George Mueller, who, as the head of manned spaceflight at NASA Headquarters in Washington, approved all flight crew assignments. At the same time, Slayton also submitted the names of John Young, Jack Swigert, and Charlie Duke as the *Apollo 13* backup crew. Every crew proposed by Deke Slayton in the past had been approved, but this time approval was not forthcoming. Mueller rightly felt that Shepard needed more training time after being grounded for so long and that a year of training was simply not enough for him and his rookie crew. Slayton did not agree, but he had no choice in the matter. At that time Jim Lovell and his crew of Ken Mattingly and Fred Haise were assigned to *Apollo 14*, and they were a far more experienced crew, with plenty of training time under their belts. Slayton asked Lovell if he and his crew would be ready to fly on *Apollo 13* instead of *14*. Lovell agreed. NASA Headquarters accepted the compromise, so Shepard and his crew gained an extra four months of training time. They also gained Lovell's original backup crew of Gene Cernan, Ron Evans, and Joe Engle. The "three rookies," as many in the Astronaut Office were now calling them, would fly to the moon aboard *Apollo 14*.

While some in the Astronaut Office still felt that Shepard should have spent time as a backup commander before receiving a prime assignment, the man himself harbored no such qualms and dove into the complexities of mission commander with vigor. His two crewmates were similarly motivated for their first spaceflight. Perhaps this was the genius of the crew selection; Shepard worked very hard to justify his newly won position to himself and others, while his two rookie crewmates worked hard to fill in any knowledge shortfalls in areas in which Shepard's interest waned. Geological study was not one of his favorite activities, reasoning that it had nothing to do with piloting after all, whereas Roosa and Mitchell both knew that they would be judged by their peers and the mission scientists in this area, at least as much as in their piloting skills.

The *Apollo 14*, or H-3, mission had gone through several schedule and destination changes since its original baseline was set. At first it had been scheduled to land in the Censorinus highlands, then Rima Bode II, and then Littrow, launching in August 1970. The mission would constitute two EVAs, similar to those undertaken on *Apollo 12* and planned for *Apollo 13* but with a landing in the lunar highlands for the first time. The Command Module pilot would carry out a photographic survey from orbit for the planned *Apollo 16* landing site at Descartes. Following the failure of *Apollo 13* to land at Fra Mauro, however, NASA mission planners and geologists felt that Fra Mauro was too important a landing site to miss altogether, so *Apollo 14* was assigned to land there instead of at Littrow, a site that could wait until a later mission.

There were other changes brought about by the investigation following the *Apollo 13* mishap. This meant that *Apollo 14* would be the last of the H missions, which were classified by NASA as short-duration stays on the moon with two EVAs. From *Apollo 15* onward Apollo missions were to be flown to the J specification that included three days' stay on the lunar surface, with three EVAS, and updated equipment. This required increased-duration lunar suits, an LM updated for longer stays on the moon, a Lunar Rover Vehicle (LRV) for covering greater distances, and a greater emphasis on science. In the meantime mission planners, engineers, and astronauts had to correct the faults that had caused the *Apollo 13* failure.

The *Apollo 13* Review Board's official report was published in June 1970. Broadly speaking, its recommendations were split into two sections. The first detailed the steps necessary to avoid such an accident occurring again, such as improvements in quality control, preflight procedures, and notices for external contractors. The second section outlined the changes that needed to be made to the hardware and flight procedures to deal with such an incident should it happen again in the future. All of these changes meant that *Apollo 14* would need another three months before it was ready to fly. A preliminary launch in January 1971 was set.

Changes to the CSM were mainly concerned with the oxygen tanks and related systems. They included removing combustible materials from future oxygen tanks and making changes to the fan system, which had been the catalyst of the explosion aboard *Apollo 13*'s SM. The addition of a third oxygen tank, sufficiently far away from the other two, with extra isolation valves to prevent it from leaking in the event of another explosion, would help to ensure the crew's survival. An additional battery was added to the CM to give an extra margin of power to the existing entry batteries. Extra storage for water was deemed necessary in the event the crew could not use the existing potable water system for any reason. These additional collapsible storage containers could also be used for urine storage if needed. The LM's power transfer system was changed to an easier method of transferring power in both directions, to and from the CM. In addition, one or two issues that had come to light during previous successful missions were also fixed.

Training for *Apollo 14* had begun as soon as the prime and backup crews were assigned but changed after the flight of *Apollo 13*. As with previous Apollo crews, the CMP and two LMPs followed separate training paths, apart from cooperative activities such as launch, reentry, midcourse corrections and the like, for which they had to train as a unit.

Roosa had a very crowded flight plan, particularly for the two days that he would be alone in the CM while Shepard and Mitchell explored the lu-

25. Ed Mitchell (*left*) and Alan Shepard in September 1970, suited up in preparation for altitude tests. Courtesy NASA.

nar surface. Beyond conducting photography of the *Apollo 16* landing site, he had a multitude of other photographic targets, including the eventual landing site of the *Apollo 14* LM, the location of the *Apollo 12* LM ascent stage, which had crashed back onto the moon after the crew had discarded it in lunar orbit, and the descent stage at the *Apollo 12* landing site. He would also photograph the location of the *Apollo 13* Saturn IV-B rocket stage, which had been deliberately sent crashing onto the surface of the moon to test the sensitive seismic devices placed on the surface by previous moon-walking astronauts. The crew of *Apollo 14* would also allow their IV-B rocket stage to impact the moon's surface as an additional test of seismic disruption. In addition, some of these pictures were to be photographed with the Hycon Lunar Topographic Camera (LTC), mounted in the hatch window during lunar orbit, which would hopefully provide high-resolution images in low orbits, down to three feet in detail.

For the LM crew of Shepard and Mitchell there was an equally crowded set of objectives. A similar set of ALSEP experiments, originally intended to be left by the crew of *Apollo 13*, would be deposited on the surface at Fra Mauro. The deployment of this equipment was planned to take up most of the first EVA. An additional piece of equipment on this mission was the Modularized Equipment Transporter (MET), a two-wheeled vehicle that the astronauts could load with tools, equipment, and soil and rock samples during their EVAS. It had a handle that allowed it to be pulled along by one of the crew and two legs that allowed it to stand freely, Which, it was thought, would be particularly useful during the crew's traverse to Cone Crater during the second EVA. This crater was of particular interest because of the possibility that the ejecta from around the rim might include some of the deepest excavated rock yet found on the moon. The MET would allow the collection of samples of hard rock from beneath the regolith that had built up on the lunar surface over thousands of years.

The mission of *Apollo 14* also meant the beginning of the end for the Apollo program. NASA announced the cancellation of *Apollo 20* in January 1970, and *Apollo 18* and *19* were similarly canceled in September 1970. With just four Apollo flights left to fly to the moon, production lines for Apollo hardware began to wind down. NASA had to make the best use of the remaining Saturn Vs from the original production run of fifteen as well as the CMS and LMS that were originally selected for those missions. Some were destined for the upcoming Skylab program and the one-off joint ASTP mission, but the rest were destined never to fly or in some cases were not even completed. It also meant that a large number of people who had found long-term employment at NASA centers and with NASA contractors because of the huge effort behind Project Apollo would be laid off. The golden era of spaceflight that had begun with the Mercury missions was coming to an end, but the man who had kick-started that era, Alan Shepard, would now unwittingly help to draw it to a close. The loss of *Apollo 20* in particular had an impact on one of the *Apollo 14* crew. Stu Roosa could reasonably have assumed that if he did a good job on his first flight then he would remain in the crew rotation and quite possibly score the role of backup commander for *Apollo 17*, which might in turn have led to the command of *Apollo 20*.

26. Mission training at KSC included planting the American flag into the moon's surface. Here Mitchell (*left*) is pushing the flag's lower staff into a patch of simulated lunar soil, while Shepard extends the telescopic crossbar. Courtesy NASA.

Ed Mitchell may also have had similar thoughts, but he was beginning to think that his future lay outside of NASA. Alan Shepard did not have this worry; *Apollo 14* would be his last spaceflight, his final hurrah, and a great way to crown his astronaut career.

The mission patch for *Apollo 14* was designed by Alan Shepard, initially from an idea of his own. He sketched a patch including an astronaut pin, the symbol of the Astronaut Office, and showed it to Roosa and Mitchell. They all pitched in to include the Earth, the moon, and their names and passed it to the NASA art center for rendering. The backup crew led by Gene Cernan decided to come up with their own patch, making use of the same basic design but with the Warner Brothers character the Road Runner standing on the moon, and the bird's cartoon nemesis, Wile E. Coyote, replacing the astronaut pin. The Road Runner represented the backup crew beating the prime crew to the moon. The Coyote had red fur, to match Roosa's red hair (he had been known as "Red Rooster" back in his smoke jumper days), and a potbelly, to represent Mitchell's general distaste for exercise. The cartoon character also had a long gray beard for the "old man" Alan Shepard. In place of the designation *Apollo 14* at the top, the backup patch said "BEEP BEEP," and it bore the names Cernan, Evans, and Engle at the bottom. The backup version of the crew patch would haunt Shepard's crew for the entire flight.

In addition, as with past Apollo flights, the crew had to name their individual spacecraft for the period when the CM and LM would be separated and it could no longer be referred to by Mission Control as simply "*Apollo 14*." Eventually, Stu Roosa named the CM "Kitty Hawk," in honor of the Wright brothers, and Ed Mitchell named the Lunar Module "Antares," a bright red star that was most visible during the pitch-over phase of the landing. Initially, Roosa and Mitchell had tried to come up with paired names such as *Apollo 10*'s Snoopy and Charlie Brown but without success. So Roosa finally told Mitchell, "You name the Lunar Module anything you want, but I'm going with Kitty Hawk."

When the time came for the launch of *Apollo 14* on 31 January 1971, it had been a long road for all of the crew, but for Shepard in particular. The rollercoaster ride in the ten years since his Mercury flight may have altered many things for him, but one thing had never changed—his desire to succeed was as strong as ever, never diminished by his illness, and if anything made even stronger by his enforced absence from spaceflight. In 1991 he was interviewed by the Academy of Achievement, and he said then that "there was the challenge to keep doing better and better, to fly the best test flight that anybody had ever flown." It was what had led him to "being recognized as one of the more experienced test pilots, and that led to the astronaut business."

After a forty-minute delay due to the weather, a period that would have no effect at all on the mission, but would have later implications for Ed Mitchell and his personal agenda, the first phase of *Apollo 14* went well. The launch and Trans-Lunar Injection (TLI) both proceeded flawlessly and to the complete satisfaction of everyone. The Saturn V, as usual, provided its own fifteen minutes of fame. Then, when the time came for Roosa to disconnect

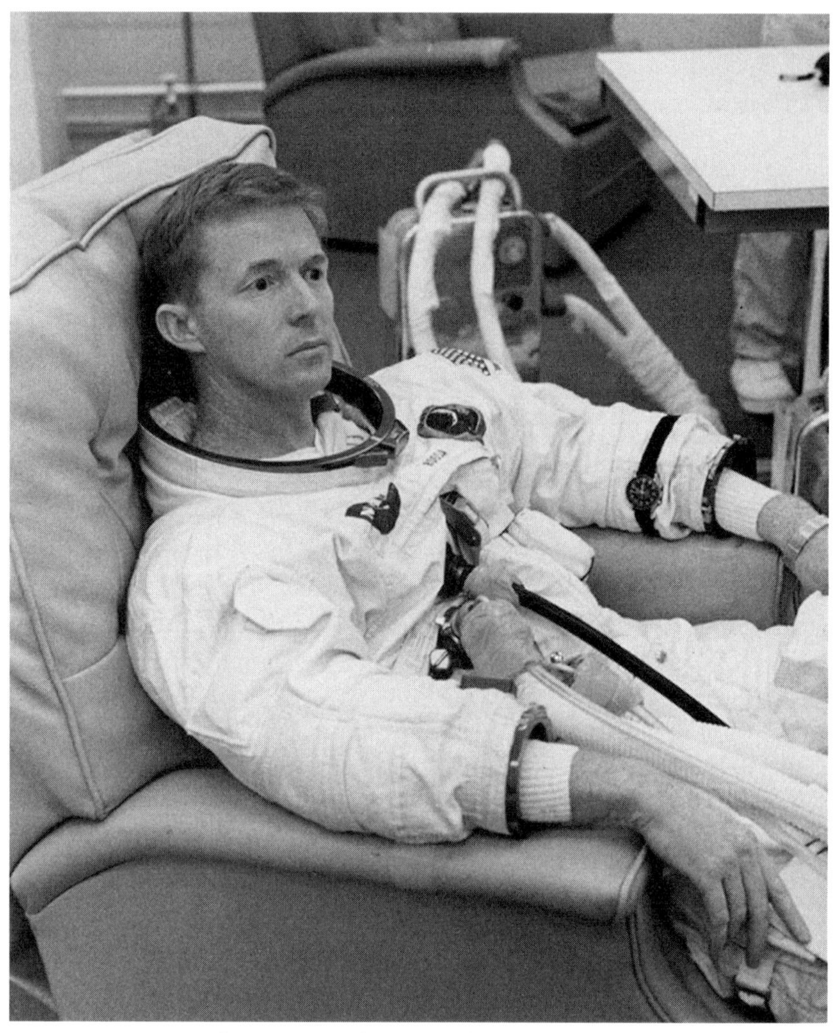

27. A contemplative Stu Roosa during suit-up on launch day. Courtesy NASA.

the CSM from the Saturn rocket's S-IVB stage and extract the LM from its berth, there was a problem. He gently eased the CSM away from the Saturn and slowly turned the craft around, careful to avoid using more fuel than necessary. Then, as he eased the nose back toward the docking port on the top of the LM, Roosa also ensured that the docking probe on the CSM was perfectly aligned. All went well until they met and then, to Roosa's bewilderment, the expected initial soft docking did not occur. The CSM bounced back from the LM and began to slowly drift away. Roosa arrested the drift but was perplexed and frustrated with the failure. Had he done something wrong? He did not believe so. Hours upon hours had been spent in the simulator practicing this very maneuver, perfecting every detail so that the minimum amount of fuel was spent and the docking system exposed to only the slightest bump. Somehow it had gone wrong. He checked with Mission Control, but they had no indication of any problem, so they cleared him for another try. All seemed well again but with the same result, a failed docking. It now appeared that there must be some kind of hardware fault, but did it lie in the docking system of the CM or the LM?

After more failed attempts, both the crew and the controllers in Mission Control were running out of ideas. It went without saying that if the two craft could not be docked, the mission was over. They would circle the moon and come straight home. The docking mechanism in the CM could be checked by the crew if they buttoned themselves into their pressure suits, dumped the cabin pressure, and brought the docking probe inside. Once that had been done they might be able to see something obvious—a loose piece of debris perhaps. They could then clear the problem and reinstall the probe. The LM's system was less easy to check, although it did occur to the crew that they could remove the probe from their own docking system and then inspect the LM from inside the CSM once Roosa brought the craft close enough together. This would be an entirely unapproved and unproven procedure, and the crew felt sure that Mission Control would deem it too risky, but it was an option. Mission Control, however, had one other idea. They instructed Roosa to try the docking again but this time to keep firing the thrusters on the CSM, holding it hard against the LM after contact. At the same time Shepard would retract the docking probe and allow the twelve capture latches to engage. This would require some particularly fine flying by Roosa, as he would be essentially aligning for the docking without the help of the probe.

By now the crew was feeling frustrated and in disbelief; to come this far and be thwarted did not seem possible, but here they were. The day was getting longer and longer, and still they had not achieved the docking that was so essential to their mission. This time it would have to work, and fortunately it did. The two craft came together and achieved a firm docking.

The question now facing everyone, and especially the crew, was would the docking system work again when the two craft attempted to lock together in lunar orbit after the landing? As the crew began to relax after a longer than planned day, they carefully shed their bulky pressure suits and packed things away, ready for their first night in space. Before they could sleep, however, Mitchell proceeded to remove the docking hatch, look at the probe and drogue, and show them to Mission Control via the television camera. Although the lighting was not good enough to see everything in great detail, it was clear that both components only had a couple of light scratches on their surface. The cause of the docking problem remained a mystery.

Much to their amusement, the crew made one unexpected discovery as they headed for a lunar rendezvous: every locker or other storage compartment they opened, they found one of the backup crew patches. Prior to launch the backup crew had stowed their alternative patches in each of the lockers and taped them to the bulkheads in both spacecraft. Now they floated free, as Stu Roosa relayed to the CapCom in Mission Control, Fred Haise.

Roosa: *We're really inundated with unauthorized objects in both spacecraft . . . I think Ed was showing you one up in there [in the LM]. If you can see this [holding the patch] . . . I don't know if any of the backup crew is in [Mission Control] tonight . . . but they've left their calling card.*

Haise: *Okay, we have a pretty good picture, Stu, and they are here.*

Roosa: *Okay, tell them we sure appreciate every compartment that we open up having one of these things come floating out . . .*

Haise: *They [the backup crew] aim to please.*

The rest of the journey to the moon was uneventful, mostly filled with discussions about what might have gone wrong during the docking procedure. Arriving in the moon's orbit was also without drama, and the crew proceeded with their preparations for the undocking sequence. All went well, but once *Antares* had separated from *Kitty Hawk*, the crew experienced some communication problems, with the S-band antenna in the LM popping its circuit breaker several times. Then, just over forty minutes after the separation, while the crew members were completing their checklists, the ABORT light came on in front of Mitchell. Eventually, it blinked off again, but at this stage of the mission it was of no real consequence or concern. Then it happened again about ten minutes later, and Houston asked Mitchell to tap the instrument panel around the area that contained the light to see if it would go off. He did as instructed, and again the light went off. Unsure of the cause, Mission Control told the crew to proceed with their checklists while controllers on the ground looked into the problem.

Everybody understood that the ABORT light anomaly would be a real problem once the *Antares* crew began their descent to the moon, when the autopilot would be in control, and the lit-up ABORT signal meant a real abort might have to be initiated. It was assumed by both the crew and Mission Control that the problem was being caused by a stray piece of solder that had worked loose from one of the many circuit boards in the panel and was now occasionally and randomly touching parts of the board and causing a short-circuit that was falsely triggering the ABORT signal. It was possible to get around the problem by completely shutting off the program, which would mean that the solder ball might continue to trigger the abort, but the signal would have no effect. The downside was that the abort program would not work at all, even if it was triggered by a genuine alarm signal.

Mission Control was not keen on leaving the crew without an abort mode. The software that ran the LM's autopilot and many other processes was conceived and written by MIT technicians. As with all suppliers to the Apollo program, MIT kept a team of experts available on site during each mission for just such an eventuality. They were now called upon to try to come up with a better solution. They needed to work fast, while Shepard and Mitchell continued to orbit the moon. The crew could not afford to wait very long as the window of opportunity for landing was limited. The man responsible for this part of the LM software was Don Eyles, and his intimate knowledge allowed him not only to come up with a way for the computer to ignore the errant switch but to do it without losing the abort mode entirely. There was no way to send this new program directly to the orbiting *Antares*. Mitchell would have to key in the changes manually, following the CapCom's instructions. Time was limited, but he was able to complete the changes without any problems. Once again, *Apollo 14* was a "go" to land on the moon.

Shepard and Mitchell had more fuel left to them on *Apollo 14* than previous missions due to a new procedure that meant the initial descent burn had been carried out by CM *Kitty Hawk* before it undocked from *Antares*. This allowed the LM crew a more relaxed descent. Both men felt a heavy sense of relief after the narrow escape with the ABORT switch, and they were now keen to get on with the landing, feeling that their problems were now behind them. Then, at the point in the descent when the LM should have begun to receive data from the landing radar, about 32,000 feet above the surface, the radar remained blank. Mission rules dictated that if the radar did not spring into life by the time they reached 10,000 feet altitude, they must abort the landing. Without the radar information the crew had no accurate means of determining their height above the lunar surface. Mission Control could also see that the radar had not engaged, and advised the crew to reset the circuit breaker that handled the landing radar. Shepard was already thinking that he could land *Antares* without the radar, no matter what the rules said. At just over 15,000 feet, data from the radar suddenly blinked to life on Mitchell's panel. The power recycle had done the trick. Shepard's voice said it all. "Okay. Go . . . go great. Great. Oh, that was close."

The rest of the descent went exactly as planned, and Shepard was finally able to announce to Mission Control, "Okay, we made a good landing." After the problems with the abort program, Ed Mitchell wanted to express his thanks. "Okay," he said, "Say, Fred [Haise at CapCom], that was really great work you did on that abort problem." Haise responded, "Yes, those guys did a lot of scratching around there, Ed." Al Shepard could not help but pitch in with "You bet. It really saved our mission."

Meanwhile, in lunar orbit Stu Roosa was delighted to hear his crewmates land on the lunar surface, but he was having problems of his own with the Hycon camera. A mysterious clacking noise had become apparent almost as soon as he had started the camera running, and now it refused to work at all. Most of his personal mission objectives involved using it, and he hoped that a workaround would be found.

On the lunar surface, after some food and housekeeping chores in the LM, Shepard and Mitchell were ready for the first EVA. Shepard began to inch his way down the ladder to the footpad, and CapCom Haise said "Okay, Al; beautiful. We can see you coming down the ladder right now. It looks like you are about on the bottom step, and on the surface. Not bad for an old man." Shepard laughed and replied, "Okay, you're right. Al is on the surface, and it's been a long way, but we're here." Shepard's determination had seen his dream come true; he became the only Mercury astronaut to walk on the moon and the oldest person to do so, at forty-seven years of age.

28. Almost a decade after becoming the first American to fly into space, Alan Shepard finally stood on the surface of the moon—the only Mercury astronaut to do so. Courtesy NASA.

The initial tasks for the first EVA followed the pattern established by earlier Apollo landings. Taking contingency samples of the lunar surface and then setting up the ALSEP package were the first orders of the day, followed by deploying the flag of the United States. All went well for a change, and both men now felt that a successful mission was within their grasp. After their first lunar EVA they clambered back into the LM, ate a rehydrated meal with cold water (all that was available in the LM), and tried to settle down to rest for a few hours. Like previous crews, they had to sleep in their suits, and it was neither comfortable nor easy, but eventually they drifted off to a restless sleep.

Meanwhile, alone in lunar orbit, Roosa was not having an easy time. The Hycon camera was still malfunctioning, and he had already spent much more time than he really wanted to troubleshooting the problem. After reconnecting every cable and cleaning every part, the problem persisted, and worse still, he was now hours past the timeline he had set for himself. He admitted to himself that the tasks for his mission were probably too much for one man to achieve, even if everything had gone perfectly, and he blamed himself for that overloaded work schedule.

When Shepard and Mitchell woke from their fitful sleep and contemplated the day and tasks ahead, their main thoughts centered on Cone Crater. On the second EVA they planned to hike to the rim of the enormous crater, peer inside, and then collect rock samples from around the edge in an effort to find deep ejecta from the crater itself. On paper it sounded simple enough, but both men knew that navigating on the lunar surface might not be as simple as it seemed. Most people navigate on Earth by using a combination of maps and landmarks to guide them. The two moon walkers each had maps attached to the cuffs of their suits, but they had been converted from photographs taken by previous Apollo CMPs from orbit and did not translate that well for someone walking on the uniformly gray surface. In addition, landmarks are not so easy to come by on the moon, where every rock looks very much like another, and the dips and undulations could easily play tricks on the eyes and minds of the moon walkers.

The previous day the two men had made preparations for this second EVA by unloading the MET from its storage bay on the LM and loading it with some of the tools they would need on their second lunar traverse. Their backup crew, they found, had also been active here; the MET was covered with the alternate cartoon patch design. It seemed there was no escape from the backup crew, even on the moon. After completing the loading of the MET, Shepard and Mitchell began the trek to the first sampling stop on their way to Cone Crater, but it soon struck them that this was not going to be an easy sojourn at all. The landmarks they were looking for were often hidden in surface depressions, which meant that the astronauts had to be almost on top of them before they could see and identify them. As a consequence, they wasted valuable time, always a limited commodity on the moon, checking and rechecking their positions. Eventually, they found the first and then the second of the scheduled sampling stops, before proceeding to their ultimate destination, Cone Crater.

The MET, which was designed to make things easier for the moon walkers by providing a mobile means of transporting tools and collection bags so they did not need to carry them, was actually becoming a pain for both men. The terrain was so rough and strewn with small rocks that the wheels of the MET were constantly being thrown into the air. The light gravity meant that the trolley, and its contents, actually flew higher than anticipated. To add to their misery Fred "Freddo" Haise, the CapCom on duty at that time in Mission Control, called to the two moon walkers, "There are two guys [the backup EVA crew] sitting next to me who kinda figured you'd end up carrying it up." It was a subtle reminder that Gene Cernan and Joe Engle had bet the prime crew a case of scotch whiskey that the MET would have to be carried if they expected to reach Cone Crater.

Both men were determined to carry on, partly because they knew how important it was to the mission and the geologists back on Earth and partly because they did not want to fail in any aspect of the mission. They knew that the flight director in Houston would eventually call a halt to their trek. They were simply taking up too much time on this single task, and it would surely affect the time available for their remaining lunar duties. At several points they thought that they must be very close to the rim of Cone Crater, only to discover that in fact there was still more slope to go. Mitchell made the comment, "Our positions are all in doubt now, Freddo. What we were looking at was a flank, but it wasn't really—the top of it wasn't the rim of Cone. We've got a ways to go yet." They were offered, and had taken, time extensions from Mission Control. The extra time, they knew, would come at the expense of later tasks.

Inevitably, Mission Control decided that enough was enough. Haise radioed to the crew, "The word from the backroom is they'd like you to consider where you are [to be] the edge of Cone Crater." After the two astronauts complained—"I think you're finks," Mitchell said, they were allowed to press on for thirty more minutes. Haise added a further inducement to continue, "Al and Ed, Deke says he'll cover the bet if you'll drop the MET." To which Mitchell replied in obvious frustration, "It's not that hard with the MET. We need those tools. No, the MET is not slowing us down, Houston. It's just a question of time. We'll get there." Shepard agreed. The distance to the crater was much greater than they had perceived from the maps.

Unfortunately, the crew did not get there. Too little time and the continued uncertainty of their position made it folly to continue. Neither Mitchell nor Shepard complained further, and they began to take samples and core tubes before heading back to the LM. As they neared *Antares*, bringing back with them almost a hundred pounds of rock and soil samples, both men were understandably tired and frustrated at not having achieved a principal mission goal, throwing a few words such as *damn* and *hell* into their transmissions. At one stage as they loped back toward *Antares*, Shepard incautiously let out a heartfelt expletive that said it all for him: "Son of a bitch!"

Once they had safely returned, their mood seemed to lighten a little. Mitchell went to collect some more rock samples, and Shepard made his way over to the ALSEP to check that it was still correctly aligned and working as it should before they left it for good. With that done, Shepard carried out a small surprise that he had planned with the permission of Bob Gilruth, director of the Manned Spacecraft Center, and the approval of Deke Slayton—but only if time allowed.

He began by pointing the TV camera at himself and then said: "You might recognize what I have in my hand as the handle for the contingency sample return; it just so happens to have a genuine six iron on the bottom of it. In my left hand, I have a little white pellet that's familiar to millions of Americans. I'll drop it down. Unfortunately, the suit is so stiff, I can't do this with two hands, but I'm going to try a little sand trap shot here." With both Mitchell and Mission Control providing advice, Alan Shepard hit several golf shots on the moon, the last of which he concluded went for "miles and miles and miles." In fact, even though he had wildly overstated the distance, that ball still traveled about two hundred yards.

As both men prepared to leave the lunar surface for the last time, a poignant exchange took place between them and Fred Haise, the *Apollo 13* LMP. "Al, you and Ed did a great job," Haise offered. "Don't think I could have done any better myself."

To which Mitchell replied, "That's debatable, isn't it, Freddo?"

"Well, I guess not now, Ed," Haise answered wistfully.

As each man left the moon's surface, neither had any memorable or regretful words; they just clambered back into the LM, shut the door, and prepared for the next stage of the mission.

Meanwhile, aboard *Kitty Hawk* Stu Roosa was still having problems with the Hycon camera. His troubleshooting had continued over two days, without much success. He had been able to take some shots with the device, but it involved him controlling each exposure manually, and consequently each planned site image took much longer than he had anticipated. Now it was time for him to stop taking pictures and prepare himself and the CM for the return of *Antares*, his two colleagues, and the rock samples they had gathered during their lunar excursions. They also needed to find some extra space to bring back the probe from the CM's docking mechanism so that they could try again to discover why it had failed earlier in the mission. Normally, the probe is left in the LM when it is jettisoned after the final docking. The liftoff from the lunar surface went smoothly, and as *Antares* lifted toward an eventual rendezvous, Roosa asked Shepard if he cared to share any profound words on his experience on the moon. "Stu, you know me better than that," came the muttered response a few moments later.

Much to the relief of the three men after their earlier docking problems, the rendezvous and linkup went just as planned. They next began transferring the rocks and equipment from the LM, which took some time. All three were very tired at this point, and they still had plenty to do before they could sleep. *Kitty Hawk*, now with all three crew back on board, was a mess from all of the extra items carried back from the surface of the moon. After the SM had fired its big engine to leave lunar orbit, Mission Control suggested that the crew might want to get out the Hycon camera and give it one last try from a fixed position. The crew, however, were having none of that; they were too tired, and the camera was now buried beneath other equipment. The idea was quickly withdrawn. After securing the CM into its temperature control roll and making some room, including placing the docking probe temporarily back into the tunnel, the crew finally grabbed some much-needed sleep.

The trip home gave the crew some time to relax. Nothing in the flight plan was too pressing for the next three days, and each man had the opportunity to look out of the windows and savor the feeling of having almost completed every assignment they undertook. It also gave one crew member the chance to conduct some tests of his own creation. Unofficially, and unknown both to Mission Control or his crewmates, Ed Mitchell had designed experiments in extrasensory perception (ESP) that he planned to carry out in space. He imagined that the idea might not be well received by his colleagues, not because it presented any danger to the mission but because it was outside of the typical activities of a military test pilot.

In truth Ed Mitchell had always been a little different from most of his astronaut colleagues. His mind had always been open to more than just the technology that allowed his moon mission, and although he loved flying as much as anyone else in the astronaut corps, it was not the be-all and end-all of his life. His religious upbringing seemed to him at odds with his later roles as engineer, scientist, and test pilot. There in space he had a unique chance to try to reconcile some of those questions, initially just for his own curiosity, but later it would provide him with a life beyond flying, the navy, and NASA.

About three weeks before the launch of *Apollo 14*, two research physicians named Boyle and Maxie had approached Mitchell. They were both very interested in the possibility of conducting an ESP experiment in space, particularly in testing the effect that great distance might have on the phenomenon. Certainly, a spacecraft traveling to and from the moon seemed to provide the ultimate test of distance. It was arranged that four Earth-based psychics, one of them the well-known Swedish medium Olof Jonsson, would try to receive information from Mitchell on board *Apollo 14*. Jonsson initially trained as an engineer, but he later became more and more involved in the psychic world, performing healings and remote readings to astounded audiences. His most noteworthy experience occurred in 1951, when he helped the Swedish police with a murder investigation, unsuccessfully. A year after the initial murder, the police officer who had assisted Jonsson was found to be the original murderer when he committed suicide after killing eight more people, including his parents and his former fiancé, leaving a note admitting his guilt. Jonsson's reputation in Sweden was ruined, and with the help of his remaining supporters, he left Sweden to live in the United States. Once there he rebuilt his reputation by allowing psychic tests to be carried out on him in a lab, which he claimed proved his "sensitivity."

Mitchell would "transmit" a number between 1 and 5 that matched one of the well-known Zener symbols (a square, a circle, a star, a cross, and a wavy line). He carried with him not the cards but, instead, a table of random numbers, also between 1 and 5, and thought about the corresponding symbol for about fifteen seconds. Each session lasted for about six minutes and involved attempting mentally to transmit twenty-four symbols. Mitchell carried out this experiment during his sleep periods when the CM was quiet and he could concentrate the most. In all he carried out four such sessions, twice on the way to the moon and twice on the return journey. The data collected from these experiments is very much open to interpretation, particularly given the forty-minute delay prior to *Apollo 14*'s launch. The delay actually meant that the listeners on Earth were tuned in at the wrong time, different from Mitchell's transmitting time as each had failed to make the appropriate adjustment. Olaf Jonsson even claimed to have received messages from Mitchell on a day when the astronaut had been unable to transmit.

During the trans-Earth coast Mitchell also had another experience; one that would shape the rest of his life. Each of the crew had spent time looking back at the receding moon and forward to the approaching Earth, marveling at the magnificence of it all, even the normally pragmatic Shepard. For Ed Mitchell it was different; he saw more than just the images out of the windows. He felt that he understood it all, how it all worked, and how everything was interconnected. The experience was to have a profound effect on him and his future.

Reentry went just as planned, with *Kitty Hawk* splashing down in the Pacific Ocean on 9 February 1971, just four miles from the recovery carrier USS *New Orleans.* The crew members were retrieved from the bobbing CM and transported by helicopter to the ship and the mobile quarantine facility. In fact, they would be the last Apollo crew to undergo these two weeks of isolation, the biological threat to Earth having by then been dismissed.

Their time in quarantine gave the crew a chance to relax after a particularly trying flight. Alan Shepard felt the satisfaction of commanding a mission and achieving all of his personal goals—at least as far as spaceflight went. For their part Ed Mitchell and Stu Roosa enjoyed a rookie's elation at having undertaken their first spaceflight.

It was during this period that the press exposed Mitchell's ESP experiment to the public, and Shepard could hardly believe it. He had been completely unaware of his LMP's intentions and said he might have put a stop to it before the flight had he been forewarned. But as far as Mitchell was concerned, his experiments in extrasensory perception were just as harmless an interest as his commander's love of golf. "He did his thing, I did my thing," he later said, shrugging it off.

Not everyone was entirely happy, however, with the crew's achievements. Some of the geologists in the Lunar Receiving Laboratory who were unpacking and classifying the rocks brought back by Shepard and Mitchell were unhappy with the lack of documentation that the crew had provided. Many of the samples had not been photographed in place before they were collected, and overall some felt that the crew had not gone out their way to be good scientists. Many scientists complained that it was wrong for geologist-astronaut Harrison Schmitt to be sitting on the sidelines without a mission assignment, when clearly the nonscientist astronauts were letting them and the rest of the scientific community down with their largely haphazard collecting of lunar regolith samples.

With the end of their mission, the crew of *Apollo 14* was once again eligible for assignment to future missions. Alan Shepard, however, had already let it be known that he would not fly in space again. He would continue in his role as chief of the Astronaut Office for three more years. Deke Slayton was now well on the way to recovering his own flight status, and Shepard was happy to take over the administration of crew assignments while Slayton pursued his eventual assignment to the ASTP flight of 1975.

In August 1974 Alan Shepard left NASA and the U.S. Navy, having achieved the rank of rear admiral. He would serve on the boards of many corporations under the umbrella of his Seven Fourteen Enterprises (named for his two flights, *Freedom 7* and *Apollo 14*). Several years earlier, through his business acumen, he had become the first astronaut to attain the status of millionaire. In 1996 he was diagnosed with leukemia, and he fought a typically stoic and mostly private two-year battle against the insidious disease, but it was a fight even he could not win.

On Tuesday, 21 July 1998, Rear Admiral Alan Bartlett Shepard Jr. passed away in his sleep at the Monterey Community Hospital in California, aged seventy-four. The world had lost a much-admired man and a truly historical figure. His loss was mourned around the world, and many paid tribute. President Bill Clinton said of the United States's first spaceman that he was "one of the great heroes of modern America." Fellow Mercury astronaut Wally Schirra openly wept as he said, "The brotherhood we have will endure forever." NASA administrator Daniel S. Goldin stated that "the entire NASA family is deeply saddened by the passing of Alan Shepard. NASA has lost one of its greatest pioneers; America has lost a shining star." Just five weeks after the death of her husband of fifty-three years, Louise Shepard died unexpectedly, during a commercial airplane flight while returning to their home.

In December 2006 the U.S. Navy launched a new ship, named the USNS *Alan Shepard* (T-AKE-3) in honor of the late rear admiral and pioneering American astronaut.

Nobody, it seemed, wanted the unenviable assignment of being backup crew for the two remaining Apollo missions. With the cancellation of the last three flights, no one expected any of the backup crew members to receive an assignment on the two remaining prime crews. As it turns out, scientist-astronaut Harrison Schmitt took the place of the LMP scheduled to fly on *Apollo 17*, bringing back a geological goldmine of samples and invaluable data. Although it was never formalized, the original backup crew for *Apollo 16* was expected to be Fred Haise, Jerry Carr (LMP), and Bill Pogue (CMP). Following the cancellation of *Apollo 19*, to which the three would normally have rotated, both Carr and Pogue requested transfers to the forthcoming Skylab space station program, on which they would later fly.

Deke Slayton was in a quandary, as everyone knew that the backup role for *Apollo 16* was a dead-end job, so he informed Stu Roosa and Ed Mitchell that—as outlined before their lunar mission—they were assigned to the task, along with Fred Haise as their commander. When Slayton had originally mentioned the assignment, before the *Apollo 14* mission, Mitchell had not been keen on doing it, but he changed his mind when Slayton pointed out that flying on *Apollo 14* was part of the deal.

After carrying out the duty of backup CMP on *Apollo 16* and again on *Apollo 17*, Roosa was assigned to work on the space shuttle program. He would leave NASA and the U.S. Air Force in 1976, some five years before the shuttle flew for the first time. He entered private business both at home and abroad. On 12 December 1994 Colonel Stuart Allen Roosa died from complications associated with pancreatitis, aged sixty-one. His wife Joan died in October 2007 and was buried at Arlington Cemetery along with her husband.

Ed Mitchell left NASA and the U.S. Navy in September 1972, having achieved the rank of captain, and changed his life dramatically. Always described by his fellow astronauts as the "smart one," he now wanted to direct his scientific mind toward investigating psychic phenomena. In 1973 he created the Institute of Noetic Sciences (IONS). The term *Noetic*, coming from a colleague's suggestion at a psychic conference, is derived from a Greek word meaning "mind." In his own words the institute was created to "research and use the tools of science to understand the nature of consciousness and the relationship between science and spirituality."

Ed Mitchell has spent the last thirty-five years carrying out investigations into all kinds of psychic phenomena and parapsychology. He is perhaps most popularly known, however, for his public statements regarding UFOs and alien visitations to Earth. In 2008, while appearing as a guest on a British radio show, he was asked about the existence of aliens. It is not an area in which he is specifically interested, but his answer seemed to shock the interviewer, and the news of his "revelations" was widely covered. Mitchell stated, "I happen to have been privileged enough to be in on the fact that we've been visited on this planet and the UFO phenomena is real." In fact, Ed Mitchell has been making similar comments for many years before this disclosure. In 1996 he published the book *The Way of the Explorer*, in which he also states: "Yes, there have been ET visitations. There have been crashed craft. There have been material and bodies recovered."

Close to forty years after walking on the moon, Edgar Mitchell is still serving on the board of directors at IONS and continues his work investigating the human mind and spirit.

Apollo 14 was certainly not the most famous of the Apollo moon missions. Apart from Alan Shepard's celebrated golf shoot on the lunar surface, the flight and its crew are largely unknown. Nevertheless, at a time when NASA and the United States desperately needed to taste success, following the high drama of *Apollo 13*, Alan Shepard, Stuart Roosa, and Edgar Mitchell delivered with competence and very little fanfare, ushering in a whole new era of scientific exploration of the moon.

7. A Whole New Focus

Colin Burgess

Oh pilot, 'tis a fearful night. There's danger on the deep.

Thomas Haynes Bayly (1797–1839)

There is little doubt that a brief but bloody episode that unfolded on the southern coast of Cuba in 1961, in a coral reef–bounded inlet of the Gulf of Cazones known as Bahía de Cochinos (Bay of Pigs), caused acute embarrassment for the administration of newly elected President John F. Kennedy. In January of that year incumbent president Dwight D. Eisenhower had broken off diplomatic relations with Cuba. Even before that, however, and with the full knowledge of the White House, the Central Intelligence Agency had been covertly training antirevolutionary Cuban expatriates for a possible invasion of their homeland to overthrow the dictatorial two-year regime of Fidel Castro. When Kennedy took over the presidency, he fully supported the invasion plan, and the rapid incursion of Cuba by a little over 1,500 fully armed exiles began on 17 April.

Hoping to establish a beachhead and then cross quickly overland into Havana to the northwest, the invading force relied heavily on spontaneous support from Americans, but sadly this did not eventuate. Nor did the planned airborne strikes by the U.S. Air Force. History records that President Kennedy, having been apprised of the rapidly faltering nature of the invasion, took a nervous step back at this critical time and canceled the air strikes. As a result, Castro's army quickly brought the invasion to a halt. Two days later, on 19 April, it ended in an ignominious defeat, with ninety exiles dead and the rest rounded up and imprisoned. It had been a botched and humiliating defeat for the new U.S. president, who took full responsibility for the disastrous raid. It was a total fiasco and one that made a furious Castro extremely wary of any future American intervention in his country. For their part Cuban exiles in the United States would never forgive the president for losing his nerve, believing he had betrayed them and their cause.

On 19 April 1971, exactly ten years to the day after the humiliating defeat at the Bay of Pigs, there was another, more passive frustration for the United States—this time on the technology front rather than on any terrestrial battlefield. It came when the Soviet Union successfully launched into orbit the world's first space station, an eighteen-ton orbital laboratory it called Salyut, meaning "Salute." Although the United States had earlier won the race to put people on the moon and continued to send crews on lunar landing expeditions, the Soviet space program had chalked up yet another significant first in space exploration.

With the grudging realization that they had lost any chance of beating the United States to the moon, the Soviet space chiefs generally diverted their attention to the long-term occupancy of space. Like NASA, the Soviet space agency had always envisioned an orbiting space station as the next logical step after a lunar landing. Plans for their first space station had actually been conceived some years earlier, in 1964, in a top-secret military space program known as Almaz (Diamond). Three years later construction would be approved.

Originally, there were three main hardware components planned: a twenty-ton space station tended by crews of three cosmonauts and equipped with an Agat optical camera for reconnaissance imaging of military installations on Earth, a reusable transport logistics spacecraft used to ferry military cosmonauts and cargo to and from the station, and Proton rockets for launching all of the components into space. Additionally, the station would be equipped with a single small capsule capable of being ejected for a rapid return to the ground, which might contain surveillance film and other small payloads.

There would be three elements to each Almaz station: habitation quarters for the crew; a large-diameter work section containing surveillance equipment and other instruments; and a transfer section that housed a docking port, the capsule ejection system, and an extravehicular activity (EVA) hatch.

Soviet engineers had completed two flight-rated Almaz station hulls by 1970, but it was then decided within the Soviet leadership to concentrate on developing a more open, civilian space station project, using elements of the Almaz design. One of the Almaz hulls was then transformed into a space station that would support genuine scientific studies, although top-secret work would continue on the military Almaz program. In order to confuse Western observers, both programs would be simultaneously developed under a generic name. Originally, the name Zarya (Sunrise), suggested by Korolev's OKB-1 design bureau, was painted on the side of the space station. Just three days before launch, however, it was decided the call sign might prove too confusing, as it was traditionally the one used since Yuri Gagarin's flight to identify ground control. An alternative was quickly sought and approved, and the station henceforth became Salyut.

At 01:39 a.m. (GMT) on 19 April 1971, a week after the tenth anniversary of Gagarin's flight, the Soviet Union launched the world's first space station, *Salyut 1* (officially *Dolgovremennaya Orbitalnaya Stantsiya* [DOS-1], or Long-Duration Orbital Station), into orbit atop a Proton rocket. It began circling the Earth every 88.5 minutes at an angle of 51.6 degrees to the equator. The station was reported to be functioning normally, leading Boris Petrov, a senior academician in the Soviet Academy of Science, to call the launching "an epoch of orbital stations and planned research work of men in conditions of space laboratories."

Speculation centered around the imminent launch of the world's first space station had been widely reported in the world's press. It was known that Soviet reporters had been leaving for the Baikonur launch complex, and specially equipped Soviet ships had been sighted in Pacific Ocean waters, presumably ready to track and communicate with forthcoming manned Soyuz flights that were expected to link up with an orbiting station. Then, following the launch of *Salyut 1*, articles began to appear in the world's press peppered with quotes from eminent scientists on the possible effects of the long-term human habitation of such stations. The last manned Soviet space flight had been *Soyuz 9*, launched the previous June, during which veteran cosmonaut Andrian Nikolayev and space rookie Vitaly Sevastyanov had spent eighteen days in orbit to study the physiological effects of prolonged weightlessness. Worryingly, the two men had returned with weakened muscles and heart troubles, casting considerable doubt on the ability of humans to remain in a weightless environment for long periods.

Four days after the launch of *Salyut 1*, at 11:54 p.m. on 23 April, Russia ushered in its second decade of human spaceflight by hurling three men into orbit aboard a spacecraft designated *Soyuz 10*, which was expected to rendezvous and dock with the massive station. "I would say that their intention is to link up and transfer men," reported Sir Bernard Lovell, a British physicist, radio astronomer, and frequent media commentator on the Soviet space program. As the director of the Jodrell Bank observatory near Manchester, England, his facility had tracked every major Soviet space flight since the first Vostok mission carrying Yuri Gagarin. "I believe the Russians are very close to carrying out procedures which have been anticipated for some time." This, he said, involved building "a large, manned space platform remaining in Earth orbit for long periods and to be visited by a sequence of cosmonauts."

Meanwhile, Soviet air force colonel Vladimir Shatalov, the twice-flown commander of this new mission, reported to ground control under the call sign Granit (Granite) that he and his fellow crew members were in excellent health and spirits and had begun "preliminary tasks." There was no confirmation from official sources that a linkup with *Salyut 1* would be attempted, merely that the main aim of the Soyuz mission was to conduct "joint experiments" with the space station. But it was noticeably a crew experienced in rendezvous techniques and crew transfer. Like his spacecraft commander, the crew's flight engineer, thirty-six-year-old Alexei Yeliseyev, was on his third space mission and was one of two cosmonauts who had transferred by EVA from the *Soyuz 5* spacecraft to *Soyuz 4* (also commanded by Shatalov) in January 1969. In fact Shatalov would later reveal the crew had been in intensive training for the final two weeks before the flight, working sixteen-hour days. "We prepared so painstakingly," he stated, "you may almost say that the ship is engraved on our eyeballs." The third member of the crew, thirty-eight-year-old test engineer Nikolai Rukavishnikov, was a space newcomer.

Vladimir Alexandrovich Shatalov, the son of a railway engineer, was born in the city of Petropavlovsk, by the Ishim River in northern Kazakhstan. When he was two years old, his parents, Alexander and Zoya, decided to move the family to Leningrad. As a boy, his heroes were aviators, particularly Valeri Chkalov, a renowned figure in aerobatics and daring test pilot who, before his death in a flying accident, was the holder of many long-

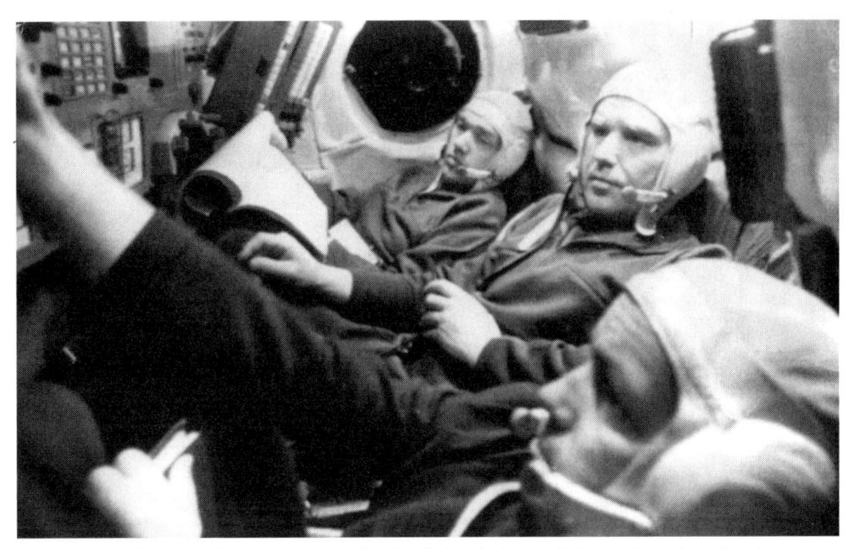

29. The crew of *Soyuz 10* in training for their mission to *Salyut 1*. *From left*: Alexei Yeliseyev, Vladimir Shatalov, and Nikolai Rukavishnikov. Courtesy Spacefacts.

distance flight records. Originally, however, there had been another influence. "I suppose it all began with my father's stories," he revealed. "During the Civil War he was a mechanic in an air unit. He told me many interesting stories about things that happened on the ground and in the air, about the first airplanes." Shatalov would also become quite expert at making model airplanes.

In 1945, after completing ten years of schooling, Shatalov was compulsorily drafted into the Soviet army, but he opted instead to take preliminary flight training at a special air force school in Lipetsk. He was then assigned to the Kachinsk Higher Air Force School for pilots, graduating with honors in 1949 and joining the Soviet air force. Later, in 1953, he undertook advanced training by further correspondence with the Red Banner Air Force Academy and graduated in 1956. By this time he and his wife, Muza Andreyevna, had a son, Igor, who was born in 1952; their daughter, Yelena, arrived six years later. By March 1962, and now a lieutenant colonel and deputy commander of his fighter regiment, he had already begun military test pilot training when he heard that recruitment was under way for a new, second group of cosmonauts. He passed all the interviews and examinations and was accepted as one of seventeen air force officers to be selected. He would commence his cosmonaut training at Zvezdny Gorodok (Star City) on 8 January 1963.

Shatalov's first assignment was as second backup commander for the first manned Voskhod mission, which flew in October 1964. He subsequently trained along with Georgi Beregovoi as one of two crews in line for a planned two-week *Voskhod 3* mission, later canceled when the Soyuz program was initiated, making any further Voskhod missions redundant. Reassigned to the Soyuz program, Shatalov was on duty as a capsule communicator (CapCom) for the first but ill-fated Soyuz mission in April 1967 and then served as Beregovoi's backup on *Soyuz 3* in October the following year. He served as the commander of *Soyuz 4* in January 1969 and was involved in the first orbital docking and transfer of crew from *Soyuz 5* to his craft; he flew again as the commander of *Soyuz 8*, launched just nine months later, on 13 October 1969. On this mission his spacecraft was part of a three-craft rendezvous-and-docking mission, which came to be known as the "troika" flight. Unfortunately, the docking maneuver could not be accomplished, and *Soyuz 8* returned on 18 October. Now a highly experienced and respected member of the cosmonaut team, he would be assigned his third spaceflight in three years when he replaced Georgi Shonin as commander of *Soyuz 10* and was teamed with Alexei Yeliseyev and Nikolai Rukavishnikov.

Alexei Stanislavovich Yeliseyev, the flight engineer on *Soyuz 10*, was born on 13 July 1934, on the outskirts of the town of Zhizdra in the Kaluga region of Russia. His father Stanislav's surname was actually Kureytis, but as he had been jailed earlier in the 1930s for anti-Soviet agitation, Alexei's parents decided it would be prudent (under prevailing Stalinist circumstances) to give him the surname Yeliseyev, his mother Valentina's maiden name. In 1941, following the invasion of the Soviet Union by German forces, the family relocated to the Soviet Far East, where Alexei began his schooling. At the end of hostilities the family moved once again, this time to Moscow, where Yeliseyev's parents found employment. Meanwhile, Alexei was given responsibility for looking after his brother Artemis, who was ten years younger than him, and tending to the household. As his mother held a doctorate in chemistry, young Alexei also became interested in science and mathematics, and his parents happily guided him along that career path.

After completing high school in 1951, Yeliseyev gained admittance to the N. E. Bauman Higher Technical School in Moscow, graduating in 1957 with a candidate of technical sciences degree. An outstanding student at the school, he had scored very highly even in such difficult subjects as automatic regulation, linear and nonlinear systems, and oscillation theory. He also excelled at the sport of fencing, twice winning the Soviet national fencing championship. Following his graduation he undertook engineering work at Sergei Korolev's OKB-1 design bureau, under the direction of Boris Rausenbakh, and had just started there when the first Sputnik was launched on 4 October 1957.

In 1963 Yeliseyev began to specialize in space engineering and the following year learned that civilian engineers might be accepted into the cosmonaut team. He applied and was one of nine civilian cosmonaut candidates personally selected by Korolev for future flights as onboard engineers for Voskhod and later Soyuz multiperson crews. When he first applied to Korolev, the chief designer said that he was four months away from selecting another group of civilian cosmonauts and to give him a call at the end of that time. He was as good as his word, as Yeliseyev recalled: "I was certain that he would forget his promise, what with all the many things he had on his mind. Four months later, to the day, I dialed Korolev's number. He said, 'You can start training. I okayed your application this morning.'"

The untimely death of Korolev in the hospital on 13 January 1966, following what was meant to be a routine operation, caused many difficulties within the Soviet space program. It also delayed the official selection of the nine men into the cosmonaut team until 23 May that year. Initially, Yeliseyev had two problems; KGB screening had revealed his father's background and his own change of last name. Additionally, his first marriage had ended in divorce, and he had subsequently married the former Larissa Komarovna, by whom he now had a daughter Yelena, born 9 March 1960. Fortunately, unlike the air force selections, these potential problems were not regarded as major issues and were dismissed as acceptable. Together with fellow candidates Valery Kubasov, Vladimir Volkov, and Georgi Grechko, Yeliseyev would subsequently report to the cosmonaut training center on 5 September.

In January 1969 Yeliseyev had been part of a three-man crew launched aboard the *Soyuz 5* spacecraft, together with mission commander Boris Volynov and Yevgeny Khrunov. The craft docked with *Soyuz 4*, piloted by Vladimir Shatalov, at which time Yeliseyev and Khrunov had carried out a dramatic one-hour extravehicular activity (EVA), making history as the first people to transfer from one spacecraft to another in orbit. They would remain aboard *Soyuz 4*, landing safely on 17 January near Karaganda in Kazakhstan. Just nine months later Yeliseyev served as a backup crew member for *Soyuz 6* and *7*, before being launched aboard *Soyuz 8* on 13 October 1969 along with mission commander Vladimir Shatalov.

Nikolai Nikolaievich Rukavishnikov, nicknamed Kolya, was born into a working family on 18 September 1932 in Tomsk, situated on the Tom River, one of the oldest cities in western Siberia. His stepfather, Mikhail Gavrilovich Mikheyev, and mother, Galina Ivanovna, were both railway surveyors. Thus, in his first years of life through to adolescence the family was constantly on the move as his parents planned and surveyed new railways.

It was a hard life growing up in many temporary tent cities and often being educated in boarding schools, but despite the hardships, it was one he says he relished. "Our frequent moving from one place to another probably influenced my view of life," he once recalled. "It may have influenced my choice of profession. My parents liked their profession. They lived like travelers and explorers. Before work in one expedition was finished they already discussed new routes, new travels. For years we did not have a permanent home and all our things fit into several suitcases. As a boy," he relates, "I had very many impressions and I met many interesting people. My favorite subject was geography [while] mathematics and physics came easy. My father was an amateur radio enthusiast and he passed on his interest in radio and technology to me."

In 1948 the family spent two years living in Mongolia, where Nikolai continued his formal education. He managed to complete high school in 1951, after the family returned to the Soviet Union. He then enrolled at the Moscow Institute of Physics and Engineering, where he studied automation and instrument making, later specializing in dielectrics and transistors. Rukavishnikov graduated from the institute on 13 May 1957 (his diploma work was published in a scientific magazine), and the following month he went to work at Sergei Korolev's OKB-1 design bureau as an engineer and later senior engineer. Here he helped to create and develop automatic control systems, and it became his ambition to test them for himself in the space environment. With this in mind, and with Korolev's blessing, he applied to join the cosmonaut team.

Now married to Nina Vasilyevna, a former senior engineer technician at an engineering plant, and with a son, Vladimir, born in 1965, Rukavishnikov was selected to join the cosmonaut team on 23 May 1966. He arrived at the cosmonaut training center in November, although his training did not commence until 8 January 1967. That same year he was assigned to the Soviet lunar landing group and began training for a possible lunar landing mission. When this program was eventually canceled, he was reassigned to the Salyut cosmonaut group.

In a 1992 interview with Dutch spaceflight researcher Bert Vis, test engineer crew member Nikolai Rukavishnikov revealed that the night launch of *Soyuz 10* had actually been delayed by a day. "They were loading it with liquid oxygen, which is extremely cold," he stated. "One of the cables that was attached to the rocket froze . . . the metal froze so the cable and the rocket froze together. It would have broken off at the time of the flight in any case, but it was an unnecessary risk, so we decided to postpone it for one day. It doesn't cost anything to postpone a flight by one day; we just wanted to minimize the risk."

Beginning on *Soyuz 10*'s fifth orbit, the crew encountered severe difficulties when their ship's automatic orientation systems would not correctly align the craft's orbit to that of the space station due to contamination of the optical surfaces. Eventually, Shatalov sought permission to override the system manually, after which he skillfully guided *Soyuz 10* down into the correct orbital path.

The following day, after the crew had rested, Shatalov switched on the Igla (Needle) radio system, which provided rendezvous data used for controlling the relative motion and attitude of the two spacecraft. The system successfully guided *Soyuz 10* to within a couple of hundred yards of *Salyut 1*, at which time Shatalov once again assumed manual control for the soft docking. "The orbital space station Salyut looked overwhelming," Yeliseyev would later recollect of the rendezvous. "I don't even know what to compare it with. It was . . . a little like a train entering a railroad terminal."

There was jubilation back on the ground when the veteran commander announced that a soft docking had taken place. Then, about ten minutes later, jubilation turned to frustration when Shatalov reported, and ground telemetry confirmed, that an indicator light signaling the final hard dock had not illuminated, which indicated a gap of just over three inches between the two vehicles. He tried firing the Soyuz's engines in an attempt to engage the docking collars forcibly, but still the gap remained. Without a pressure-tight docking the connection was not considered secure enough to allow the hatches on both craft to be opened. It was then suggested to Shatalov that he disengage and try again, but after a few minutes the crew reported "incredible difficulty" in trying to undock from *Salyut 1*. It was a very disturbing turn of events.

A little over five and a half hours after launch, Shatalov tried to undock again, and this time he was successful. The *Soyuz 10* craft disengaged from *Salyut 1*, and following considerable discussion by leading experts on the ground, the order came through to abandon any further docking attempts and return to Earth. After taking photographs of Salyut's docking node for engineers on the ground, Shatalov instituted reentry procedures. During the landing (the first ever night landing by a manned Soviet spacecraft) an air inlet valve was prematurely activated, and toxic fumes permeated the air supply. Rukavishnikov was overcome by the fumes and passed out during the descent through the atmosphere. *Soyuz 10* touched down in darkness perilously close to a lake some seventy-five miles northwest of the Kazakh town of Karaganda. Rukavishnikov was treated and quickly recovered, and none of the three men suffered any serious aftereffects.

It was immediately assumed in the West that there had been technical problems preventing a docking and crew transfer. The Soviets emphatically denied it, issuing statements claiming the flight had been a complete success and that the Soyuz crew had never intended to enter the station. Not surprisingly, Western observers were unconvinced. After a number of years the Soviets admitted that the original goal of the mission had indeed been to link up *Soyuz 10* with *Salyut 1*, but the hard docking necessary to allow a crew transfer had not been achieved, and the mission had been aborted. The cause of this failure was traced to a slightly wrong angle of approach and resultant damage to what would be described as a "weak" docking unit in the Soyuz craft. On the positive side photographic images returned by the crew indicated that *Salyut 1*'s docking system was intact and undamaged, clearing the way for the flight of *Soyuz 11* to proceed once engineering changes had been made.

Initially, Georgi Dobrovolsky, Viktor Patsayev, and Vladislav Volkov had not even been together as one crew. They were cosmonauts in an era in the Soviet manned space program when four crews would routinely be assigned to a flight program, in this case the flights of *Soyuz 10* and *11*. Ordinarily, once the prime crew had flown, their backups then trained for an upcoming mission, each crew generally moving up one step as successive prime crews were launched into orbit.

At one time Dobrovolsky had been the commander of the third crew in training for a future mission to the Salyut space station, while Patsayev and Volkov had been assigned to the second, backup crew for *Soyuz 11*. When Shonin, the original commander of the *Soyuz 10* prime crew, was forced to stand down "for medical reasons," the hierarchy at the Gagarin Cosmonaut Training Center moved the commander of the backup crew, Vladimir Shatalov, into command of the prime crew for that mission. Consequently, Dobrovolsky was also promoted in the order, becoming the new backup commander for *Soyuz 11*. The crew was now made up of Dobrovolsky, Patsayev, and Volkov.

When asked why Shonin had been disqualified as commander of *Soyuz 10*, early cosmonauts Nikolai Rukavishnikov and Vitaly Sevastyanov both shied away from the question, declaring that it had been decided on medical grounds and not for rumored disciplinary reasons. When asked the question directly in 1993, Shonin was at first a little coy but then sighed and tapped himself on the chest. "This was my fault," he admitted. "Only mine." When pressed, he laughed and said "We were very young men, and my best friend was [Mars] Rafikov. . . . I was like Rafikov." The cosmonaut colleague he referred to had been dismissed from the cosmonaut team for excessive drinking and womanizing. When asked if this meant too many parties, he smiled and responded, "Yes . . . maybe!"

Although he had performed well on *Soyuz 6*, Shonin's drinking had subsequently become well known to his superiors, who mistakenly tolerated his excesses. After being replaced as commander of *Soyuz 10*, he was hospitalized in Moscow suffering from depression and alcoholism. The well-liked cosmonaut would recover and straighten out his life, but his glory days were over, and he would never fly in space again.

Once the major problems associated with the *Soyuz 10* docking mishap had been identified and remedied (the docking apparatus had been damaged during the maneuver), the prime crew assigned to fly to Salyut with *Soyuz 11* was formally approved by the State Commission. Commanding this crew would be Alexei Leonov, renowned as the first person to walk in space on his *Voskhod 2* mission six years earlier. His flight engineer was Valery Kubasov, a spaceflight veteran of the *Soyuz 6* mission, and space rookie Pyotr Kolodin was set to make his first flight.

Preparations and training for the linkup flight went smoothly, until shortly before the scheduled launch date. During a routine medical examination Kubasov was unexpectedly diagnosed as having a minor lung ailment. "They saw a spot on my lung," Kubasov recalled in a 1992 interview with Bert Vis. "They couldn't find any other biochemical markers . . . so as a result we all three got knocked out."

According to a central figure in the Soviet space program, Lt. Gen. Kerim Aliyevichs Kerimov, in a 2002 interview with Vis: "The medical commission . . . just three days before the launch they x-rayed all the crew members. Kubasov, after observations, they noticed coin-size dark spots on his lungs. He was disqualified by the medical commission." Although it later turned out that the spot on Kubasov's lung was in fact the result of an allergic reaction to a pesticide commonly used to spray trees at the Baikonur Cosmodrome, Kubasov was forced to step down from the prime crew.

Initially, Kubasov was simply replaced on the crew by Vladimir Volkov. Then, astonishingly—just "eleven hours before the launch," according to Leonov—the prime crew was replaced in its entirety, with an unambiguous direction that the entire backup crew would now fly *Soyuz 11*. "We had a special rule written down on paper concerning crews," Kerimov added. "It was prohibited to exchange one person—only an entire crew. That's why the entire crew was changed for Dobrovolsky, Volkov and Patsayev." It was a precaution that in hindsight probably saved three men's lives, but it would have a dire impact on the replacement crew.

Leonov was both astounded and furious at the decision so close to launch. He protested to Vasily Mishin, Korolev's successor at OKB-1, arguing that he and the crew had not only been involved in the design of the Soyuz spacecraft, but they had been training long and hard for the mission since their assignment just a month earlier, whereas the backup crew had only trained lightly, fully expecting to fly on the subsequent *Soyuz 12* mission some months away. Mishin tried to soften the blow by promising Leonov's crew a two-week vacation to the Black Sea before they went back into training for *Soyuz 12*, but Leonov stayed at Baikonur long enough to witness the launch of *Soyuz 11*.

Liftoff from the remote steppes of Kazakhstan took place at 4:55 a.m. Moscow time on 6 June 1971 and proceeded with very few problems as the spacecraft thundered into orbit atop its rocket, fitted with a new docking mechanism. In order to save precious space aboard the cramped craft, a decision had been made that the operating crew would not wear protective space suits and helmets on their journey to and from the space station. It was a dangerous practice that had gained acceptance in the Soviet manned space program after the flight of the first Voskhod seven years earlier. Dispensing with the bulky suits created more room for the cosmonauts and saved weight and effort, but it was a risky resolution that would soon end in tragedy.

In Northamptonshire, England, young male students at the Kettering Grammar School picked up the first reported signals from *Soyuz 11* an hour after it was launched. Physics teacher and leader of the school's satellite tracking team, Geoffrey Perry, was able to announce the launching well ahead of the first reports emanating from Moscow. The boys had been patiently monitoring radio signals from the space station since its launch, and they began to notice increased activity in early June, indicating that another craft might soon be joining the station in orbit. Then, on 6 June, they heard what they were listening for—a signal from a spacecraft carrying three cosmonauts. As one of the amateur trackers explained: "A good three-man satellite signal makes a kind of purring noise, followed by fifteen bleeps. The fourth bleep is someone's heartbeat, the eighth identifies the seat the cosmonaut is sitting in, and when we put a name to him we can tell when they move from their seats or go to sleep. Occasionally we even hear voice casts from the crew to their ground control."

Later that day British newspapers, using the information supplied by the Kettering group, released front-page news of three cosmonauts in orbit about to link up with the Salyut station—a full half-hour before the news was announced to the world by the Soviet Union. The following day banner headlines in Soviet newspapers proclaimed yet another triumph in the nation's manned space program and offered photographs of the three cosmonauts. Vladislav Volkov was a name already well recognized for his ear-

30. Relaxing prior to their *Soyuz 11* mission. *From left*: Vladislav Volkov, CDR Georgi Dobrovolsky, and Viktor Patsayev. Courtesy Bert Vis.

lier flight aboard *Soyuz 7* and something of an idol to teenage Russian girls due to his striking good looks. His companions, making their first flights, would now become equally well known to the adoring Soviet public.

Using the replacement docking collar now fitted to his Soyuz craft, Dobrovolsky, using the call sign Yantar (Amber), made a flawless hard docking with *Salyut 1*, followed by the successful linking of their hydraulic and electrical systems. Seal and pressurization checks were carried out and found to be satisfactory, which meant that the station's hatch could be opened. Leonov recalls everyone in the Mission Control Center standing and applauding in undisguised relief as this took place. Patsayev floated in first and switched on the air regeneration system as the others made their way into the station. A delighted Dobrovolsky announced to ground control, "This place is tremendous. There seems to be no end to it." Soon after, however, the crew reported a strong burning smell, and they were advised to return to the confines of their spacecraft for their first night's sleep period. The following day they reentered *Salyut 1* and were relieved to find that the smell had dissipated. They reported to the ground that with the odor problem resolved they were ready to begin occupying the station.

Following this reassuring message, Leonov and his crew headed off for the Black Sea and a well-deserved break. Despite their minimal training, it seemed that all was now going smoothly for the Soviet Union's latest three-man crew as they settled in for a lengthy occupancy of *Salyut 1*.

Mission commander Georgi Timofeyevich Dobrovolsky was born in the seaside city of Odessa on 1 June 1928, but when he was just two years old, his father, Timofey Trofimovich, walked out on the family, leaving Maria Alekseyevna to raise the youngster. Georgi would attend Odessa School No. 99 for six years from 1935 and was thirteen years old when World War II broke out. Like other local youngsters, he actively took part in minor but nevertheless dangerous acts of sabotage against the Germans. When Odessa fell in 1941, he joined an underground partisan group but was captured by the Germans in February 1944. Found to be in possession of a firearm, which he had never used, he was savagely beaten, tortured using electric shocks, and thrown in jail, sentenced to twenty-five years hard labor. The following month, on 19 March, some false documents were smuggled in to him after a relative bribed a guard, after which he managed to bluff his way free. Odessa was liberated on 10 April, and Dobrovolsky remained on the run until the Nazis were completely driven out of the area.

The immediate postwar months were a struggle. The young man studied by day and at night helped to unload ships in the port. Having grown up living by the Black Sea and with a love for everything nautical, he applied to enter the Odessa Nautical School, but to his dismay he was advised that his application had been handed in too late. Shattered but undaunted, he decided instead to enroll in the Odessa Special Air Force School, which trained young men for service in the Soviet air force, and graduated in 1946. "The moment I got there I could think of nothing but the sky," he once told reporters. "I am crazy about flying." He then transferred under his trainers' recommendations to the Chuguyev Air Force Pilots School, receiving his lieutenant's wings at his graduation on 7 September 1950.

Following this training, he was assigned as a navy fighter pilot in the Sevastopol region. After serving in a number of fighter regiments flying MiG, Yak, and Lavochkin aircraft, during which time he graduated to senior pilot, flight commander, and finally deputy squadron commander, Dobrovolsky decided to further his aviation knowledge. Without interrupting his flying career, he completed an extensive correspondence course with the Red Banner Air Force Academy, graduating in 1961. Four years earlier he had wed Lyudmila Steblyova, a mathematics teacher he had met at a dance while posted to the town of Valga in Estonia. After his graduation from the air force academy Dobrovolsky became a political worker while continuing to fly. He also made a total of 111 parachute jumps and became an instructor in this discipline. He would later become a parachute instructor for his fellow cosmonauts.

In January 1963, with the rank of major in the Soviet air force, Dobrovolsky joined the Soviet Union's cosmonaut team, reporting for training at Zvezdny Gorodok, outside of Moscow. Lyudmila and their four-year-old daughter, Marina, joined him. The couple would later have another daughter, Natalya, born in 1967.

Viktor Ivanovich Patsayev was born on 19 June 1933 in the city of Aktyubinsk (now Aqtobe) in northwestern Kazakhstan, near the Russian border, although they would move in 1937 to the nearby town of Alga. His father, Ivan Panteleyevich, a bakery director, was called into action and killed in the defense of Moscow in October 1941 and buried in a mass grave. A year after the end of hostilities, Viktor, his mother, Mariya Sergeyevna, and his sister Galina moved away from Alga. By this time his mother had remarried. Her new husband, Ivan Volkov, was a widower banker with four children. After living for a time in the village of Kos-Isteko, the family finally settled near Penza in the Kaliningrad region in 1948. "I was an average pupil," Patsayev once told a biographer. "I finished school in the town of Nesterov [and then] I entered the Penza Polytechnical Institute. After graduation [in 1955] I was sent to work at the Central Aerological Observatory." The CAO was located in Dolgoprudny, just three miles north of Moscow, and Patsayev was employed there as a design engineer.

In his youth Patsayev had discovered a passion for reading, with a preference for science fiction novels, although his tastes were very wide-ranging, and he enjoyed the works of many poets and authors from Mikhail Lermontov to Jack London. In later years he would also develop a keen interest in advanced physics and mathematics. For recreation he involved himself in a variety of activities such as skiing, cycling, fencing, and shooting.

Not long after he started working at the CAO, Patsayev began publishing scientific articles and demonstrated exceptional talent in creating and developing miniaturized meteorological instruments. Soon after, and following the first Sputnik launches in 1957, he left the observatory and began working with precision instruments at the OKB-1 design bureau outside of Moscow, headed by Sergei Korolev. Here he befriended future cosmonaut Vladislav Volkov, a fellow member of a local flying club, and their friendship deepened when they were recruited into recovery teams for manned space missions.

Unlike the gregarious Volkov, Patsayev had a serious and studious nature. Russian journalist Evgeny Riabchikov would describe him as "unhurried, calm, rarely given to jokes and witticisms. He weighed his words carefully, and when he did speak, spoke in the clear, precise words of the engineer."

Patsayev's decision to become a cosmonaut was apparently gradual, but he was probably spurred on by his friend Volkov's acceptance into the cosmonaut team in 1966. He finally told Korolev that he would like to do design work related to spaceflight and as part of this undertaking was keen to apply for cosmonaut training. Korolev questioned Patsayev on what drew him to contemplate such a bold move, but he had already been keeping an eye on the young man's excellent progress at the design bureau and liked resolute and persistent people. He felt Patsayev would fit in well as a civilian cosmonaut and later wrote out a letter of introduction to accompany his personal recommendation. "Brief testimonial of Patsayev, Viktor Ivanovich. [He] has shown himself to be a good engineer and specialist capable of solving tasks. He has carried out many technically complex works, published several scientific articles, taken his candidate's examinations. Performs all his work well and on time."

In order to further his qualifications, Patsayev began parachute training with other cosmonauts in 1968 and was accepted into the cosmonaut team on 27 May 1969. By this time he and his wife, Vera Alexandrovna, a research worker, had a son, Dmitri, born in 1957, and a daughter, Svetlana, born in 1962.

Vladislav Nikolayevich Volkov was born in Moscow on 23 November 1935 and as a youth derived a lot of early career influences from his parents. His father, Nikolai, was an aeronautical engineer, while his mother, Olga, had worked in aircraft plants for many years. He once revealed that his family background had definitely set him on a course to the stars. "I made my choice long ago, as a boy. Partly, probably, because my father and mother worked in the aircraft industry and spent many hours building airplanes. I probably inherited their enthusiasm. I dreamed of becoming a test pilot and testing all the newest planes. I think I was fortunate."

Essentially, Volkov's story was similar to that of his *Soyuz 11* crewmate Viktor Patsayev. Like Patsayev, he entered school during the war, taking his early schooling at the Moscow High School, and at fourteen (again like Patsayev) joined the Komsomol, the junior branch of the Communist Party. In 1953 he attended the Moscow Aviation Institute's aero club, graduating the following year. He then reenrolled at the institute in order to gain his engineering degree but maintained his flying lessons and also took up parachute jumping. Volkov graduated from high school in 1953 and then studied at technical institutions, including the Moscow Aviation Institute, where he worked as a specialist engineer-electrical technician. While studying at the institute, he also met his future wife, Lyudmila Biryukova. They married early in 1957. Their son, Vladimir, was born on 14 February the following year, and in early 1959 Vladislav and Lyudmila both graduated from the institute. In April that year Volkov began work at Korolev's OKB-1 bureau as an aviation designer-assembler in Department 4. By this time he had met and befriended Viktor Patsayev, with whom he worked at the design bureau, and together they took advanced lessons at the same Kolomna flying club, usually in the same airplane. They completed the course in 1968.

From 1961 to 1966 Volkov participated in the design, construction, and testing of the Vostok and Voskhod spacecraft as deputy leading designer, and in May 1966 he was selected to join the cosmonaut team. When asked in a later press interview about his decision to become a cosmonaut, Volkov responded, "I want to master space equipment, to test it and help in improving it, to do something resembling what test pilots now do in aircraft design offices."

Evgeny Riabchikov characterized Volkov as an amiable person, saying he was "an outgoing man with the build of an athlete. His constant inclination to humor, his feeling for the right word, his enthusiasm for sports, his knowledge of song, his eagerness to be of help to his comrades . . . made him the most popular man on the cosmonaut team." Tennis and ice hockey were favorite sporting activities. He also spent hours sketching and wrote a book for young people on his life and space travel.

Volkov's first trip to space came as flight engineer aboard the three-man *Soyuz 7* in October 1969. During this flight different rendezvous and docking techniques were carried out (albeit unsuccessfully) with *Soyuz 8*. On his return he was awarded the title of Hero of the Soviet Union.

Following standard media release practices then in place, very little information was given on the progress of the work being carried out aboard *Salyut 1*, other than daily announcements stating that the mission was "proceeding smoothly" and the crew was "fulfilling their scientific research program." They would organize their work in shifts around the clock, combining it with a regimen of physical exercises. This included the use of a moving "caterpillar" track to simulate the exertion of walking and help tone up their muscles, which tend to atrophy in the weightlessness of space. Six days into their flight aboard the orbiting laboratory, Dobrovolsky revealed that he and Patsayev had decided out of convenience to grow beards, even though battery shavers had been provided on board.

Two weeks into the mission there was still no public indication from official Soviet sources about how long they would remain in orbit, when the mission might end, or how soon the crew would board *Soyuz 11* for the return journey home. The crew even cast the first votes from space, as the Russian population was electing members of the Supreme Soviets, the Parliaments, of the republics of the USSR. "We give our votes to the inviolable bloc of communists and non-party members," the three men formally transmitted to the ground.

One potentially catastrophic problem faced by the crew on 16 June has been described as a small electrical fire in one of the station's systems, filling the station with acrid smoke. Vitaly Sevastyanov was on the second backup crew, and he downplayed the incident. "Yes, there was an electrical fire in one piece of equipment," he confirmed, but then pulled back just a little from that statement. "Well, there was just smoke. It wasn't really a fire. It was then disconnected from the energy supply system and the smoke stopped. It wasn't a real fire."

Valery Kubasov partially confirmed Sevastyanov's story. "It's true . . . that there was a small fire aboard the space station. One of the electronic devices started to smoke a little and they simply disconnected it. They had a fire extinguisher which they sprayed it with, that's all."

The fire, however big or small, did not prevent the trio from continuing their work program. The Soviet space hierarchy certainly considered it an event not appropriate for public release or comment, and like other problems, it was quickly covered up, only to be revealed many years later. Subsequently, however, strong rumors surfaced that the flight was far from trouble-free in other areas. While Soviet television showed images on the news of jovial, bearded crew members apparently enjoying themselves and in good spirits, there was said to have been a significant degree of tension and personality clashes aboard the station. The normally jocular Volkov, in particular, is said to have had ongoing problems in taking a subordinate role to Dobrovolsky, and it took lengthy discussions between the crew and the ground before the matter was resolved to a satisfactory degree. In his masterful book *Challenge to Apollo*, space historian Asif Siddiqi wrote that during the fire drama Volkov "had become extremely nervous and had tried to resolve the situation himself, ignoring the assistance of his crewmates." Siddiqi further wrote that in a 1989 interview OKB chief designer Vasily Mishin revealed that Volkov, "the only spaceflight veteran on board, declared himself the commander of the mission, usurping Dobrovolsky's role," and that there were several "complicated conversations" between Mishin and Volkov after the incident. Despite the apparent tension and their limited mission training, the crew managed to perform a raft of complex scientific experiments with great competence.

On 19 June, three days after the fire incident, Viktor Patsayev woke to his thirty-eighth birthday, becoming the first person ever to spend a birthday in space. Television viewers across the Soviet Union were able to share in the happy occasion, watching as Patsayev's crew members, in an impromptu celebration, toasted him with tubes of prune juice.

Patsayev had also become the first space gardener, carefully squeezing a rubber bulb attached to a small water reservoir and applying moisture to some seeds. The hydroponic vegetable patch soon took to life; the seeds sprouted, and the little garden was soon a pleasing green in the otherwise sterile environment of the station. An automatic camera taking periodic photos would record the progress of Patsayev's little garden in space. In one of its bulletins the Soviet news agency Tass reported that "experiments are continuing aboard Salyut to study the influence of conditions of weightlessness on the development of higher vegetation. Grown for this purpose are Chinese cabbage, flax and bulb onions, cultivated by the hydroponics methods."

With nightly television broadcasts from *Salyut 1* allowing the Soviet people to follow the progress and spectacle of their flight through words and images, the three cosmonauts not only became well known but greatly adored across their nation, leading the authoritative space historian James Oberg to observe: "As week followed week that June in 1971, the excitement and exultation mounted. A glorious welcome home was in store for the cosmic heroes. The humiliation of the Apollo moon triumphs had been exorcised. Soviet space supremacy had been restored by three personable, heroic cosmonauts."

Meanwhile, the crew that was scheduled to take over from them had finished their enforced break and resumed training. "For twenty-three days the mission went remarkably well," Leonov later reflected, "despite the fact that the crew had had so little time to prepare for their packed schedule of biomedical and scientific experiments. When we returned from our two-week vacation to the Black Sea, the mission was nearing completion."

On 27 June, two days before the crew was scheduled to vacate *Salyut 1*, the Soviet space program would suffer the first of two major setbacks. At Baikonur a third attempt was being made to launch a massive N1 rocket, this one carrying as payload a mock-up of the Soviet lunar orbiter and lander. But as Russian space historian I. B. Afanasiev observed:

Soon after liftoff, unexpected gas-dynamic moments [eddies and countercurrents] at the base caused the rocket to roll. The rate and angle of roll grew steadily. At 39 seconds, the gyroscope-stabilized platform of the launcher control system hit its stops, and at 48 seconds, the large amount of torque started the destruction of Block B [second stage]. The emergency rescue system was only a mockup and, naturally, did not fire. At 51 seconds, when the angle of turn reached 20 degrees, the KORD system [Kontrolia Raketnykh Dvigatelei, the system for controlling the rocket engines] issued a command that cut off all engines of the first stage.

The gigantic N1, like its two predecessors, was destroyed. Interestingly enough, space researcher Bart Hendrickx points out that an article on *Salyut 1* published in a 1996 issue of *Novosti Kosmonavtiki* claims that the launch of the third N1 was at one point scheduled for 20 June and that "according to some data" the orbiting cosmonauts would have been able to observe it had it taken place that day. Hendrickx explains:

> *They would have used an instrument called Svinets, an infrared device for detecting missile launches that was originally developed for the 7K-VI [military version Soyuz] in the 1960s and was also supposed to have flown on* Voskhod 3. *However, the article goes on to say that the experiment was cancelled when a decision was made on 18 June to delay the N1 launch until 22 June, presumably because the ground track of* Salyut 1 *no longer carried it over Baikonur at the right moment. Of course eventually the N1 didn't fly until 27 June.*

Having completed nearly twenty-four days in orbit, and almost doubling *Gemini 7*'s former endurance record, the crew of *Soyuz 11* reactivated their powered-down Soyuz spacecraft, ready for the return journey home. Once this had been completed, they transferred all their scientific data records, exposed film, and logbooks into the descent module of the craft, before strapping themselves in. After the three men had carefully checked all the spacecraft's systems, Dobrovolsky initiated the undocking from *Salyut 1* and then flew alongside the station for an hour while his crew took photos. He then allowed their spacecraft to drift slowly away from the orbiting laboratory, and over the next three orbits they made final preparations for the reentry phase.

As commander of the backup crew, and having originally trained at length for this mission, Alexei Leonov was monitoring events at the Mission Control Center in Kaliningrad, outside of Moscow. According to Leonov, he was also in radio communication with the crew, going through a checklist of instructions with them. "As the crew went through the control of positioning the air vents located between the landing and orbital modules," he recalled in his memoirs, "I advised them to close the vents and not to forget to re-open them once the parachute had deployed." He suggested they write it in a logbook. The vents were meant to close and then automatically open once the parachute had deployed following reentry, but Leonov said he harbored concerns that the vents might open prematurely, placing the crew (who were only clad in lightweight tracksuits and without helmets) in immediate peril from oxygen starvation. "It seems," he would reflect sadly, "that the crew did not follow my advice."

As the *Soyuz 11* crew prepared for the seven-minute burn of their braking rockets, ground control radioed a brief message to the crew, "Goodbye Yantar; till we see you soon on Mother Earth." To which Dobrovolsky responded, "Thank you; be seeing you. I am starting orientation." It would be the final transmission received from the spacecraft. Next, following a successful burn as they passed over Madeira, the spacecraft dipped back into the atmosphere, and contact was lost with the ground. This was a normal occurrence generally lasting about four minutes, as radio waves are unable to penetrate the ferocious turbulence and ionization associated with reentry. Several minutes passed, by which time radio contact should have been reestablished with *Soyuz 11*, now approaching the ground under its main parachute. There was only an ominous silence, and the longer it went on, the more ground controllers became concerned. Soon after the radio blackout ended, the crew should have been in touch with flight control. They had no way of knowing that the crew had earlier switched off all the radio transmitters in an attempt to locate the source of an air leak from their spacecraft.

Meanwhile, on the steppes of Kazakhstan everything seemed quite normal to the recovery crews, waiting in their helicopters. To everyone's jubilation they reported seeing the descent module gracefully drifting down to a wheat field beneath its massive red and white parachute. Then, as planned, braking rockets automatically fired moments before the craft touched down, slowing the craft and absorbing the shock of impact. Moments later the charred module had come to rest sideways on the ground as recovery and medical crews landed alongside in helicopters then hastened over to it, ready to open the hatch and greet the record-breaking crew. On hearing that the spacecraft had successfully touched down in the designated area, relieved ground controllers concluded that the spacecraft's radio must have failed in the final phases of reentry. Their celebrations would be short-lived.

What happened in the next few minutes would be lost in a disturbing shroud of official secrecy for many years. Once the rescue team had reached the heat-scarred descent module, they playfully knocked on its side but received no response. They quickly unfastened the bulky hatch and swung it open, excitedly shouting congratulations and messages of welcome to the three crew members still strapped in their couches. Alarmingly, they were greeted by an eerie stillness and silence. Dobrovolsky, Patsayev, and Volkov were in an apparent state of repose, as if asleep. Their faces were tranquil, but blood had formed sticky trails from their ears and noses.

Over the next few frantic minutes the cosmonauts, their bodies still warm, were unbuckled with great difficulty from their seats and dragged out onto stretchers on the hard ground, where CPR was immediately applied by two doctors working on each of the cosmonauts. But it was a futile effort, even when electroshock fibrillation was attempted. None of them registered a pulse—all three were dead. After working on the bodies for almost an hour, the doctors finally gave up. The horrifying news was quickly relayed to Moscow, where, after hurried meetings and decision making, government officials could do little more than issue a statement to the Russian media containing the catastrophic news that the lives of three Soviet cosmonauts had been lost.

Russians were left in a state of shock upon hearing the news. People broke down in tears when they heard it on their radios, and the pace of Moscow traffic slowed as drivers and their passengers listened for further updates about the tragedy. One parked car with a radio attracted so many passersby in Moscow that it caused a massive traffic jam.

"I made no attempt to hide it," Evgeny Riabchikov recalled. "I wept. There were tears in every eye I saw that day. Even the bravest and the manliest betrayed their grief. The grim news of the death of the brave crew of *Soyuz 11* went straight to the heart of all humanity. They had been children of this earth and their fate had become a personal loss to all of us who live on this blue planet."

In Houston Robert Gilruth, director of NASA's Manned Spacecraft Center, issued a statement saying, "I am very shocked and distressed at the news. It is completely unexpected after what would appear to have been a successful mission." George Low, deputy administrator of NASA, said: "The loss of the three cosmonauts is a terrible tragedy. I extend my deepest sympathies to their families and to their colleagues. We have the greatest respect for their achievements in space and our hearts go out to them in their loss."

Soon after the tragic news came through, fellow cosmonauts Leonov and Sevastyanov were immediately dispatched to the landing site as members of a hastily formed investigative committee dealing with the aftermath of the tragedy. It was their task to secure the spacecraft and take photographs.

Three hours after the *Soyuz 11* descent module had touched down, the two men, still in shock, were at the scene. "Their blood-soaked seats, and signs that attempts had been made to resuscitate them, were the only evidence of the tragedy," Leonov would later reflect.

With very few facts to go on, the loss of the *Soyuz 11* crew caused immediate and profound concern among Western scientists and space officials. Soviet space officials attempted to get information about how the men had died, but those overseeing the matter remained frustratingly tight-lipped, only saying that a thorough investigation was under way to determine the probable cause. An autopsy conducted on the three cosmonauts revealed they had suffered brain hemorrhages and bleeding from the middle ear. Their blood contained ten times the normal amount of lactic acid, indicating that the men had died not only of anoxia but under conditions of sheer panic and terror.

In the aftermath of the tragedy there was speculation around the globe that the mysterious deaths might have been the result of a complete breakdown of the three men's vital organs when confronted with the sudden stress of gravity during reentry, after being accustomed to working at a reduced pace. It was known that during Project Gemini there had been worrying signs of the human heart growing lazy in the absence of gravity. In July 1969 a strapped-down research monkey named Bonnie had died of heart failure after his recovery from a nine-day orbital Biosatellite flight, launched to investigate how prolonged periods of weightlessness might affect vital organs. Following that incident the principal investigator of the mission, W. Ross Adey, had suggested that no human being should be permitted to orbit more than twenty-eight days without more animal test flights.

For many perhaps the greatest fear was that if indeed long-duration spaceflight caused a breakdown of vital organs, then the program might need to be halted, at least temporarily. More research on the human body's ability to adapt to long periods in space would need to be carried out. American space officials implored Russia to resolve the question quickly and provide some answers for the safety of all future human space explorers. "We all got a bit worried," said NASA astronaut chief Deke Slayton. "Given that Nikolayev and Sevastyanov had been pretty weak [after *Soyuz 9*] . . . here you had this crew dead. Was there something about being weightless that could kill you?"

A top Soviet physicist, Ivan Vorobyov, issued a statement of concern that was quickly picked up by Western journalists. "From today," he wrote, "every new day of [a human space] flight is a step into the unknown—no man has flown in space so long. Do the negative body processes stop at some time in weightlessness? Is there a limit to the body falling out of earth habits? Is there a fringe which should not be overstepped? We cannot give a final answer to these questions."

A column in the *Sydney Morning Herald* newspaper at that time summed up the fears that were resonating throughout the scientific community.

If it was the sudden effect of gravity that killed the cosmonauts, the Russian and American space exploration programs have received a stunning blow. Mechanical failure can be corrected for future spaceflight. But if the human body cannot survive gravity after prolonged weightlessness man may become a prisoner on earth—when Neil Armstrong's lunar footprints have put him on the threshold of the universe. Before these three cosmonauts, no man had ever spent so long in a state of weightlessness, living 24 hours a day unshackled by gravitational pull.

Did those three weeks in space so weaken the cosmonauts' cardio-vascular systems that, when Soyuz 11 *hit the massive pressures on re-entry, their bodies broke? A tentative theory at the Manned Spacecraft Center in Houston is that the cosmonauts' vital organs, made sluggish by protracted weightlessness, could not respond sufficiently to counteract the return to gravity. When a man is under the influence of earth gravity, the heart must strive to pump blood through the body and keep it circulating. In space, blood can "wash" through the body with little effort. The cardio-vascular system becomes lazy. Then, suddenly, it is called on for almost superhuman performance during re-entry to the atmosphere. It may well be that these were the first fatal cases of "space bends."*

The truth has to be found quickly if space exploration is to continue.

Speculation, some of it wildly inaccurate, continued in the West as people sought answers. Meanwhile, NASA pressed ahead with plans for the continuation of the Apollo program and the forthcoming Skylab missions. George Low conceded that some changes in Skylab were "conceivable, but unlikely." Charles Berry, NASA's chief physician, commented that American astronauts had shown some body changes during flight, including a temporary weakening of the heart as well as general deconditioning and a loss of red blood cells. "But none of these changes are believed to be life threatening," he emphasized. Veteran astronaut Wally Schirra, when pressed for comment, said he felt sure the cause would turn out to be a "hardware failure" rather a physiological breakdown of the heart and circulatory system after twenty-three days of weightlessness.

Twelve days after the triple fatality, the Soviets issued a surprisingly brief and mostly unsatisfactory statement that the three cosmonauts had died of "depressurization from a loss of the ship's sealing" and that an inspection of the descent module "showed that there are no failures in its structure." What they failed to disclose, and would keep close to their chests for the next three years, was the actual cause of the tragedy. They also knew from cardiogram records that Dobrovolsky had died two minutes after the depressurization in the descent module, Patsayev after a hundred seconds, and Volkov after just eighty seconds. One immediate determination was made—in the future all cosmonauts would wear spacesuits during launch and reentry. With the extra space this would require, each craft would now be restricted to just two crew members.

Finally, with the signing of an agreement for a future joint Apollo-Soyuz spaceflight, NASA demanded answers and got them. The undocking of *Soyuz 11* from *Salyut 1* had apparently gone exactly as planned. The problem occurred when the spacecraft separated into three distinct components right on schedule, ten minutes after the reentry burn. The orbital and service modules on either end of the descent module were explosively discarded and would burn up in the atmosphere, while the cosmonauts would reenter in the bell-shaped descent module. Twelve pyrotechnic bolts had fired to separate this occupied module, but investigators believed there was an overload at this time, the resultant jolt displacing a ball joint from its seating, which in turn caused a pressure equalization valve—designed to activate after landing—to pop open, effectively venting the cabin's atmosphere. The crew would have realized almost immediately that there was a rapid drop in cabin pressure and tried to determine the source of the leak.

It is thought that Dobrovolsky assumed it was a problem with the hatch and quickly loosened his restraint straps to investigate, but the hatch would have been properly sealed and secure. The outflow of cabin pressure certainly would have created an insistent whistling noise that, adding to the crew's confusion, would mingle with a similar noise emanating from the onboard radio transmitters and receivers. Knowing that precious air was quickly bleeding from their spacecraft and that they only had a few seconds to locate the source, Volkov and Patsayev also unfastened their belts and switched off the communications system. They would then have pinpointed the noise as coming from beneath the commander's seat—the location of the ventilation valve. "The cosmonauts heard the air whistling and Patsayev unbuckled and tried to block it with his finger, but he failed," remarked Vasily Mishin. "There was a manual drive—they could have protected the capsule; but they forgot or did not know about it. Perhaps it had been omitted from their training."

Having traced the source of the noise, Dobrovolsky and Patsayev would likely have known where and what the problem was, but acute aerodynamic forces had built up to such an extent that it would have been impossible to do anything about it, and they had little option but to struggle back into their seats and try to refasten their belts. They did this clumsily in haste (and probably in great fear), and as a result their straps were found tangled by the ground recovery crew.

Asif Siddiqi wrote that the air pressure in the module would have dropped and been lost very quickly.

Just four seconds after the ventilation valve failure, Dobrovolsky's breath rate shot up from sixteen (normal) to forty-eight per minute. After the beginning of the pressure loss, the cosmonauts lost the capacity to work in ten to fifteen seconds and were dead in forty-eight to forty-nine seconds. They were apparently "in agony" [he is quoting from Kamanin's diaries] three to five seconds after separation until about twenty to thirty seconds before death. All the pressure in the capsule dropped from a normal level of 920 millimeters to zero in a matter of 112 seconds. As one Russian journalist [Mikhail Rebrov] put it, the cosmonauts "passed away fully aware of the tragic consequences of what had happened."

It was later said by Mishin and Nikolai Kamanin that something as simple as placing a finger over the leak would likely have saved the cosmonauts. Tragically, the three men died only minutes away from the triumphant conclusion of one of the greatest Soviet achievements up to that time in the Space Age.

As a member of the investigative team, Leonov would later state that the tragedy had affected and saddened the entire cosmonaut corps, but he also felt a very personal guilt and frustration at what had happened. "Had I been allowed to fly in their place I am sure my crew would have survived," he reflected in his memoirs. "I never told anyone that the crew had failed to follow my instructions and that this had led directly to their deaths." He added that Patsayev's widow, Vera, worked at the OKB-1 design bureau and was in a high enough position to gain access to the radio exchanges between Leonov and the crew and uncovered what she felt was the sad truth about their failure to heed his advice about ensuring the vents were fully closed. Most of all, according to Leonov, he could not bear to look into the eyes of his dead colleagues' children. "Even though it was not my fault, I blamed myself for what had happened. It was not until much later that the children learnt how desperately I had tried to avert the tragedy."

When specifically asked about Leonov's assertions, former chairman of the State Commission Kerim Kerimov shook his head vigorously. "That's rumors," he told interviewer Bert Vis. "That's false. Everything is false. The reason is, that valve had a small ball inside of it. And when the spacecraft hit the atmosphere it came loose. You know they were buckled up; they heard the hissing sound of the air escaping. If they had pushed it . . . but they were unable to do that because they were buckled up."

Also sitting in at the interview was Zakhar Fedorovich Brodskiy, a former OKB engineer and the aide to Lieutenant General Kerimov. "I'm a tester myself," he added to the conversation.

We tried to find out the reason of that event. While jettisoning the service module from the command module, the jettisoning takes place with the help of pyrotechnics. All the bolts must be blown one by one, within a second. Not all at the same time, but one by one. Normally, that valve was opened to level the pressure inside and outside. The moment was an explosive one. . . . Under the great force of the explosion the valve opened in vacuum, at forty kilometers altitude. They were buckled up. If they had been in spacesuits . . . maybe one of them realized the reason and tried something, but nobody could move. We duplicated the events in the low pressure chamber at Chkalovskaya. We practiced more than a million such jettisons. Only once did we get the same results. Only once, in a million attempts. Now it has been modernized and it has a special reserve system. A backup system.

To which Kerimov added: "They have a backup system in the form of spacesuits. You can't rely on a single system. It may fail you."

Dobrovolsky, Patsayev, and Volkov did not get the hero's welcome that everyone envisaged once they had undocked from the space station. Instead, their bodies, with shaved faces and dressed in civilian clothing, were laid in open coffins for several hours in the Central House of the Soviet army in Moscow on 1 July to allow tens of thousands of dignitaries and other mourners to pay their last respects. Patsayev was the only one to show any sign of injury—he had a large dark bruise high on his right cheek. Each of them had a posthumous Gold Star of the Hero of the Soviet Union pinned to their chest; for Volkov it was the second such honor to be bestowed. Early the following morning the three bodies were cremated.

On a day declared one of official mourning—a hot, humid day in Moscow—their remains were solemnly transported to Red Square in a long cortege attended by Soviet dignitaries and several cosmonauts. Astronaut Tom Stafford was also there, having been flown over in haste to represent President Nixon. Following a number of speeches and eulogies made before a massive, grieving crowd, the urns containing the three men's ashes were placed with honor in a niche in Moscow's Kremlin Wall, behind the Lenin Mausoleum and close to those of their fellow cosmonauts Yuri Gagarin and Vladimir Komarov. They were the third, fourth, and fifth cosmonauts to have their ashes consigned to the historic wall and also the last. With the later demise of the Soviet Union, further inurnments in the wall would cease. Along with the other posthumous honors, three new tracking ships used to support the space program would be named after the three cosmonauts.

Many years later numerous details about the Soviet space program would be gradually declassified. Far more reliable information on its history slowly but surely found its way to the media and to sleuths in the West who had tried to unearth and report on the program's secrets since the very early days of Soviet cosmonautics. As part of this declassification, some amazing film footage was released, which had been shot immediately after the landing of *Soyuz 11*. It showed the bearded bodies of the three cosmonauts lying on stretchers next to the descent module, while members of the recovery team frantically tried to resuscitate them.

The *Salyut 1* station would never be occupied again. All future crews had been halted nine days after the tragedy, while ongoing efforts were made to keep the space station in orbit. The two-man *Soyuz 12* crew, which would be launched on 27 September 1973, conducted a test flight designed to check out improvements made to their Soyuz spacecraft, but it was not scheduled to approach or dock with the station. As with every crew subsequent to *Soyuz 11*, this crew of Vasily Lazarev and Oleg Makarov was wearing full space suits and helmets for liftoff and reentry. By the time a new resident crew could have flown to and linked up with *Salyut 1*, the station had already been decommissioned. On 11 October that same year it was deliberately brought back down by radio command and burned up in a spectacular display over the Pacific Ocean.

Superstition is endemic in Russia's space program, and every person—regardless of nationality—launched aboard one of its Soyuz variant spacecraft has to observe certain traditions prior to the flight. For instance, they have to visit the Kremlin Wall and pay homage there to Gagarin, Komarov, and the crew members of *Soyuz 11*; sign the door in their living quarters; and visit the still-intact office of Yuri Gagarin and place their signature in a guest book on his desk. On their way to the launchpad the males are supposed to urinate on the left rear wheel of their bus, as Gagarin allegedly did before his Vostok mission (although unedited film evidence of the transfer bus on the way to the launch suggests this never occurred).

Cosmonaut superstitions were never more evident, however, than in the aftermath of the triple fatality in 1971. The failed *Soyuz 10* mission, the replacement of the original *Soyuz 11* crew, the small fire on board the Salyut station, and finally the deaths of three cosmonauts had a major impact on the remaining members of the cosmonaut team. To them *Salyut 1* was a vehicle cursed at worst and unlucky at best, and as a result there was open relief when *Salyut 1* was decommissioned and deliberately destroyed and a replacement station sent into orbit.

The cosmonaut corps had their own newspaper, a copy of which was pinned on a notice board at the Star City training facility following the loss of the crew of *Soyuz 11*. In it there was a tribute to the crew: "The conquest of space is a difficult journey, but it is one upon which mankind has already embarked. It is an inevitable step in the logic of world progress."

8. On a Roll at Hadley

Geoffrey Bowman

The mind is not a vessel to be filled, but a fire to be lighted.

Lucius Mestrius Plutarch (c. 46–120 BCE)

As the light aircraft climbed into the hazy blue sky above the Mojave Desert, Jim Irwin was perplexed. This was his trainee's fifth instructional flight of the day, but the harder Sam Wyman tried, the worse his flying seemed to get. Irwin himself was an exceptional flier, recently appointed as sole test pilot for the U.S. Air Force's top-secret YF-12A Interceptor, but as he waited for his dream assignment, he added to his flying hours by teaching men like Master Sergeant Wyman.

Sam was a nervous flyer who tended to overreact. Irwin was hoping if he could just calm his trainee down, a solo flight might be possible later that day. But as the aircraft made a steep climbing turn, its nose rose sharply, and Irwin sensed the unmistakable shudder of a developing stall. He never quite managed to piece together the events of the next few seconds, as the nose dropped and the wings canted sickeningly into a flat spin. He tried to recover, but the controls seemed to have frozen. Perhaps Sam was trying to pull the doomed aircraft out of its death plummet, but the laws of aerodynamics would not be cheated.

The impact with the desert floor was brutal. Irwin's head slammed against the back of his pupil's seat, which then collapsed onto his feet, shattering the right ankle and leaving bone fragments protruding through flesh. Miraculously, the aviation fuel did not ignite, so the rescuers from Edwards Air Force Base (AFB) were able to extract the badly injured men from the wreckage and take them to hospital.

Wyman, who was severely concussed, was treated and given the last rites, but he managed to pull back from the brink. Irwin suffered two broken legs, a broken jaw, head injuries, and numerous cuts and bruises. Writhing in pain, he was flown to the intensive care unit of March AFB Hospital, east of Los Angeles, where doctors contemplated amputating his right foot because of badly impaired circulation.

Irwin's wife, Mary, visited him every day in hospital, but his young daughters were horrified by the sight of their father wrapped in bandages like an Egyptian mummy. Recuperation was long and painful. With both legs in plaster and his jaw wired, he was bedridden, unable to eat solid food, and barely able to communicate. The doctors saved his foot but were sure he would never fly again and uncertain if he would ever walk properly.

As the summer of 1961 wore on, helped by Mary's devoted daily nursing care, Irwin's body slowly healed. When the plaster casts finally came off, his legs were shriveled and horribly scarred, particularly the right leg, which was now slightly shorter than the left. His muscles were so weak he had to undertake a grueling series of exercises to recover his strength, but gradually he regained the ability to walk and even dared to think about flying again. Then came the hammerblow: the air force doctors broke the news to him that because he had suffered a concussion, there was a mandatory grounding for at least a year. Instead of testing the YF-12A, a precursor of the legendary SR-71 Blackbird, Jim Irwin could see his flying career plunging into its own flat spin.

James Benson Irwin was born on St Patrick's Day, 17 March 1930, in Pittsburgh, Pennsylvania, of Scots-Irish descent. His grandparents had emigrated from Pomeroy, County Tyrone, in what is now Northern Ireland. When he was a boy, looking up at the moon, he felt that he would somehow be able to go there one day. He made the mistake of telling his family and neighbors, who all laughed at him. His mother was a little more direct, telling him: "Son, that's foolishness. Men will never be able to go to the moon!" At the age of seventy-one, Elsa Irwin might have recalled those chiding words ruefully as she watched live television images of her astronaut son kicking up lunar dust.

Irwin showed an early interest in science, scoring 100 percent in a high school exam, but he had no obvious interest in flying. He wanted to enroll at West Point Military Academy but fell short in his exam grades and instead entered the U.S. Naval Academy in Annapolis, graduating with a degree in naval sciences in June 1951. He transferred to the recently created U.S. Air Force, not so much out of enthusiasm for a career in aviation but because he did not care for the idea of a life in the navy. It was not until he made his first solo flight that he felt the "great joy" and "complete relief" of being alone in the air, the master of his own fate.

There is a sad chapter in Jim Irwin's military life that does not appear in any NASA biography: he was briefly married to and then divorced from his first wife, Mary Etta Wehling, the daughter of an officer at Reese AFB in Lubbock, Texas. By his own admission they were both young and immature, but ironically it was religious differences (his family was Protestant, hers Catholic) that finally drove them apart—perhaps in part a legacy of his ancestry.

Now bitten by the flying bug, Irwin realized that progress in his new career required further education. He applied to the University of Michigan, where future astronaut colleagues Jim McDivitt and Ed White were also studying, and earned master's degrees in both aeronautical engineering and instrumentation.

Returning to flying duties, Irwin met and married his second wife, coincidentally also called Mary, and in spring 1960 he was accepted into test pilot school. The couple moved to a new home in the stifling heat of Edwards AFB, and shortly after graduating as a test pilot, Irwin was given the top-secret assignment of being the first test pilot of the YF-12A. He could not even tell Mary about this great honor. It was while he was biding his time, waiting for the assignment to begin, that he crossed paths with Sam Wyman.

Fourteen months after his near-fatal crash, Jim Irwin was restored to flying status. He applied successfully to the Aerospace Research Pilot School (ARPS) at Edwards, a nursery for budding air force astronauts, and graduated in 1963, now looking to the burgeoning space program as a future career path. But it was a path strewn with obstacles. Irwin was one of 770 applicants who met NASA's qualifications for its third intake of astronauts. The 14 selected in October 1963 included future moon walkers Buzz Aldrin, Alan Bean, Gene Cernan, and Dave Scott . . . but not Jim Irwin. He was convinced it was due to lingering concerns about his accident.

When NASA began recruiting again in September 1965, Irwin was approaching his thirty-sixth birthday, the age limit for astronaut applicants. He realized that it was his last chance. He again faced tough competition—there were 351 applicants—but this time he was successful, becoming one of 19 men selected in April 1966. With Mary and their four young children in tow, Jim Irwin set off for a new life in Houston, Texas.

Just as Project Apollo was the product of a common effort by some 400,000 individuals striving toward the goal of putting men on the moon, so each individual Apollo mission was the culmination of a process that interwove the lives of three often disparate human beings, finally bringing them together at the tip of a Saturn V rocket. One of them, Alfred Merrill Worden, was born in Jackson, Michigan, on 7 February 1932 and grew up on a small farm, one of seven children. From childhood he worked in the fields and milked the cows, and as a teenager he virtually ran the farm, but he quickly realized that farming was not the life he wanted. He made up his mind to go to college and earned a place at West Point Military Academy.

Like Jim Irwin, Worden was not thinking about flying as he grew up. He never looked up in the sky and thought, "Gee, I want to be a pilot." What he most recalled about West Point was the teaching of military leadership, but then in his final year he was impressed and influenced by two tactical officers who were in the U.S. Air Force. Worden decided to join the air force, not so much out of a desire to fly but because he mistakenly thought that promotions there happened more quickly than in the army. Once he took the plunge, he soon realized that flying was his kind of game.

In another interesting parallel with Irwin's career, Worden attended the University of Michigan on a guided missiles course. In a 1989 documentary Worden referred to astronauts as "overachievers," perhaps a self-deprecatory swipe at his own clutch of degrees: bachelor of military science from West Point; and master's degrees in astronautical and aeronautical engineering and in instrumentation engineering from the University of Michigan. He spent a year at the Empire Test Pilots' School (ETPS) in England and had been assigned for a three-year stint at a research flight test facility in England when the commandant of the test pilot school at Edwards Air Force Base asked him to come back to California as an instructor. As the commandant had come to England in person, and was none other than the legendary Gen. Chuck Yeager, first man to fly faster than sound, Worden was inclined to oblige—although there was a gentlemanly dispute between the two NATO allies about the transfer, which was eventually resolved by the U.S. secretary of state.

Six weeks after graduating from the ETPS, Worden was an instructor at Edwards, where he trained hot-shot pilots including future NASA astronauts Charlie Duke, Hank Hartsfield, and Stu Roosa. When NASA sought applications for its fifth group of astronauts, it was a foregone conclusion that Worden would apply and a surprise to no one that he was accepted.

When a boy shares his middle name with an airfield, there must be a fair chance he will develop an interest in aviation. David Randolph Scott was born at Randolph AFB near San Antonio, Texas, on 6 June 1932 and was hooked on flying from the age of three, as he watched his father piloting biplanes in tight formation across the Texas sky. As a ten year old, he spent his spare time building model aircraft and had absolutely no doubt that he wanted to be a pilot like his father. Two years later he had his first flight. When his father threw the aircraft into a series of aerobatic maneuvers, Scott was overjoyed. "It was the most exciting thing I had ever experienced," he later wrote.

In retrospect Scott's trajectory after that was predictable. He was accepted into West Point in 1950 and graduated fifth out of a class of 633. The four years of discipline, education, and grueling exercises proved to be "probably the most valuable and formative" of his life. Applying successfully for the U.S. Air Force, he was posted to Marana Air Base, Tucson, Arizona, where he felt the "enormous exhilaration" of achieving his childhood ambition by learning to fly.

As the Cold War cast its dark shadow over an already war-ravaged world, Scott trained to drop nuclear weapons from single-seat fighter-bombers. Posted to Europe, he flew fighters over Holland. After the Soviet invasion of Hungary, his squadron patrolled the interface between East and West, on high alert.

Determined to become a test pilot, Scott was advised to obtain a degree in aeronautics, so he applied successfully to the Massachusetts Institute of Technology (MIT) in 1960 and made the difficult transition to a civilian life of study. Even before earning his degree (which combined both aeronautics and astronautics), he had been invited to Edwards AFB for test pilot school flight evaluation. Once there, he flew with Chuck Yeager and graduated as the best pilot in his class. He felt on top of the world: "To win the best pilot award, at such a hard school to get into, and from the best pilot in the world, was a more cherished achievement than anything else I had ever done."

He progressed to the newly opened ARPS at Edwards, before selection by NASA in March 1966, into the third astronaut class. Even among such an elite bunch of flyers, Scott was recognized as an obvious future commander. Another member of the group, Walt Cunningham, saw in Scott "the controlled arrogance of those who are born to lead."

With three years' seniority over Worden and Irwin, Scott would clearly be the first of the *Apollo 15* crew to fly in space. *Gemini 8*, launched on 16 March 1966, was a test pilot's dream. Scott, along with mission commander Neil Armstrong, made the first docking in orbit with a target spacecraft. Although the mission's primary goal was achieved, a faulty thruster made the spacecraft gyrate wildly and forced an abort. Instead of preparing to perform the United States's second spacewalk, Scott found himself floating in the South China Sea less than eleven hours after launch. Their coolness and skill under extreme pressure marked both men out as key assets for NASA's forthcoming Project Apollo.

Scott flew again on *Apollo 9*, the first launch of all three modules required for a lunar mission. While Jim McDivitt and Rusty Schweickart tested the new LM's spaceworthiness, Scott remained at the helm of the CSM. Although overshadowed in the public perception by *Apollo 8*'s Christmas lunar spectacle, *Apollo 9* was another dream assignment for a test pilot and a key step on the road to the moon.

It was the *Apollo 9* mission that began to weave together the careers of the men who would become the crew of *Apollo 15*. While Scott trained as a member of the *Apollo 9* prime crew, Worden was selected as a member of the support crew. The assignment offered no guarantees of a future flight but was an opportunity to get noticed. While the other support crew members, Fred Haise and Ed Mitchell, spent a lot of time supervising the preparation of the LM, Worden flew regularly to California, where he became well acquainted with Scott while working on the CM. His engineering skills helped him become an expert in the nuts, bolts, and wiring of the CM, and along with future *Apollo 13* astronaut Jack Swigert, he wrote the malfunction procedures for the spacecraft.

Meanwhile, Tom Stafford, who would command the *Apollo 10* mission in a dress rehearsal for the first lunar landing, asked Irwin to join his mission's support crew. Irwin had previously been in charge of thermal testing of LM systems in a vacuum chamber, a task he found thoroughly satisfying. He mistakenly believed Stafford wanted him to be on the *Apollo 10* backup crew and was initially disappointed to be on the support crew. To make matters worse, one of his contemporaries, Stuart Roosa, *was* on the backup crew. But Irwin swallowed his pride and worked long hours to gain a well-earned reputation as an LM expert.

Shortly after the successful flight of *Apollo 9*, with Project Apollo in overdrive and launching a mission every two months, Scott, Worden, and Irwin were named as the backup crew for *Apollo 12*. Irwin was delighted: his hard work on the LM meant that he would now be the understudy for Alan Bean on what would be a lunar landing flight. In 2008 Worden still had vivid memories of how he felt about the selection: "I was in awe of Dave Scott. He was the 'poster-boy' for astronauts, and considered the 'boy scout' of the program. I was absolutely thrilled to be on a crew with Dave because I believed him to be the best of the corps. Jim was my buddy and we shared an office. I was happy to have him on the crew. He had been through a lot and bounced back quickly to get in the program. He was a very thorough and detail-oriented guy, which I knew would be great for our crew."

Conventional wisdom during the heady days of Apollo stated that if you did a sound job as backup, you would skip two flights then get a prime-crew seat on the next flight. Dave Scott was determined to command a lunar landing mission, and he pushed his team relentlessly. Worden described him as a "slave-driver" but also "a very professional, very no-nonsense kind of commander."

In theory Scott could have become the first man on the moon, if Armstrong and Aldrin had failed to land the LM *Eagle* and if *Apollo 12* commander Pete Conrad had been injured in training. Ironically, this unlikely sequence of events became even more unlikely as the result of a lunar landing simulation run by Scott and Irwin on 5 July 1969. The computer simulation team decided to test the skills of Gene Kranz's Mission Control team. As Scott and Irwin made their landing approach, the computer sounded a "1201 alarm."

Steve Bales, Mission Control's LM computer expert, knew this meant that the computer was overloaded with data, but no procedures had been worked out to deal with such a problem. Scott and Irwin were instructed to abort the landing, which they achieved. Crucial lessons were learned in that simulation. When a real 1201 alarm sounded fifteen days later as Armstrong and Aldrin descended toward the Sea of Tranquility, Bales knew they could safely ignore it and similar 1202 alarms. The rest is history.

On 26 March 1970, four months after the *Apollo 12* crew returned from their successful lunar landing mission, Dave Scott's arduous training regime paid off when the prime crew for *Apollo 15* was announced as Scott, Worden, and Irwin. Each Apollo crew was a different blend of personalities, from the close buddies of *Apollo 12* to what Michael Collins described as the "amiable strangers" of *Apollo 11*. Worden says of the *Apollo 15* crew: "We were a very business-like crew, and did not socialize a lot. In fact, I was the one who did not always agree with Dave so there was a little tension between us, but I think that was for the best, as we were ultimately seen as the most scientific crew, that accomplished the most on our flight. In retrospect I believe our functioning as a crew was the most efficient because we never had to cover for each other. We just did everything on the flight plan, plus extras."

After the euphoria of the first landing on the moon came the financial hangover for NASA and Project Apollo, and diminishing budgets carved a swathe through the later, more scientifically challenging missions. The cuts had a dramatic impact on the plans for *Apollo 15*. In January 1970 the final planned landing mission, *Apollo 20*, was canceled, thereby providing a Saturn V to launch the Skylab space station. That left three "H" missions (*Apollo 13*, *14*, and *15*) and four advanced "J" missions, which would carry lunar rovers to allow extended geological excursions. Then, in September 1970, as NASA recovered from the effects of the *Apollo 13* near-disaster, the purse strings tightened further, and two more flights were canceled. The final three landing missions would all be "J" missions, and the first astronauts to use the Lunar Rover would be Dave Scott and Jim Irwin on a new, beefed-up *Apollo 15* mission. Once again, Scott would get to be a test pilot, even if on this occasion the test vehicle had no wings and rolled across the ground on wheels made of wire mesh. Instead of covering a few miles on foot, Scott and Irwin would be able to cover a total distance of about

31. The crew of *Apollo 15* in front of a training Lunar Rover holding a model of the subsatellite they would release in lunar orbit. *From left:* CDR Dave Scott, CMP Al Worden, and LMP Jim Irwin. Courtesy NASA.

twenty miles on three extended extravehicular activities (EVAs). The mission opened up golden opportunities for the NASA geology team, who were salivating at the prospect of a real voyage of discovery.

Among the thousands of individuals who worked on making *Apollo 15* a success, several names stand out. Harrison "Jack" Schmitt, a qualified geologist selected as an astronaut in 1965 who would himself walk on the moon on *Apollo 17*, was a driving force in helping to turn test pilots into competent field geologists able to pinpoint the most significant rock samples. He realized that the traditional classroom lectures and "learning by rote" methods were boring his colleagues rigid. As a student at California Institute of Technology (Caltech), Schmitt had been particularly impressed by one of his teachers, Lee Silver, who later helped train *Apollo 13* astronauts Jim Lovell and Fred Haise in geology. Silver readily agreed to take over the geological training of Scott and Irwin for *Apollo 15* and from May 1970 to May 1971 became teacher, guru, and mentor, accompanying the two men and their backups (Dick Gordon and Jack Schmitt) on at least sixteen field trips. Scott has described Silver as "an inspiring teacher and a really nice guy [who] knew how to fill us with enthusiasm for all that could be achieved in the geological study of the Moon." To Irwin, Lee Silver was "the greatest guy, just a ball of fire."

Then of course there were the three crew members. Unlike most astronauts, David Scott developed—and has retained—a genuine love of geology and was determined to learn all that he could in order to make his three days on the moon as productive as possible. One of the Apollo science team, Don Wilhelms, has written of Scott that "geologists who worked with him are unstinting in their praise of his interest and ability in their subject [which] blossomed into excitement and total commitment."

Wilhelms believes that Scott transmitted his enthusiasm to Irwin and acknowledges Irwin's devotion to his geological work on the mission. Irwin is a good example of an astronaut who lacked Scott's innate love of geology but was canny enough to know that if he wanted to *go* to the moon, he had to *understand* the moon and be able to select the best rock samples to shed light on the mystery of its origin.

More than on any previous mission, the CMP on *Apollo 15* would be conducting his own detailed survey of the moon. Al Worden, praised by Wilhelms as "an enthusiastic and staunch observer from orbit," had at his disposal a new battery of scientific instruments (known as the SIM bay) lodged in the spacecraft's Service Module. They included a camera powerful enough to spot the LM (and even the Lunar Rover) on the surface, sixty miles below. Scott and Irwin had Silver, but Worden had a teacher of his own—Farouk El-Baz, an irrepressible Egyptian geologist affectionately known as "The King." While Scott and Irwin prepared to study a relatively small patch of lunar real estate, Worden learned to understand the "big picture" as the mountains, craters, and plains rolled beneath the spacecraft.

But where to land? With only three J missions available, NASA scientists had to make each site count. After much debate the choice was between Marius Hills, a relatively accessible area of possibly volcanic rock, and Hadley-Apennine, a spectacular site on the eastern edge of the Sea of Rains (the Mare Imbrium to astronomers). Almost four billion years ago a huge asteroid struck the moon, gouging out the Imbrium Basin (the right eye of the "man in the moon") and throwing up the 15,000-foot Apennine Mountains. Lava bled from the depths of the wounded moon, washing up against the lower slopes of the mountains. Hundreds of millions of years later, an immense river of lava carved a sinuous valley deep into the petrified floodplain. Where it meanders past the foot of the Apennines, Hadley Rille is three-quarters of a mile wide and a thousand feet deep, providing a window through which Imbrium's history can be viewed.

Selecting Hadley-Apennine presented a bold choice: the landing approach over the mountains would be steep, and the presence of the rille all but ruled out "landing long," as on *Apollo 11*. But the rewards to be unearthed in this geological wonderland outweighed the difficulties. In particular Scott and Irwin were encouraged to seek out samples of anorthosite: crystalline rock from the moon's original crust that might unlock the secret of the battered world's origins. The area was just too good to ignore. In September 1970 Hadley-Apennine was selected as the landing site for *Apollo 15*.

Monday, 26 July 1971, 9:34 a.m.: The Saturn V rocket stood poised on Pad 39A, wreathed in icy vapors and silhouetted against a hazy morning sun. According to the local press, a million people in and around Kennedy Space Center were waiting for the launch of *Apollo 15*, braving summer temperatures that had already exceeded 85°F. Launch director Walt Kapryan reported "the most nominal countdown we have ever had."

Three hundred and thirty feet above the launchpad, the crew members were alone with their thoughts and the voices sounding over their headsets. Reclining in the right-hand couch, Jim Irwin recalled the hatch closing, like a dungeon door clanging shut, but welcomed his "imprisonment." Having little to do during the uneventful countdown, he nodded off several times. Forty feet below him lay the enhanced Lunar Module, in which he would ride down to the surface of the moon with Dave Scott. The all–air force crew had named the landing craft "Falcon" after the official mascot of the Air Force Academy. In keeping with the scientific flavor of the mission, they had named their Command Module "Endeavour" after the sturdy little ship that had carried the great explorer James Cook on the first of his three epic voyages of discovery. Two centuries before *Apollo 15*, Captain Cook had charted the South Pacific, New Zealand, and parts of Australia. How might Cook have felt had he known that one day the crew of a very different kind of ship of exploration, also bearing the name *Endeavour*, would honor his memory by exploring another world?

For the fortunate watchers in sight of launch complex 39A, and for untold millions of Apollo addicts glued to TV screens all over the world, the moment of truth had arrived. For anyone too young to have watched an Apollo launch, even vicariously on live television, it is almost impossible to appreciate the buildup of tension as the familiar backward litany of numbers dwindled toward zero or the sheer adrenaline rush as the rocket lifted off the pad on an incandescent tail of fire. For Jim Irwin it was a moment of pure joy, as he recalled to me in 1983: "It was one of the happiest moments of my life. I heard the word *ignition*, and I sensed, and felt and heard, all that tremendous power being released under the rocket. I realized that after all those years of training, preparation, and education, at last it was my turn, at last I was leaving Earth and going into the heavens."

For Dave Scott this was the second time a Saturn V had accelerated him from zero to five miles per second in the time it takes most people to drink a cup of coffee, and there were differences: some subtle, some not so subtle. "I would suspect from what I heard from other guys that every launch was different. The configurations are only slightly different, but you have different times of day, different wind conditions, who knows? But from what I can recall the *Apollo 9* launch was really pretty violent, especially at staging. Fifteen wasn't quite that violent, but it was still pretty violent."

To this day Al Worden delights in telling audiences that he and Jim Irwin were caught unawares by the sudden jolt at the separation of the first and second stages, only to be told later by their grinning commander, "Sorry, I forgot to tell you about that."

But otherwise, the flight into orbit was as trouble-free as the countdown. The crew was calm, almost taciturn, as they accelerated into the hazy blue Florida sky, and when the third-stage engine finally shut down, they were riding at the tip of the heaviest man-made object that had ever orbited the Earth.

Less than three hours after launch, the crew fired up the S-IVB engine again and accelerated rapidly toward their appointed rendezvous with the moon in three days' time. In a moment of empathy with the crowds back in Florida, Irwin reflected on how far he had traveled compared with all of the motorists returning through traffic jams to their hotels.

The journey to the moon was not without incident. Shortly after Worden had deftly extracted *Falcon* from its housing on top of the Saturn booster, a warning light on the instrument panel suggested a serious fault in the mother ship's main Service Propulsion System (SPS) engine. If it was a genuine fault, the LM would have to stay attached to the CSM to provide backup engine capability, and the landing would be scrubbed. Fortunately, a laborious process involving the crew and Mission Control identified a short circuit in a switch. Isolating the switch, Scott performed a brief midcourse correction burn of the engine by manual control. It worked perfectly. "That burn was exactly what we wanted to see," remarked CapCom astronaut Joe Allen. "We'll proceed with a nominal mission." In space and on the ground many sighs of relief were heaved.

Around sixty-one hours after launch, the astronauts also had to contend with a leak from *Endeavour*'s drinking water plumbing. An uncontrolled flow of water in the weightlessness of space could have been a serious problem, with a high risk of electrical short-circuits, but a procedure was communicated to the crew within fifteen minutes, and Scott was able to use an on-board tool to tighten the leaking joint. With the danger behind them, the astronauts saw the funny side of the incident. "We've got a bunch of towels hanging up," Scott reported. "Looks like somebody's laundry."

Thirty-seven years later Scott points out that this potentially catastrophic problem was ultimately fixed by a tool held in a human hand. He is concerned that NASA's current plans to return to the moon involve leaving an unmanned Orion spacecraft in lunar orbit while all four astronauts descend to the surface, perhaps for several weeks, in the new version of the Lunar Module.

When they have a problem "upstairs," what they will tell you is: "We can solve it from Mission Control robotically." They're wonderful guys at Mission Control, they're brilliant engineers, but if they make a mistake they say: "Gee, I'm sorry!" But the guy up there, the Al Worden flying up there, if he makes a mistake then he's in the same position the rest of us are in. So if I'm on the moon I want one of my colleagues, in whom I have ultimate trust or I wouldn't fly with him; I want him up there making the decisions. So NASA of today, they don't think that through, and they put four guys on the moon, and now they've got a surprise up there—like a water leak. And you don't have anyone there to fix the water leak. And if you leak all the water out, you're all dead!

On Thursday, 29 July, nearly seventy-nine hours after launch, *Apollo 15* slipped into lunar orbit. As the spacecraft reappeared around the moon's eastern edge, Scott reported, "Hello, Houston, the *Endeavour*'s on station with cargo, and what a fantastic sight!" Just over four hours later another burn of the SPS engine dropped the spacecraft into an elliptical orbit with a low point of only 56,000 feet. Scott was jubilant. "Hello, Houston, *Apollo 15*. The *Falcon* is on its perch!"

Eighty-six hours into the mission the crew settled down for a sleep period. At the time Mission Control was predicting that the spacecraft's orbit would have a low point of around 50,000 feet the next day when the two modules were due to separate. But the moon's gravity is "lumpy" and tends to pull a spacecraft's orbit out of shape. By the time the crew woke up on landing day, Houston was predicting an altitude of 33,000 feet, "plus or minus 9,000 feet," which left open the possibility that *Endeavour* would shortly be clearing the Apennine Mountains by only 9,000 feet at an orbital velocity of over 3,500 miles per hour. To Worden this was "close enough to count rocks," so before making final landing preparations, the astronauts had to fire *Endeavour*'s reaction control thrusters to raise their low point back to 50,000 feet, the planned height from which *Falcon* would begin its descent.

Friday, 30 July 1971: In Houston it was just after five o'clock in the afternoon, but the mission controllers were not about to head home for a relaxing weekend. A few miles above the moon *Falcon* was arcing down toward Hadley-Apennine, legs forward and windows facing the sky. To the millions following this most crucial phase of the mission on TV, everything seemed to be progressing smoothly, with neither the nail-biting drama of *Apollo 11* nor the exuberant dialogue of *Apollo 12*. Communications from Scott and Irwin were sparse, businesslike, and laced with technical jargon. Like the graceful movement of a swan across a pond, *Falcon*'s descent was outwardly sedate, masking the considerable activity beneath the surface.

For a start the best maps of the landing site showed nothing smaller than sixty feet across, and the geologists thought that the area might be strewn with rocks and boulders. In order to clear the barrier of the Apennine Mountains, *Falcon* had to descend at an angle of 26 degrees, almost twice as steep as the approaches of the three earlier missions, and Scott was painfully aware that if they came in too high, they could end up landing on the wrong side of the rille, denying them access to the mountains. In fact, the radar was telling Scott they *were* a little high. And then, shortly before they were due to pitch forward to gain their first view of where the computer was steering them, Scott was warned by Houston that they were coming in 3,000 feet too far to the south.

In 1983 Jim Irwin described what thoughts were going through his head as *Falcon* descended toward the moon:

Most of the time I was concentrating—it was deep concentration—on the instruments, and I kept telling myself: "Jim, you're just in a simulator," because I thought that if I really did *realize I was coming in for a landing on the moon, I might get so excited that I would forget something I was supposed to do. So I kept trying to convince myself it was just a simulation. It proved to be a very smooth simulation. The only exception would be when we pitched forward about 30° at six thousand feet, and then we could see the surface where we hoped to land, and it was a great surprise there because we could look out the windows and we could see mountains that went up another seven thousand feet above us, and our first inclination was, "We're coming into the wrong place!" because we'd never seen the mountains that high around us in our simulations. But then it was very reassuring to look out and see the canyon, Hadley Rille, and in doing so we convinced ourselves we were coming into the right place, so we continued the approach and I went back to concentrating on all of the instruments and all of the readings that had to be made.*

But it was Scott who was actually flying *Falcon*, and two problems were combining to confuse him. First, the landing site was barely recognizable compared with the views projected on the screen of the LM simulator back in Houston. It turned out that the mapmakers had taken the medium-resolution pictures of the landing site and stretched the contrast so much that every little depression on the rolling plain looked like a sharply defined crater. On top of that, it later emerged that the southerly deviation in *Falcon*'s trajectory had been corrected before the landing commenced, so Houston's warning that they were too far south was misleading. Unable to spot craters that were so obvious in the simulator, Scott adjusted the descent northward.

What goes through the mind of an Apollo commander as he brings his spacecraft down toward the moon? According to Dave Scott:

Well, you have to get it down; you have to land it successfully within a certain amount of propellant, so what you are thinking about—totally focused on—is where to land, where to touch down, and to keep going until you get there. Previous flights had flown a "stair-step" descent, and I think it's a natural tendency. It uses up propellant, so we focused on taking it straight on in, and getting it on the ground some place that was reasonable, *and we didn't know until we got there because the photography only had a twenty-meter resolution, so there wasn't any real way to tell exactly where you wanted to land. Mission control thought we were south, and we couldn't really tell, so we tried to go north. The thing that goes through your mind is, "I want to land as close as possible to the predicted touch-down point, but I want to get down* soon *to save propellant." It's a constant focus. It's what they now call "zoning."*

Having selected a relatively flat area to land, Scott encountered a final problem: the lunar dust. By the time *Falcon* had dropped to within sixty feet of the surface, billowing clouds of dust had completely obscured his view. All he could see was the approaching shadow of the LM cast on a gray smear, so, while continuing to look outside, he was also concentrating on Irwin's voice reading out altitude and descent rates.

Shortly after 5:16 p.m., Houston time, a blue light flashed on the instrument panel. Irwin called out, "Contact!" and Scott instantly cut the engine, determined (with good reason, as it turned out) not to let the newly elongated engine bell contact the surface while still firing, in case the back pressure caused an explosion. *Falcon* dropped the last few feet and struck the uneven surface with a solid thump that rattled every piece of equipment on board.

"Bam!" exclaimed Irwin. Scott reported, "Okay, Houston, the *Falcon* is on the plain at Hadley." Irwin added: "No denying that. We had contact!" A relieved CapCom in Houston said, "Roger, roger, *Falcon*!" and a burst of applause from those around him was clearly heard on television sets all over the world. It was a testament to the quality of their training and their coolness under pressure that Scott and Irwin were able to make landing on the moon seem almost routine. Viewers did not realize at the time how much concentrated effort it had actually required.

The landing had not only been a hard one, but *Falcon* had come down partially inside a small crater, with the engine bell contacting the crater's rim. The spacecraft pitched backward 6.9 degrees and rolled 8.6 degrees to the left, producing a tilt angle of 11 degrees from the horizontal. This was well within safe landing parameters but produced a good example of how two people, both superbly trained observers, can remember the same event in rather different ways. In 1983 Jim Irwin discussed the landing with the author: "I said 'Bam!' when we hit, because it was a very hard landing, and I'm surprised that I wasn't prepared for it. It's the only landing I had ever made standing up, so it was a great surprise. When we landed, we hit on the side of a crater and we started pitching up and rolling off to the side, but then we reached about 20 degrees and it rolled back and came to rest just on the rim of the crater, at about 10 degrees."

Scott, as the man doing the flying, had a somewhat different perspective when discussing the landing in 2008:

We landed on a slope, because one pad was in a little crater that tilted us, I don't know what it was, six or seven degrees, something like that. But no, we didn't roll and pitch 20 degrees. One of the differences in our outlook between Jim and me was that he was looking at the gauges on the inside. I'm looking out the window, so he was very surprised when we got there, because he wasn't seeing the ground come up. So it was a surprise to Jim, so he may have overstated the pitch and roll, but it came right on down pretty steady.

There is a possible explanation for Irwin's comment in the *Apollo 15 Mission Report*, which concludes that "the landing was very stable in spite of the relatively high lunar surface slope" but that the vehicle briefly tilted at a pitch rate of 17 degrees per second and a roll rate to the left of 15 degrees per second.

With the landing successfully achieved, the clipped, technical talk from *Falcon* occasionally gave way to excitement and anticipation. "Hadley Base here," Scott reported. "Tell those geologists in the Backroom to get ready, because we've *really* got something for them!" This was a message to Lee Silver that his students were ready to show him how much they had learned.

It had been a long day, and the crew would sleep before stepping out onto the moon for their first EVA, but two hours after the landing came an appetizer: for the first and only time in Project Apollo the astronauts removed the docking hatch above their heads, and the commander stood on the ascent engine cover to look out over the unexplored landscape that sur-

32. This composite image shows the dramatically tilted LM *Falcon* with Mount Hadley Delta looming up in the background. Courtesy NASA.

rounded them. Scott provided an enthusiastic commentary as he took a series of photographs, some with a telephoto lens. For international space enthusiasts listening live to a *Voice of America* short-wave radio broadcast, a picture of the mountains of the moon began to emerge.

"All of the features around here are very smooth," Scott reported. "The tops of the mountains are rounded off. There are no sharp, jagged peaks or no large boulders apparent anywhere." So much for the drawings of artists like Chesley Bonestell, who had invariably given the moon sharp, jagged peaks. Inside *Falcon* Irwin could only look through the windows, but he was a patient man. His time would come.

Saturday, 31 July, 8:29 a.m.: In Houston it was breakfast time. It was also morning on the moon, and the LM cast a long shadow across the hummocky plain. David Scott eased himself backward through the hatch and carefully descended the ladder toward the beckoning surface and into what Houston justifiably called an "extraordinary television picture." I was one of millions of viewers all over the world who vividly recalls the thrill and excitement of that live color view of Scott dropping onto the footpad, gripping the ladder with his right hand, and turning to face the TV camera. The sky above was deep black, his spacesuit bright white with blue and red details even in the shadow, and the gold-tinted visor of his helmet reflected parts of *Falcon* and the lunar horizon. Clearly, TV technology had advanced in leaps and bounds since the disappointing *Apollo 14* images only six months earlier.

Stepping onto the lunar dust, Scott spoke from the heart. "As I stand out here in the wonders of the unknown at Hadley, I sort of realize there's a fundamental truth to our nature: Man *must* explore. And this is exploration at its greatest!" He did not mention that the front footpad seemed to have little or no contact with the surface, but as he moved around *Falcon*, checking the vehicle's condition, he reported that the left rear pad was in a small crater and the big engine bell had contacted the rim of the crater and was visibly buckled.

A few minutes later Jim Irwin followed his commander down the ladder. He had not lowered his reflecting gold visor, and his face was clearly visible through the clear plastic of the bubble helmet. Dropping straight onto the footpad, right hand gripping the ladder, he appeared to lurch backward and pivot almost out of sight to his left. "Boy, that front pad is really loose, isn't it?" he commented. In 1983 he explained what had happened: "My weight came to bear on the lip of the footpad. The footpad wasn't resting securely on the surface, and so when I put my weight on it, like a big dish it rotated out from under me throwing me off balance. I would never have lived it down if I had fallen on my backside in full view of millions of television viewers!"

It seemed implausible that a LM could be resting solidly on the moon with one of its four landing legs off the ground, yet both astronauts agreed during post-mission analysis that the front footpad was, at best, only very lightly in contact with the lunar dust. In a conversation I had with Scott in 2008 he further analyzed what had happened, agreeing with Irwin that touching down on the rim of a small crater (obscured by the dust during the landing) meant that *Falcon* probably had made surface contact with the engine bell about the same time as the footpads, hence a "five-point landing." If so, the LM struts would not have absorbed as much of the impact as intended, explaining the jarring bump that had surprised Jim Irwin so much. Photographs taken by Scott of the underside of *Falcon* show the front edge of the engine bell in contact with the rising rim of the crater, so that it partly supports the weight of the LM. The visible crumpling of the bell seems to confirm this conclusion, and Scott also agrees with it. It is all

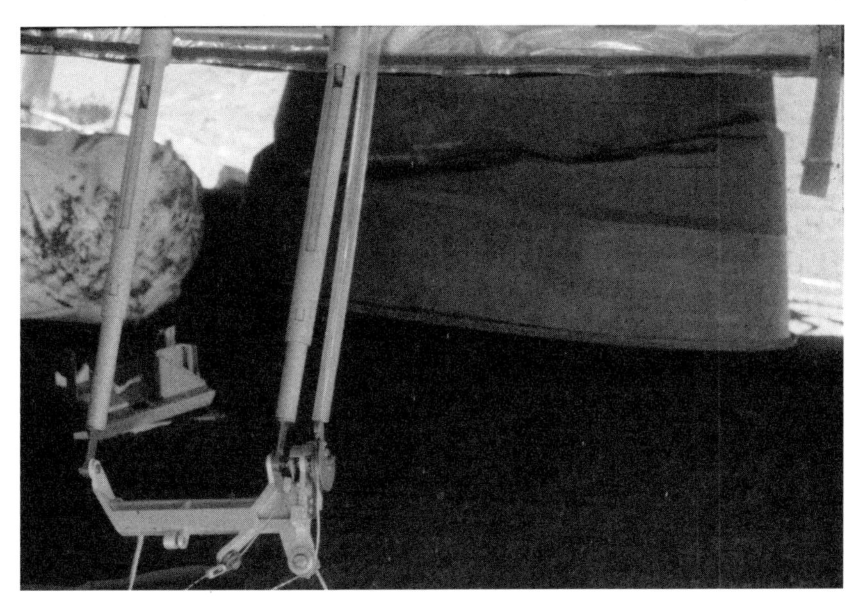

33. The damage to the engine bell of *Falcon*'s descent stage is clearly visible in this close-up image. Courtesy NASA; enhanced by Kipp Teague, Project Apollo Archive.

too clear from the photographs that switching the engine off promptly had been the right thing to do.

Shortly after descending to the surface, the astronauts set up the TV camera on a tripod facing the shadowed front of the LM and provided Project Apollo's only visible record of the deployment of a Lunar Rover. The efficiently designed "moon buggy" was folded into one of the compartments of *Falcon*'s octagonal descent stage. Slowly and jerkily, as the astronauts pulled on a series of lanyards, the Rover tilted down toward the ground and opened up like some mechanical butterfly struggling from its chrysalis and spreading its wings. Finally, it was resting on the lunar dust, and the astronauts were able to lock the seats and controls into place. A miracle of 1970s technology, the Rover was just over ten feet long and six feet wide. Each wheel unit contained a quarter-horsepower, battery-operated electric motor, permitting speeds up to eight miles per hour. Both the front and rear pairs of wheels had independent steering, a degree of redundancy that proved fortuitous.

When the Rover was ready, Scott climbed aboard for an initial test-drive, only to be cautioned by Houston to "buckle up for safety." The first lunar motorist obediently attached his seatbelt—not an easy task in a bulky spacesuit. As he drove slowly around and behind *Falcon*, the TV camera picked him up, transmitting the first and only live image of a motorized vehicle being driven on the moon.

As he briefly put the Rover through its paces, Scott realized that the front steering was inoperative. Fortunately, the rear steering was working, which gave him sufficient control to maneuver the vehicle. He parked beside *Falcon*, and the two men began loading up the Rover with all the necessary tools and equipment for their first expedition. But it was the scenery in the background that captured the attention of TV viewers. There were the mountains of the moon in all their glory, backlit by the low sun, a panorama of gold-edged mounds and inky shadows. Looming out of the darkness on the edge of the picture crouched the brooding bulk of Mount Hadley.

The live pictures showed Scott and Irwin mounting a newly designed TV remote-control unit on the front of the Rover, topped off with a delicate gold mesh antenna that had to be pointed precisely at Earth. This system was the key to public interest in the mission: if it could provide high-quality color TV at every stop, the astronauts would always be visible to the audience back home.

There was a brief loss of picture as Scott transferred the TV camera to its mounting on the Rover, then viewers were treated to a glorious color picture under the control of a Houston-based operator named Ed "Captain Video" Fendell. He could make the camera point up, down, right, or left and could zoom in on views of scientific and public interest. Astronaut Joe Allen, serving as CapCom in Houston, was clearly impressed. "And the TV scenery for us is breathtaking," he commented.

"Good!" replied Scott. "Can't be half as breathtaking as the real thing though, Joe, I tell you. Wish we had time to just stand here and *look*." But time passes relentlessly on the moon. Less than two hours after Scott climbed down the ladder, the first lunar motorists set off for the southwest toward Hadley Rille.

"Man, this is really a rocking, rolling ride isn't it?" enthused Scott.

"Never been on a ride like this before," Irwin responded.

"Boy, oh boy!" Scott added. "I'm glad they've got this great suspension system on this thing!"

Interviewed in 2008, Scott retains fond memories of driving the Rover:

It was great, mainly because it was such a wonderful machine. It was hard to drive, in the sense that the moon is so irregular—you know, undulations and craters and rocks and shadows—that you're constantly "on the stick" driving it. You have to watch every moment, and that's why Jim would do the discussion about where we were going and I would watch the "road" ahead of us, and with all the bouncing there is feed-back through the suit to the hand-controller. You want to keep the stick straight, but it's bouncing so you get that feedback into the system. But the Rover itself was such a great design, it's just remarkable to me that it responded as well as it did. It was a remarkable piece of equipment. They should make another dozen of them for the next set of missions!

After a brief stop at Elbow Crater—appropriately located at a point where the rille makes a sharp turn to the north—Scott and Irwin headed up the lower slopes of Mount Hadley Delta toward an immense smooth-edged pit gouged out of the mountainside, St. George Crater. They were hoping to sample boulders thrown up by the impact that had made the crater, but the rim of St. George and the slopes below it were surprisingly bare. Finally, Irwin spotted a single boulder, at five feet across the largest they had yet seen, and they decided to sample it. Dismounting from the Rover, Scott stole a moment to look back down the slope toward the rille and the plain.

"Oh! Look back there, Jim. Look at that. Isn't that something? We're up on a slope, and we're looking back down into the valley and—"

"That's beautiful!" Irwin interjected.

"That is *spectacular*!" Scott went on, a little later adding, "The most beautiful thing I've ever seen."

For viewers back on Earth this was like describing the Mona Lisa to a blind man. Finally, after equipment had been unloaded and Scott had aligned the high-gain antenna, a picture appeared on terrestrial TV sets. Scott and Irwin were sampling the boulder. When Joe Allen in Houston pointed out that it was "not older than three and a half billion years," Scott put the comment into context.

"Can you imagine that, Joe? Here sits this rock, and it's been here since before creatures roamed the sea on our little Earth."

But what about the rille? Ed Fendell in Houston must have had the same thought because at that moment the camera began to pan downslope to the right, revealing a breathtaking vista. Scott's enthusiastic description had

34. In this sublime panoramic photo Dave Scott is working alongside the LRV at Station 2 against the backdrop of Hadley Rille. Courtesy NASA.

been no exaggeration. There on the screen was an enormous V-shaped gash snaking across the moon's surface, its bright sunlit western slope contrasting sharply with the deep shadows engulfing most of the eastern side. Features that would have been called meanders and interlocking spurs in a river valley on Earth were clearly visible. As the camera zoomed in on the rounded floor of the rille, angular boulders as big as houses swam into view. It was a truly, uniquely spectacular sight, supremely well worth the wait.

Scott had picked the perfect spot to park, allowing the camera to view right along the length of the rille from above the point where it turned sharply to the north at the foot of Hadley Delta. In Houston Joe Allen reacted like most TV viewers: "And we have a view of the rille that is absolutely unearthly."

"Yeah, didn't we tell you . . . ," Scott replied.

Irwin added, "Glad you can enjoy it with us."

Scott then remarked, as if challenging the politicians who had canceled the last three Apollo missions: "Yes, sir, Joe. Tell me this isn't worth doing, boy!"

In 1983 Jim Irwin discussed the rille with me. "We could have walked down, particularly where we were on our first EVA near St. George Crater. That was a very gentle slope, I think maybe a maximum of a 20-degree slope. Walking down would have been very easy, walking up would have been much more difficult!"

He was grinning as he spoke. Both astronauts had mused playfully on whether the Rover, or their legs, could have tackled the rille at the point where its slope was little greater than the lower slopes of Hadley Delta. Of course, the mission plan was sufficiently conservative to rule out any such adventure, and the astronauts had to make do with photographs of the boulders on the floor of the rille taken with their 500-mm telephoto lens.

The Apollo explorations of the moon were planned meticulously, with the crew intimately involved in the establishment of the mission rules. The Rover was never driven so far from the LM that the astronauts could not walk back. Safety margins were rigidly adhered to. Compared with Captain Cook's voyage in the original *Endeavour*, it may sometimes have seemed that the exploration of the moon was forever hamstrung by too much attention to safety and absolutely no indulgence in risk taking. But Cook was not sailing in a vacuum a quarter-million miles from home.

After their geological sampling on the lower slopes of Hadley Delta, Scott and Irwin headed back toward *Falcon*, crossing their outbound tracks several times and reaching speeds of almost eight miles an hour. They had a magnificent view of Mount Hadley to the northeast, rising fifteen thousand feet above the plain. They arrived back at the LM after an absence of two and a quarter hours, but the EVA was far from over. They still had to set up their Apollo Lunar Surface Experiment Package (ALSEP) scientific instruments, now including a battery-powered drill with which Scott hoped to extract a ten-foot deep sample of the rock beneath the landing site. He would also try to drill two ten-foot holes to accommodate temperature probes to measure the escape of heat from the interior of the moon. Determining whether the moon has, or once had, a molten core like the Earth is a key to understanding what makes the moon tick.

Unfortunately, the lunar drill would prove to be Scott's nemesis. At first everything seemed to go well. Live TV showed it boring down into the ground, but as it met more and more resistance, Scott had to lean his full weight on it (only sixty pounds, fully suited, in the moon's gravity). He was audibly grunting, but because of a flaw in the drill's design, no amount of effort could persuade it to penetrate more than halfway to the planned depth. He started to drill the second hole nearby but had to stop barely a third of the way to the ten-foot target. This was a task to which he would have to return.

The first EVA lasted six hours and thirty-two minutes, an Apollo record. Apart from the drilling it had been a great success, but both astronauts were exhausted and needed rest. For Irwin it went beyond exhaustion. Each space suit incorporated a water bag to quench the occupant's thirst during the long excursion, but a supply tube in Irwin's suit was kinked and failed to deliver any water. After almost seven hours of strenuous activity, he was badly dehydrated, which would later have serious implications.

Sunday, 1 August, 9:24 a.m.: Delivered by their indispensable four-wheeled friend to a vantage point on the lower slopes of Mount Hadley Delta, Scott and Irwin gazed down over the stark beauty of their small corner of the moon. Over three hundred feet below them and three miles away, a tiny speck of gold glinting against gray revealed the presence of *Falcon*. Scott was mesmerized by the view from the mountain, his favorite of the entire mission, and he took several pictures with the telephoto lens to capture his fragile home away from home nestling among the craters and ridges of the plain below.

The astronauts were about to depart from Station 6, their first sampling point on this second EVA. The excursion had started well, with the Rover's front steering inexplicably choosing to operate, giving Scott even better control on the run to the south. The plan was to drive part of the way up the flank of Hadley Delta and to take rock samples at various stops. Ideally, they would locate a crater that had penetrated through the surface material, throwing out ancient highland material, but Station 6 had been something of a letdown, with no obvious evidence of the elusive anorthosite.

It was time to move on to Spur Crater, an impressive three hundred–foot hole drilled more than sixty feet into the flank of the mountain. On the way they made a short stop, designated Station 6a, at a ten-foot boulder farther up the slope. Irwin had spotted the boulder in the distance on the first EVA, and it was Irwin who announced a fascinating discovery about the structure of the big rock. "Come up and look at this," he cried, "because this rock has got *green* in it, a light green color . . . the first green rock I've seen." Scott still recalls the incident with relish:

The Rover was having difficulty traversing the soft regolith, and walking in the soft stuff was a real challenge. But because we had a feeling the boulder might be significant we decided to give it a go and attempt to sample and photograph it quickly. We were as usual behind time and mission control was pressing us to keep moving. But this looked too good to miss. I maneuvered the Rover to a position below the boulder and parallel to the contour and we both hopped off. As I moved toward the boulder, I saw the Rover begin to slide down the side of the mountain: our return to the LM was slipping away! I alerted Jim and he hopped down to the downside of the Rover to hold it steady and prevent the slide—the photo I took when I sampled the boulder clearly shows Jim holding the Rover with one wheel off the ground—a pretty sporty situation, but certainly worth it. Again, training and education had paid off, along with curiosity and innovation. The relatively high risk was worth the gain. This observation tuned our perceptions into recognizing a color on the moon amongst the vast landscape of shades of grey only.

Having bagged their green rock, Scott and Irwin drove a short distance downhill to Spur Crater, where they established Station 7. In the space of five minutes, they made two of the most important finds of Project Apollo, one of which would only emerge from the other's shadow decades later.

If the "green boulder" was the overture, the astronauts were about to perform the symphony, in two triumphant movements. As Scott aligned the TV antenna, Irwin spotted what appeared to be more green material scuffed up by their boots. At first Scott was dubious: "I've got to admit it really looks green to me, too, Jim, but I can't believe it's green." In spite of the green material already sampled, he still wondered if his visual impressions were being compromised by the helmet's reflective gold visor, designed to protect their eyes from the sun's intense glare.

Irwin persisted. "Oh, it's a good story. I hope it's green when we get it home." And then came realization, vindication, and a burst of dialogue that gladdened the hearts of everyone watching.

"Oh my!"

"It is green!"

"It *is* green!!"

"I *told* you it was green!"

"You're right! Ooh! Fantastic!"

The green rock samples were a serendipitous find that the astronauts had certainly not been expecting. The subtle color came from tiny spheres of green glass hurled from the depths of the ancient moon by volcanic "fire fountains." They would come to prominence again many years later.

Jim Irwin's skillful observation once again demonstrated that he was no mere drone working in his commander's shadow. As Scott told me: "Jim was a great guy to work with. For me, he was the perfect guy to be on the moon with. Steady; knew what he was doing; always offered suggestions; on top of everything. People asked, 'Gee, did Jim have a physical problem because of his legs?' [from the 1961 crash]. No, he didn't have any problem, he did it all. Jim was a great guy and did a great job."

Scott admits to feeling an "almost child-like delight" as the treasures of Spur Crater were unearthed. But the best was yet to come. A few minutes after securing samples of the green material, the two men almost simultaneously spotted something intriguing. Perched on a pedestal of rock, as if waiting to be found, was a fist-sized lump of white material. As Scott detached the rock from the pedestal, he saw it glinting in the sunlight.

"Guess what we just found!" he exulted. "*Guess* what we just found! I think we found what we came for!"

"Crystalline rock, huh?" Irwin added, for Houston's benefit.

"Yes, sir! You'd better believe it!"

What they had found was a chunk of anorthosite, the "holy grail" that Lee Silver had trained them to locate. Back in Mission Control, Silver and the rest of the geology team were ecstatic. The press almost immediately called it the "Genesis Rock." It was not entirely appropriate, but the name stuck. Later found to be around 4.15 billion years old, it was not the oldest Apollo sample, nor even the oldest *Apollo 15* sample, but it was almost certainly a fragment of the ancient lunar crust, thrown to the surface by the impact that had gouged out Spur Crater.

Ask Scott today if he straightaway recognized the significance of the crystalline rock, and if that was because of his geological training, and his response is immediate and succinct: "Yes! Yes!" He called Spur a "gold mine," prompting Joe Allen in Houston to add, "And there might be diamonds in the next one." Further exploration on the mountainside would have been desirable, but other considerations took precedence. Scott and Irwin were up against an implacable enemy—the passage of time. With each breath their oxygen supplies dwindled, and if they remained much longer on Hadley Delta, they would be unable to walk back to the safety of *Falcon* in the event of a Rover failure. So they mounted up and headed for their distant base, hugely satisfied with their discoveries but frustrated at the thought of what else could have been achieved if only they had had more time.

If Spur Crater had been the high point of the second EVA, undoubtedly the low point for Scott was his return to the heat flow experiment and the lunar drill. Twenty-two hours after leaving the site, he was again tackling the second hole. After fifteen minutes of strenuous effort, he had drilled down about five feet, matching the first hole. Unfortunately, two of the drill stems had become detached, with the result that the temperature probes ended up only a few feet below the surface. It was a testament to Scott's character that he endured these difficulties with good humor, and the heat flow experiment was at least partially successful. Important lessons were learned for the final two Apollo missions.

The third and final drilling chore was the extraction of a ten-foot-deep rock sample. At first the drill cut rapidly through multiple layers of lunar geological history, but soon Scott was grunting with the effort of penetrating harder bands of rock. This time he reached the target depth of ten feet, but the removal of the core would have to wait until the start of the third EVA.

Monday, 2 August, 4:41 a.m.: As the rover arrived at the ALSEP science station, Scott was contemplating the task that loomed ahead. "Okay, now to the drill. We last left our friend— "

"Now it's *our* friend, huh?" Irwin responded warily.

"Yes it is!"

The drill, with its precious core sample, was firmly embedded, and no amount of hauling by Scott alone was going to free it. This clearly called for teamwork: one man for each drill handle. First, they hooked their elbows under the handles and pulled. It gave a little, then a little more. The dialogue was coming from the moon but sounded for all the world like two Earthbound furniture removal men at work.

"Okay, you say when. One, two— "

"When!"

". . . three! One, two, three. A little bit."

Then they put their shoulders under the handles and pushed upward, heaving and straining, their heavy breathing as obvious to the TV audience as their elevated heart rates were to the nervous doctors in Mission Control. The reluctant drill core finally relinquished its grip and emerged from the ground, clearly visible on TV. It was one of the most important geological samples ever returned from the moon, containing fifty-eight distinct layers of lunar history, but the price paid for it was a high one. Scott suffered painful bruising to his fingertips and torn shoulder muscles. Irwin's exertions were adding to the impact of dehydration on his heart, and the accumulation of drill-related delays would ultimately mean scrapping a planned visit to a cluster of craters named North Complex, which were believed to be volcanic. This would be a bitter blow to both astronauts and to the geologists in Houston.

Preliminary tasks completed, Scott and Irwin drove westward toward Hadley Rille. On the first EVA they had enjoyed an elevated view into the rille, but this time they would be able to take samples on its very edge. The rille provided a cross-sectional study a thousand feet down through the geological history of the Imbrium Basin, created all those eons ago. The astronauts were hoping to determine whether the basin was the result of a single massive outpouring of lava or whether the rock had been laid down by repeated flows like the skin of an onion. Orbital photography by unmanned spacecraft had hinted at layering, but nothing beats on-the-spot observation. They were also hoping to find bedrock exposed at the edge of the rille, where repeated cratering had removed the layer of dust. After the frustrations with the drill, this task was much more to Scott's liking.

TV viewers expecting the Rover to be parked at the very edge of the rille were initially disappointed. Instead of a giddying view down into a deep chasm, all they could see was a gentle slope edged by a line of boulders, with the far wall of the rille visible beyond. The Rover was parked on a gradual decline that gave way to a 25-degree slope. But for Scott the view was stunning, and almost immediately he had the answers to the big questions about the rille and the Imbrium Basin: "It was one of the geological excitements for us. We were hoping to find an indication of layering. Or not. And as soon as we saw the rille we could see the layering. Spectacular! The 500-mm lens brought back some great pictures. And it was about thirteen hundred meters from the point of the camera to the far side of the rille, yet we got these great pictures."

Meanwhile, Irwin had spotted an accessible layer of rock that certainly appeared to be the bedrock they had hoped to sample. The two men bounced confidently down the slope and began chipping samples off some of the boulders. Zooming in on them, Ed Fendell in Houston treated Earthbound observers to a stunning view of the white-suited figures hard at work, with the far side of the rille as a backdrop. In fact, the view was so spectacular that Joe Allen cautiously asked how close they were to the lip of the rille. On TV it looked like the two men were teetering on the edge of a precipice. Scott had to reassure him they were not about to slip to their doom.

Despite his reassurances, Scott did suffer one of Project Apollo's more memorable falls at this site, sprawling full length right in front of the TV camera. Irwin rushed to help, but even Mission Control seemed more concerned about the camera that Scott had dropped into the dust. In 1983 Jim Irwin described falling on the moon:

We had great confidence in the suit; that it could withstand any falls we would make. But there were a few times when Dave fell that he didn't want to waste the energy of trying to turn over or get into a better position to push himself up, so he asked me to help him. So I would just grasp him with a couple of fingers and just flip him up, and it made it much easier because that way we wouldn't get quite as dirty and we wouldn't be expending more energy than we had to, because towards the end of the EVA we were just exhausted from our exertion.

With their successful foray to Hadley Rille completed, Scott and Irwin headed back toward *Falcon* on the ever-reliable Rover, their time at Hadley-Apennine almost at an end. Gazing up at the soaring mountains and rounded hills that surrounded them on three sides, they were deeply impressed in their own ways.

"Oh, look at the mountains today, Jim," Scott enthused. "When they are all sunlit, isn't that beautiful? By golly, that's just super! It's, you know, unreal." Ahead of them reared Mount Hadley, drowned in shadows when they first stepped onto the moon but now glowing in the sunlight, painted in hues of tan and gold and dominating the landscape.

Sitting beside his ebullient, enthusiastic commander, the equally enthusiastic but more reflective Irwin seemed lost in thought. A quietly religious man during his years of astronaut training, Irwin later wrote that when he blasted off to the moon, he was absolutely dedicated to the challenge of achieving a perfect flight. He was thinking only of the science and had no inkling of the spiritual journey on which he would soon embark. Here on the moon, far away from his home, he felt an overwhelming sense of the presence of God, far more strongly than he had ever experienced on Earth. He wanted to say something appropriate. Gazing up at the unearthly grandeur of Mount Hadley, he sensed that this was a fitting moment to quote from the Bible. "Dave, I'm reminded of a favorite biblical passage from Psalms: 'I look unto the hills, from whence cometh my help . . .' But of course we get quite a bit from Houston, too!" He later joked that he had felt obliged to give Houston equal billing or they might switch off the radio link, but he was content that in a small way he had given voice to his inner feelings, not something usually associated with "right stuff" astronauts.

The final five hours on the moon were mostly hectic, with a few interludes. Scott and Irwin returned—yet again—to the drill site, this time to disassemble the ten-foot core tube. Right to the last, the core was a pain, but they finally reduced it to sections short enough to fit inside *Falcon*.

Back at base for the last time, they transferred equipment, film, and precious rock samples to the spacecraft cabin. Before taking leave of Hadley Base, however, Scott had a surprise for his Earthbound viewers: a small demonstration of gravity that recalled a famous experiment on the Leaning Tower of Pisa attributed to Galileo. In his right hand Scott flourished a geological hammer. In his left hand was a genuine falcon feather, plucked from the Air Force Academy's mascot. He released the two objects at the same instant, and the TV audience watched in delight as they fell at the same lazy pace in the low gravity, striking the lunar dust simultaneously. On the airless moon there was nothing to slow the feather's descent. It was a wonderful piece of showmanship, perhaps David Scott's more cerebral response to Alan Shepard's golf shots, and the brief sequence was shown again and again on TV news bulletins all over the world.

Scott then drove the Rover to its final resting place, some three hundred feet to the east, where the TV camera would transmit the first live coverage of a launch from the surface of the moon. While there, and unknown to Mission Control, he conducted a brief, private ceremony on behalf of the whole crew to pay tribute to fourteen individuals who had lost their lives in their nations' space programs, either in training or in flight. With profound respect for their sacrifice he slipped into the dust of a small, subdued crater a card bearing the names of eight astronauts and six cosmonauts known to have died before the flight of *Apollo 15*, including, most recently, the crew of *Soyuz 11*. In front of the card he placed a small, deliberately toppled tin figurine that the sculptor Paul Van Hoeydonck had called "The Fallen Astronaut." While photographing the small memorial site, Scott was asked by CapCom Joe Allen what he was doing, and he replied noncommittally, "Oh, just cleaning up the back of the Rover here a little, Joe."

While Scott was away, Irwin had a few minutes with nothing to do, which he spent running around the LM and bouncing from foot to foot. He knew he would never be coming this way again and wanted to savor his last moments on another world. He later wrote, "It was the most relaxed time I had on the moon."

His work done, Scott now stood alone on the lunar dust at Hadley Base. The first three Apollo landing crews had made comparatively brief stays. Scott felt that he and Irwin were the first to "set up camp" and actually live and work on the moon. Now they had to leave. "This was a somewhat melancholy time. We had completed most of our planned tasks, we were still on a roll, but consumables had essentially expired and if we were to return to Earth, now was the time. But we felt like we had settled into a comfortable routine and could have continued for many days given adequate consumables. This, plus the results of the exploration, of course, were the main factors in satisfaction. But, being on a roll, it was somewhat sad to leave our amazing and friendly lunar home."

As Scott climbed back into *Falcon*, Joe Allen in Houston passed on the final verdict on the performances of Lee Silver's geology students. "Dave and Jim, I've noticed a very slight smile on the face of the professor. I think you may very well have passed your final exam."

Monday, 2 August, 2:36 a.m.: On board *Endeavour* in lunar orbit Al Worden was coaxed from his sleep by the rousing sound of Herb Alpert's "Tijuana Taxi" played from Houston. It was the lot of a CMP to plow his lonely furrow, all but forgotten by the Earth, with only the voices from Mission Control to keep him company. At one point on 30 July Worden became (according to the *Guinness Book of Records*) the most isolated human being in history, as his orbit carried him almost 2,235 miles away from the two nearest members of his species. He still holds the record, but in truth Worden did not have time to feel lonely and positively relished having the spacecraft to himself. He had spent three days conducting experiments, observing the moon through the windows, and tending the battery of high-powered cameras and surface sensors mounted in the belly of his spacecraft. The wealth of data flowing from these instruments would revolutionize knowledge of the moon's structure, but it would be overshadowed by the spectacular results from Hadley Base, particularly such geological treasures as the Genesis Rock. With the possible exception of Michael Collins on *Apollo 11*, the CMPs were always underrated by the public, who rarely looked beyond the moon walkers and seemed unaware that the man who stayed in lunar orbit was second in command on the mission.

Far below Scott and Irwin had enthused about the beauty of their mountainous landing site, but Worden saw a different world.

The moon is a pretty desolate body. It looked like it had been used for target practice for a long time: lots of craters and large meteor impact basins, very wild looking but also very dead looking. My job was to spend much time just looking at the surface, and to describe that surface to the geologists back home. In fact, I looked very hard for signs of volcanic activity and I found it in the Taurus-Littrow area. I saw and described some small cone-shaped objects that appeared to be the results of volcanic eruptions. The Apollo 17 *landing site was changed to this area based on my observations.*

Beauty, of course, is in the eye of the beholder, and when *National Geographic* showcased many of his best images in February 1972, a wider audience had the opportunity to decide for itself whether the rugged face of the moon deserves to be called beautiful. But there was no doubting the delicate beauty of another heavenly body that Worden observed many times during his solo voyage. No matter what task he was performing, he set it aside long enough to watch the Earth rising above the lunar horizon. It was the high point of his two-hour orbital "day," and it always blew him away.

Monday, 2 August, 12:11 p.m.: Back on the moon Scott and Irwin were ready to leave Hadley Base. This liftoff would be dramatically different from those of the previous missions: in Houston and all over the world TV screens were showing a live color image of *Falcon* transmitted from the TV camera on the parked Rover. Slipping gears in the TV control mechanism meant that Ed Fendell would be unable to follow the spacecraft as it rose into the black sky, but that was a minor disappointment. It was still a spectacle not to be missed.

There was no audible countdown. Dave Scott pressed the PROCEED button, and the spacecraft computer ignited the ascent stage engine. On the TV screen the static image of the tilted *Falcon* seemed to explode in a silent shower of sparks and debris as the rocket exhaust tore away shards of gold and silver insulation foil from the descent stage and blasted it in all directions. The ascent stage rose rapidly, at a noticeable angle, and disappeared from view. It was all over in three seconds, but they were seconds of pure magic.

No sooner had the ascent stage disappeared from view than a loud blast of music caught everyone's attention. Scott and Irwin were riding into orbit to the strains of the U.S. Air Force favorite, "Off We Go, into the Wild Blue Yonder." British TV commentator James Burke, sitting in the BBC Apollo Studio, jumped to the same conclusion as most viewers around the world: "This is coming from the spacecraft! What a crew!"

In fact, the music was being played by Al Worden in the CM. It was intended solely for the benefit of Houston but was inadvertently piped through the main radio loop. There was a certain amount of bewilderment in Mission Control, and Scott and Irwin were initially irritated by the distraction at a crucial time in the mission, but in retrospect the *Apollo 15* musical liftoff is one of the most memorable TV moments in Project Apollo. It seemed to epitomize a mission conducted with a blend of scientific precision and swashbuckling flair.

A key factor in the success of any Apollo mission was a harmonious working relationship between the experts in Mission Control and the astronauts in the spacecraft. Unfortunately, with *Falcon* and *Endeavour* safely reunited in lunar orbit, discordant notes began to disturb this harmony.

Shortly after the docking, the spacecraft maneuvered into an "orbital rate" attitude that allowed the SIM bay instruments to observe the moon. Apollo was flying like an aircraft circling the Earth, with the "ground" always "below" the spacecraft. Then, a little over two hours after the docking, the flight plan called for the spacecraft to be maneuvered into "inertial attitude" for jettisoning *Falcon*. This placed the spacecraft in a fixed attitude relative to the stars, and the NASA public affairs announcer pointed out that in this attitude the crew could no longer use the SIM bay instruments to observe the moon.

By this time Scott and Irwin had been without sleep for twenty hours and had not eaten for eight hours. Before they could eat and rest, they had to don their space suits for the undocking from the now-redundant LM, a safety precaution added to the flight plan after the deaths of the *Soyuz 11* cosmonauts only five weeks earlier. There was a delay when leaks were detected in both moon walkers' suits, a problem caused by lunar dust in the seals. After the mission NASA doctors briefed the Mission Control team about the potential effects of dehydration and resultant loss of elements such as potassium on Scott and Irwin. They talked about heart irregularities and "possible disorientation and memory loss." Writing about the period *Apollo 15* spent in lunar orbit in his book *Failure Is Not an Option*, flight director Gene Kranz seems to have put two and two together to make five, recalling the post-mission medical briefing and mistakenly attributing problems such as the suit leaks to crew errors caused by exhaustion.

There was a further problem with pressure in the tunnel between the LM and the CSM, possibly again caused by dirt or debris in a seal. The combination of problems delayed *Falcon*'s release by two hours and ten minutes (slightly more than a full orbit). Following the release of the faithful *Falcon*, a small engine burn was needed to change *Endeavour*'s orbit and prevent a later collision. The story of the separation is an unfortunate tale of confusion and lack of understanding in Houston. The engine burn was supposed to take place with *Endeavour* perpendicular to the moon's surface and *Falcon* "overhead." The accumulation of delays meant that when the crew was ready for the engine burn, the spacecraft's orbital motion had carried it almost 90 degrees around the orbit, placing it almost horizontal to the moon's surface, orbiting tail-first with *Falcon* following behind.

Mission Control ordered a brief engine burn that would take *Endeavour* "forward," mistakenly believing this would separate the two vehicles. Looking out of the window, Scott could see *Falcon* floating fifty yards in front of his nose. Carrying out Houston's instruction was not an appealing prospect. He responded, "And 'forward' takes us right back to the LM."

The CapCom in Houston, Bob Parker, eventually replied, "Ah, hold the burn, Dave," and several minutes later added: "Stand by, guys. Confusion still reigns, I think." CapCom then instructed the crew to maneuver to a "trailing position" and thrust against the direction of their orbital path, adding unhelpfully, "Right now we can't give you any sound advice down here."

Scott pointed out that *Endeavour* was not trailing but leading Falcon at that point in the orbit. Finally, Houston figured out that the solution lay in simply pointing *Endeavour* toward *Falcon* and thrusting away from the LM, a solution that had been obvious to the crew from the start. Even today Scott and Worden—who had slept and eaten normally in orbit—both feel disappointed that these difficulties have been attributed to crew exhaustion rather than a failure of mission controllers to remain aware of the spacecraft's correct attitude in orbit. Kranz and his team lacked what today would be referred to as "situational awareness." As Scott points out, "Mission Control is an outstanding organization and is *almost* always right—but not always, as the record will show."

Shortly after the separation burn, Dr. Charles Berry, NASA's chief flight surgeon, approached the flight director then on duty, Glynn Lunney, with disturbing news. The account is recorded in almost identical terms in books written by NASA management legends Gene Kranz and Chris Kraft. Berry referred Lunney to irregularities in Jim Irwin's heartbeat while on the moon. More ominously, back in lunar orbit, Irwin suffered episodes of "bigeminal rhythm," in which both chambers of the heart try to contract at the same time. According to Kraft's account, Berry said it was a "serious situation." On Earth he would have sent Irwin to an intensive care unit and treated him for a heart attack. The only consolation was that Irwin was already in the equivalent of an ICU, breathing pure oxygen and having his heart monitored. Furthermore, he was weightless, relieving the stress on his heart.

Scott's heart also displayed irregularities, to a lesser degree, on the return to Earth. The fact that Worden had no such problems led to the conclusion that dehydration during EVA training on Earth and exacting work on the moon had depleted the potassium in Irwin's and Scott's bodies, resulting in irregular heartbeats.

Irwin's problems led to one of the great unanswered questions of Project Apollo: why did Mission Control withhold from the mission commander the important information that one of his crew had, in effect, suffered a mild heart attack? Although Irwin had felt very tired around the time of the bigeminal episodes, he had no chest pain or breathlessness, and Scott therefore had no inkling of the problem.

Before launch a private radio channel had been arranged (to be activated by the use of the code words *West Point*) for confidential discussions with the crew out of earshot of press and public. It was never used by Mission Control, but in retrospect it is obvious from the mission transcript that the heart issue had been discussed privately in Houston. Scott and Irwin were advised to take Seconal, a sedative, before beginning their sleep period. Having not been given a reason, Scott thought this was ridiculous. He and Irwin were tired and needed no barbiturates to go to sleep. The pills stayed in the bottle.

The decision not to tell Scott assumed far greater significance on the way back to Earth. On Wednesday, 4 August, after seventy-four orbits, the crew fired Apollo's main engine and started the three-day journey home. The cameras in the SIM bay would not survive reentry into Earth's atmosphere, so the flight plan called for an EVA in deep space by Al Worden to retrieve the precious film cassettes. The plan also called for Irwin to assist Worden by standing in the open hatch of the spacecraft and passing the cassettes back into the cabin. Was this really a suitable activity for a man who had just suffered what his doctor would have treated as a mild heart attack?

Here perhaps lies the answer to why Houston kept quiet about Irwin's problems. What might Scott have done if he *had* been told? Might he have decided that recovery of the film, however important, was not worth even a modest risk to Irwin's life? Scott is in no doubt he should have been told.

Had we been informed of Jim's situation, the Flight Director and I would have discussed the options in detail and would jointly have made a decision, the ultimate responsibility for which would have been mine as field commander. But I would certainly have made such a decision only after very carefully discussing, even debating, the situation with "Flight" (who of course represents all of NASA *management). In hindsight, my inclination is that I would have minimized Jim's activities but would have proceeded with the* EVA. *Al was very good at his task and could have performed it successfully without the* EVA *by Jim. Bottom line: based on what we know now and assuming that the discussion with Flight at the time provided the same information, I would have discussed it with Jim and Al and—based on how Jim felt as well as Al's assessment—if Jim felt generally okay and Al concurred, we would have had Jim suited in the* LMP *couch and Al do the* EVA, *whereby from the commander's couch I could have grabbed the film canisters as Al passed them in. Then Al could have entered and closed the hatch [which] would have been an acceptable workaround. I doubt that we would have cancelled the* EVA; *only if Jim had felt that he could not handle a minimum exercise such as that.*

This is, of course, conjecture, because Scott (for reasons never explained to him) was not told, so the EVA proceeded as per the flight plan, with Irwin standing up in the hatch. Had he suffered a full heart attack at that moment, there would have been what Scott delicately describes as "a really serious situation." In any event, floating in the blackness of deep space 196,000 miles above the Earth, Al Worden made the spacewalk look easy, completing his tasks in only thirty-nine minutes. He had trained so hard for his moment in the spotlight that he simply carried out the EVA "by rote." Only later did he fully appreciate the glorious view of the Earth and the moon. He had no camera, and it was left to a *National Geographic* artist to reproduce a stunning view of Jim Irwin as Worden saw him, halfway out of the spacecraft and framed by the huge orb of the nearly full moon.

On Saturday, 7 August, spaceship *Endeavour* splashed down safely in the Pacific Ocean, a little earlier and a little faster than planned. One of the three huge parachutes collapsed during the descent, proving graphically (and prophetically) to a worldwide audience that spaceflight has no guaranteed safe returns, but that built-in redundancy improves the odds. The mission had lasted 295 hours and 12 minutes. (Half a world away, with an excess of teenage hyperbole that overlooked Cook, Darwin, and a few others, I noted in my diary that "the most important and ambitious scientific exploration in the history of the world has ended happily.")

Scott, Worden, and Irwin returned from the moon as heroes. After the tragedy of the *Soyuz 11* cosmonauts, *Apollo 15* was a reaffirmation that man could explore space and return safely. After an exhausting series of medical tests and debriefings by NASA officials and scientists, the crew went on a tour of the United States, which included an open-top motorcade through the streets of Manhattan with the mayor of New York, dinner at the White House, and the honor of addressing a joint session of Congress.

But fame is a fickle mistress. Less than a year after appearing before Congress as heroes, the three men had to return to testify before the Senate Space Committee about a sad episode that would forever dog their footsteps. It had emerged after the flight that the crew had taken hundreds of lightweight postal covers to the moon's surface. Two hundred and fifty had been officially recorded in the flight manifest. A further four hundred had not been, although all had been checked for fire safety purposes. After the mission one hundred of the covers had been given to a German stamp dealer, who in return set up a six thousand dollar trust fund for each astronaut for the education of their children. But the dealer broke the terms of the agreement by selling his covers before the end of Project Apollo, and the astronauts returned the trust fund money to him. In an era when it was impossible for astronauts to take out life insurance, however, the crew members had each kept one hundred covers for their own use, and it was widely assumed that some would be given to friends and family as souvenirs and the rest kept as an investment for the proverbial future "rainy day." NASA mounted an internal investigation, and although postal covers, medallions, banknotes, and other souvenirs had been carried on most previous flights, the *Apollo 15* crew was singled out for what was portrayed as poor judgment, and all three were officially reprimanded.

James Hansen, writing in *First Man*, the authorized biography of Neil Armstrong, refers to "a sad and complicated affair, with the three astronauts . . . unfairly bearing the brunt of the blame for a questionable practice involving personal mementos that had been going on ever since the early days of Mercury." The affair could never destroy the scientific legacy of *Apollo 15*, but it did distract public attention from the treasure trove of rock samples and photographs the astronauts had brought back to Earth.

After *Apollo 15* David Scott served as director of NASA's Dryden Flight Research Center, before retiring from the space agency in 1977. In the 1990s film director Ron Howard engaged Scott as a consultant on the Hollywood blockbuster *Apollo 13*. The star of that film, Tom Hanks, also called upon Scott's expertise in the making of his epic miniseries *From the Earth to the*

Moon. Now well into his eighth decade, Scott is a popular guest at space conventions and is a gifted and enthusiastic speaker. He retains a keen interest in geology and a fierce pride in the achievements of the Apollo astronauts and the thousands of dedicated workers who made Apollo possible.

What does it mean to him today to have been one of only twelve human beings to walk on the moon? "I think probably we were in the right place at the right time working with the right team, and I got to do something that was exceptional, and hopefully we made a contribution to knowledge, science, and the manner in which you explore the Moon, or a planet. It's very personally satisfying to me that I was on that team, and that I got to carry the tongs and pick up the rocks. So it was really 'right place, right time' and I thoroughly enjoyed it."

Like his former commander, Al Worden went into NASA management, working at Ames Research Center, before pursuing a business career in the aerospace industry. Uniquely among the Apollo astronauts, he published a book of poetry about his experiences in space. Like NASA's early rockets, some are misfires, while others soar gracefully, providing a unique insight into the thoughts of one lunar explorer. Today, as chairman of the board of the Astronaut Scholarship Foundation, Worden attends numerous space conventions and delights audiences with his eloquently delivered anecdotes and wicked sense of humor. Recalling his flight to the moon all those years ago, Worden says: "Apollo 15 was the most successful thing I ever did. It was the epitome of what can be accomplished on a flight such as ours. I am very proud of the fact that I was part of that crew."

For Jim Irwin the return to Earth marked the end of one incredible journey and the beginning of another. Before the flight his religious faith had been eclipsed by his career ambitions and his single-minded concentration on preparing to go to the moon. But the overwhelming sense of God's presence that he felt on the moon revitalized and renewed his faith.

After returning to Earth, it took time for Irwin to comprehend fully what had happened to him. Medically, all was not well. Tested shortly after the mission, he had seriously elevated blood pressure and took some time to shake off the after-effects of weightlessness and potassium deficiency. He

35. Author Geoffrey Bowman (*right*) discusses a pyramid-encased fragment of moon rock with Jim Irwin during a 1986 visit by the *Apollo 15* astronaut to Bangor, Northern Ireland. Courtesy Geoffrey Bowman.

had to come to terms with public adulation and hero worship. Then came the pain of the stamps affair and the official reprimand. The popular astronaut began to speak at religious meetings, overcoming his natural reserve and talking with increasing enthusiasm about his epiphany on the moon. He retired from NASA and the air force and established a nonprofit missionary foundation called High Flight, after the poem written by fighter pilot John Gillespie Magee, killed in 1941 at just nineteen years old.

For seven months Irwin toured the world preaching his message of God's love to large audiences. Then on 4 April 1973, while playing handball in Denver, he suffered a heart attack, probably caused by blocked coronary arteries. He wondered if his time had come, but he gradually recovered and resumed his ministry. As Al Worden recalls: "He continued his work for High Flight as if nothing had happened and he continued to run way too hard. He dropped a lot of weight, but he would not slow down. We were all a little worried about him."

Irwin made several visits to the land of his ancestors, meeting relatives in County Tyrone and addressing schools, public meetings, and religious gatherings. In March 1983 he completed a half-marathon in Downpatrick, County Down, the final resting place of Saint Patrick. I interviewed Irwin for an article in a local newspaper and recall a small, dapper, slightly gaunt man who radiated warmth and good humor. At a time of great turmoil in Northern Ireland his message of hope struck a chord with thousands. Asked what it meant to him to have been one of only twelve human beings to have walked on the moon, he answered in the context of his new life and career: "I feel very fortunate and very blessed, and I sometimes wonder: why was I so lucky? Why was I chosen? And I don't know the answer to that. I do know that I have a new responsibility to the people on the Earth, to share the adventure as widely as I can, so that they might be enriched, that they might share this enlightenment, that they might have a new perspective on life, a new appreciation of the Earth, and for life, and for God. Certainly, the flight prepared me for my future life."

In the 1980s Irwin also mounted several expeditions to eastern Turkey to search for the biblical resting place of Noah's Ark on Mount Ararat. Faith can sustain a man and drive him to the limits of his endurance. Irwin found no evidence of the presence of the Ark, but in August 1982 he suffered an accident near the summit of Ararat, resulting in severe lacerations to his face and legs. He was carried down the mountain on horseback by a team of Turkish commandos. Dave Scott recalls initial surprise when he heard about his fellow moon walker's expeditions to Ararat: "Jim chose High Flight as his path and good that he did—he was peaceful and satisfied in his calling, and he did good things for many people. Searching Mount Ararat was a surprise, but again probably should not have been: Jim was an explorer, and what better place to explore for something meaningful in one's life!"

Irwin continued to deliver his High Flight message but suffered another, more serious heart attack in 1986 that nearly killed him. He told audiences that his time was short and that he was prepared for the highest flight of all. He had always loved mountains. On 8 August 1991, while walking in the mountains of Colorado, Jim Irwin suffered a final heart attack and died, age sixty-one.

At the funeral service in Colorado Springs his former crewmates Dave Scott and Al Worden, together with fellow moon walkers Buzz Aldrin, Alan Bean, and Gene Cernan, acted as pallbearers. Later, when the body of (Ret.) Col. James Irwin was laid to rest with full military honors at Arlington National Cemetery, the eulogy was given by Dave Scott, who many years later dedicated his autobiography to "my partner on the moon, the very best."

The story of *Apollo 15* is told best by Dave Scott and Al Worden, two gifted speakers who survive, and by the words and writings left behind by the late Jim Irwin. But the rocks collected on the moon can also talk, and even today they have new things to tell us.

Every scientist who originally examined the Apollo samples agreed that they contained no water: the moon was bone dry. Then in July 2008, using new analytical techniques, a team led by Alberto Saal of Brown University, Rhode Island, detected water in samples of volcanic glass brought back by *Apollo 15* and *17*, including the green glass material spotted by Jim Irwin on the edge of Spur Crater. The quantities were surprisingly high (forty-six parts per million), and the finding may require new thinking about how the moon was formed. As sampling techniques improve, the green glass may have more secrets to whisper to us.

The fire that was lighted in 1971 still burns brightly.

9. Worth the Wait

Simon A. Vaughan

History is concerned with time, space, and change.
It is concerned with the unique person, with the
unique event, and with their combination.

James C. Malin (1893–1979)

"At that point the most compelling fear was that I messed something up. You go over it time and again, trying to retrace your steps. What could I possibly have done? I can't find any answers. As the clock ran down I just had this gnawing feeling that . . . we had so much confidence in our hardware that it has got to be . . . not the spacecraft. What have I done? But the clock's running out and there's no more time, and I think the rule says it's a wave-off." It was 20 April 1972, and Lt. Cdr. Thomas Kenneth Mattingly II of the U.S. Navy, a balding, thirty-six-year-old spaceflight rookie, was Command Module pilot (CMP) aboard the good ship *Casper*, flying solo around the moon, and he was in a quandary. He was passing behind the moon and therefore out of radio communication with Mission Control in Houston, while his fellow crew members John Young and Charlie Duke had already undocked the spindly Lunar Module *Orion* and were holding off at a safe distance. They were only minutes away from firing their engine for the final descent to the moon. But then Mattingly noticed a warning light suggesting a problem with *Casper*'s main engine. Rather than risk the chance that the engine might be pointing at the wrong angle when it fired, which could prove calamitous, he had immediately called off the burn. It was a move later fully endorsed by mission controllers, who said he had "done exactly the right thing" by balking and not firing the main engine as scheduled. The crew's immediate prospects for a successful lunar landing were not looking good, however, and any recovery of the situation now rested with Mattingly and the collective wisdom and experience of the people back in Houston.

Appropriately, hanging against a wall outside NASA's Mission Control Center that day was a small mirror, beneath which a framed inscription had been attached that read: "This mirror flew on *Aquarius* to the moon, April 11–17, 1970. Returned by a grateful *Apollo 13* crew to 'reflect the image' of the people of Mission Control who got us back." The inscription was signed by the three astronauts who had nursed their crippled CM to a safe return to Earth, following the most serious in-flight drama ever faced by the space agency. But for now the only thing reflected on the faces of the decision makers in Mission Control was a deep concern, as the fate of *Apollo 16* rested firmly in their hands.

Apollo 16 was something of an enigma. Of the six eventual Apollo landing missions, this one was unique in aiming for the lunar highlands, which are the brighter regions of the moon as viewed from Earth. The darker areas are the relatively flat "seas," or mares, where ancient lava flows spread out and solidified into vast smooth plains. With the exception of *Apollo 16*, all of the landings were on these flat lunar seas or—in the case of *Apollo 14*—on low hills adjacent to the seas. Even the crew of *Apollo 15* had set down on the edge of one of the great lunar plains, where once a surge of molten magma had lapped against the lower slopes of the towering Apennine Mountains before freezing into a sheet of solid rock.

The task for the crew of *Apollo 16* was to make a landing high in the mountains, nine degrees south of the lunar equator near an ancient crater known as Descartes, named after the French mathematician and philosopher. It would prove to be the most southerly landing site in the Apollo program. Lunar scientists were expecting the rock samples to be mainly volcanic and hoped to fill some of the gaps in their knowledge of the moon's history. They were soon to find that many of their preconceived theories and notions about the lunar highlands were very wide of the mark.

They were a good crew. The veteran mission commander was forty-one-year-old navy captain John Young, a test pilot with two world altitude climb records who had joined NASA as a member of its second nine-man astronaut group in 1962. He had subsequently flown the first Gemini mission with

36. *From left*, the crew of *Apollo 16*: CDR John Young, CMP Ken Mattingly, and LMP Charlie Duke. Courtesy NASA.

Gus Grissom in March 1965 and the following year had commanded a second two-man mission aboard *Gemini 10* along with Mike Collins. On his third space flight Young had served as CMP on *Apollo 10* together with Tom Stafford and Gene Cernan. While Young remained in lunar orbit aboard the CM, nicknamed *Charlie Brown*, Stafford and Cernan had flown their LM *Snoopy* to within eight nautical miles of the moon's surface to test the LM's descent and ascent systems while at the same time locating and photographing the proposed landing site for *Apollo 11*. There was no doubting Young's flying skills, his focus and enthusiasm. It would have been hard to find an astronaut more dedicated to the space program. Indeed, seven years of almost continuous flight training had led to the breakdown of his marriage, and he had since remarried. *Apollo 16* would be his fourth mission for NASA, making him the world's most experienced spaceman.

The LMP on *Apollo 16* was thirty-six-year-old Air Force Lt. Col. Charles Duke, from Charlotte, North Carolina. He had achieved a degree of prominence three years earlier as the *Apollo 11* capsule communicator (CapCom) who had spoken for millions when he told Neil Armstrong and Buzz Aldrin that their troubled landing had "got a bunch of guys about to turn blue." Duke could sometimes prove to be the joker in the pack, but his wisecracking amiability, delivered in a southern drawl, could not disguise his keen mind.

NASA must have been satisfied that Thomas K. Mattingly II bore no grudges. Two years earlier Ken Mattingly (as he prefers to be known) was meant to have flown to the moon on the ill-fated *Apollo 13* mission, but to his undisguised chagrin he was hauled off the crew when NASA thought he had been exposed to German measles, and he was replaced by backup CMP Jack Swigert barely two days before launch. It had been an unexpected development, traced back to a couple of weeks before launch, when the crew and their backups were given permission to spend a last weekend with their families. For backup LMP Charlie Duke part of this weekend involved going on a picnic with some family friends, but it was later revealed that one of the friend's children had measles, thereby exposing Duke to the disease. It was not considered a big deal, but the NASA doctors decided to conduct some precautionary tests to determine the six men's susceptibility.

The only problem was with Mattingly, and the first inkling he had of any issue occurred when the doctors asked him to give another sample of his blood. When he questioned them about it, the response was that his blood chemistry did not offer any indication of previously having measles nor any immunity to the disease. Mattingly could not recall having had the illness, so he called his mother, who ventured that she did not think so either, unless it was a very mild dose.

After that the medical staff continued to monitor Mattingly, drawing fresh blood samples every morning and evening and asking him if he felt ill. He argued that he felt just fine, but the medical staff was concerned that he had been infected and would still be in the incubation period while flying solo in lunar orbit. Then came the first indication that it could be a bigger problem than initially indicated; he was told his backup CMP, Swigert, was going to be given more time in the simulator. When he asked if a crew switch was being considered, the doctors hastened to assure him that this would be a last resort, and their caution was meant merely to demonstrate to the American public that they had done everything by the book. Even Deke Slayton could not give him solid news; he told Mattingly no decision had been made and empathized with him for having to sit around watching Swigert practice in the simulator.

Two days before the launch, Dr. Charles Berry arrived at the Cape and personally oversaw tests on a blood sample taken from Mattingly that morning. Slayton could see that his astronaut colleague was concerned and tense and suggested he drive over to Patrick Air Force Base (AFB) and do a bit of flying to relax himself. Mattingly took Slayton's advice and did a little unwinding in the sky. "Sure enough, felt better," he reflected. "Got in the car and drove back. I'm driving up the road, turned the radio on, and they interrupt [with a bulletin] that this afternoon NASA has announced that they have changed and substituted Jack Swigert for me. I just kind of pulled over to the side of the road and sat there for a while. . . . I went back out there, and nobody knew what to say. I know Deke well enough to know that he never said anything about it, but he must have been ready to kill somebody for letting that happen."

Mattingly was understandably devastated, and Slayton could appreciate his frustration better than anyone. Ten years earlier Slayton had been scheduled to become the fourth American in space aboard *Mercury-Atlas 7* and had spent four months training for the orbital flight. Just ten weeks before launch, Slayton was at the Cape going through the simulator one evening when he received a call to report to Washington immediately. He had no idea what was going on but headed straight to Patrick AFB and jumped in a T-33 for the flight to the nation's capital.

Shortly after reporting as an astronaut, Slayton had undergone a routine physical before a centrifuge test when the doctors discovered a slight heart flutter on his EKG. Further tests concluded that he had an idiopathic atrial fibrillation, but no one seemed particularly concerned, and he remained on flight status. "I realized that every couple of weeks I would go through a period of a day or two when my pulse would act up," he recounted in his autobiography. "I had no other symptoms. It certainly didn't stop me from working. And I found that if I did some heavy exercise, like running a couple of miles, the thing just went away. That should have been the end of it."

Slayton was completely unaware that his condition was again a hot topic of conversation at NASA's highest levels just weeks before the flight. After an examination by a panel of doctors at the office of the surgeon general of the U.S. Air Force, Slayton was again passed fit for flight. NASA administrator

James Webb still was not satisfied, however, and ordered the astronaut to report to NASA headquarters to face another board of doctors.

"I was sent into the next room and told to take my shirt off," Slayton recalled.

One by one these guys came in there and poked and prodded me, listening to my heart with their stethoscopes—nothing like a serious medical exam, which was something I knew a lot about by now. I came back in, and just like that, [deputy administrator] Hugh Dryden told me I was off the flight. These guys didn't find any medical reason to keep me from flying—what they said was as long as NASA has other pilots without this condition, why not fly them instead. I hadn't expected anything like this. I was just devastated.

By coincidence one of the three doctors who had grounded Slayton was a Thomas W. Mattingly of the Washington Hospital Center. Slayton now found himself standing before an astronaut with virtually the same name, stripping him of his inaugural flight for a tenuous medical reason even closer to launch than his own bitter experience.

Slayton asked Ken Mattingly if he wanted to stay at the Cape for the launch, fly down to the Mission Operations Control Room (MOCR) in Houston, or simply go home. He said he would rather not stay at the Cape and MOCR would be a good alternative, so a sympathetic Slayton phoned and told them to expect the replaced astronaut. That night Mattingly arranged for another aircraft and flew to Houston. The 1995 Ron Howard movie *Apollo 13* would depict a dejected Mattingly standing beside a gold Corvette on Cocoa Beach watching the launch of what was once his mission, but he was actually more than nine hundred miles away in Texas.

The *Apollo 13* mission had almost ended in tragedy, but Mattingly was subsequently handed a second chance to go to the moon as CMP on *Apollo 16*, ironically with Charlie Duke—the man who had originally come down with the German measles. In fact his boss Deke Slayton had given Mattingly a choice: CMP on *Apollo 16* or LMP on *Apollo 18*—an assignment that could have seen Mattingly walk on the moon—but Slayton had broadly hinted that the first flight would be the better option, saying: "It's your call. Just think about a bird in the hand." It turned out to be sound advice when *Apollo 18* was subsequently canceled.

Now Mattingly was aboard an Apollo spacecraft orbiting the moon, on a mission that might also have to be aborted, and wrestling with what could very well prove to be nothing more than a minor technical malfunction or an erroneous signal. Or something far more serious. The unfolding drama emphasized yet again that humans were still fragile interlopers in the hostile environment of space.

A bachelor like his *Apollo 13* replacement Jack Swigert, Mattingly was born in Chicago, Illinois, on 17 March 1936. Shortly afterward his father had taken up a position with Eastern Airlines, so his parents had packed together their belongings and new child and moved to Hialeah, Florida, home of the world-renowned Hialeah Park horse track.

Founded in 1925, and nowadays one of the largest employers in Dade County, Hialeah was an exciting and progressive place to live and grow up, located on a large prairie between Biscayne Bay and the Everglades. One of the principal developers of Hialeah Park, and indeed the city of Hialeah, was famed aviator Glenn Curtiss. His was not the only famous name in aviation that is today synonymous with the area. In the same year that Ken Mattingly was born, Hialeah would also be the departure point for Amelia Earhart's ill-fated circumnavigation flight around the world, along with her navigator, Fred Noonan. Having made their final good-byes, they took off in a Lockheed 10E Electra on the first leg of their highly publicized journey on 1 July 1937, later disappearing without trace over the central Pacific Ocean near Howland Island. Little wonder that with so many influences the young boy quickly developed a lingering fascination for aviation and airplanes.

"I think it's in the genes," Mattingly recalled in 2001. "As a kid, my earliest memories . . . all had to do with airplanes. My dad worked for Eastern Airlines. Before I had any idea what that was, my toys were all some kind of airplane, and any picture that you could glean from when I was a child, they always had an airplane in it. As I got to be an older child, I built every model airplane that I could find, ate every box of cereal that had a cut-out paper airplane on the back, all that sort of stuff. It was just a way of life." As the years passed and the young aviation enthusiast grew more capable of taking care of himself, his father would often reward good behavior by presenting him with employee passes on Eastern Airlines. "I would just fly to the end of the route and back. Never got off the airplane, just go one place to the other. In those days the [Douglas] DC-3 was the only transport airplane, and so if you went up the East Coast from Miami up north and back, that was a long day. That was my big thrill, to sit there and just look out the window. Airplanes didn't fly so high in those days, so you could see a lot."

Mattingly would attend elementary school in Hialeah and later took his high school studies at Miami Edison High School. He was also active in the Boy Scouts movement and would achieve their second-highest ranking of Life Scout. Earlier, when he was just nine years old, his father had allowed him to convert part of the family garage into a workshop and unofficial hobby store. Here he would happily design and build model airplanes, selling many in order to build ever more sophisticated models. In fact, one model was so well designed and constructed that it won a world speed record, attracting so much business to the family garage and away from other local shopkeepers that they petitioned to have the little unlicensed store closed down.

Meanwhile, on most weekends the family would drive over to the airport just to watch airplanes. "So when it came time to pick a career, I don't think there was ever any choice," Mattingly reflected. "It was just, well, I want to be in aviation. I wanted to fly."

After graduating from high school, Mattingly was enrolled at Auburn University in Alabama on a navy scholarship, training in the Reserve Officer Training Corps (ROTC), initially studying engineering physics before dropping the subject and taking on a preferred course in aeronautical engineering. It was while studying there that he had his first ride in what he would describe as a "real airplane," something other than a commercial airliner. One of the family's neighbors was a marine colonel at the Opa-Locka Marine Corps Station in Miami, whose unit was flying the Douglas AD Skyraider, normally a single-seat, propeller-driven attack airplane. However this particular marine unit was also equipped with a variant that had a widened fuselage, allowing two pilots to sit side by side, and when the colonel found out that Mattingly would be spending his first Christmas away from his family, he decided to give the young man a special treat. His plan involved landing a two-seat Skyraider in a field near the school, picking him up, and flying back to Miami in time to celebrate Christmas. "I still remember that," Mattingly recalled later with obvious fondness.

After that the urge to fly—perhaps to become a test pilot—grew even stronger within him. Then, on graduating with his bachelor of science degree in aeronautical engineering in 1958, he was brought down to earth with a thud:

When I graduated from college and got my commission, I volunteered for flight training and wasn't selected. I was crushed. Other people in my class had grades that weren't as good, were not engineers—and I could never figure it out. I ended up being sent to a pre-commissioning detail on a ship called the USS Galveston. *It was in dry dock in Philadelphia. So here I was, a brand new ensign, wanted to be in flight training, and instead I was up in this shipyard, a ship that didn't even go to sea. You walked across a gangplank, but there was no water underneath you; it was just concrete.*

The *Galveston* was a ship still being fitted out and yet to be commissioned, so Mattingly, with very few funds available while paying off his college debt, was feeling quite dejected. He now found himself as one of only two young ensigns actually living on the ship and felt he had nowhere else to turn. "Everybody else lived ashore, but we couldn't afford that," he mused. Fortunately, he had struck up an immediate friendship with the other ensign, who came from West Virginia, and they were of tremendous support to each other as they lived a frugal life aboard the ship. Still, he remained deeply puzzled about why he had failed to be selected for flight training.

One day his immediate superior in the gunnery department happened to ask Mattingly about all the aviation books he devoured in his spare time and why, if he was so interested in airplanes, he had not applied for flight training. Mattingly despondently replied that he had tried but had not been selected, adding that he had no idea how that had happened—particularly when others with lesser grades had been accepted. The NCO was similarly puzzled and placed a call to the Navy Bureau of Personnel in Washington DC. To his astonishment he was told that they too were disappointed when young Mattingly had seemingly decided against flight training, as he had excellent grades and other skills. It turned out that he had simply not been told to complete an application form in order to qualify. His rejection was due to nothing more than a bureaucratic oversight. The Navy Bureau officer hastily added that if Ensign Mattingly still wanted to fill out the appropriate application form, then they would be only too glad to have him; he could drive down the following day and be given a set of orders.

Understandably, Mattingly was overjoyed to hear this news and desperately wanted to make up for lost time. Before anything else could happen, he had to have a form officially requesting his transfer signed by each of his superiors. That proved to be no real problem, but the last person on his list was Lt. Col. Glenwood Park, the crusty head of his department and a man certainly unimpressed when the young ensign meekly entered his office armed with an application for a transfer. Mattingly had been learning all about the innovative Talos missile system and what made it function, and his excellent work had already come to the attention of the gunnery officer. He did not want to let such a good and promising person go without a fight and stressed that missile technology was not only the way of the future, but it would also revolutionize naval warfare, whereas aviation was destined to become a thing of the past. Mattingly was literally quaking in his boots but persisted; he wanted to transfer to flight training. Park glared long and hard at the young man, who was becoming increasingly nervous, and asked him in clipped terms if he definitely wanted to do this. When Mattingly confirmed his intentions, the colonel grabbed the form, scrawled his signature along the bottom, and dismissively tossed it back across his desk. "You are the dumbest ensign I have ever met," he growled. "Out!"

There is a humorous sequel to this story. Twenty years later, after leaving NASA and returning to the navy, Mattingly was working at the Naval Space and Warfare Systems Command when he and his colleagues were notified of an impending visit by their new commander, Vice Adm. Glenwood Park. Mattingly knew the name well but was quite sure he would not be recognized by his former gunnery officer after all those years. Having briefed Admiral Park on the command's programs, they retired to his office and relaxed for a few minutes before resuming the tour of inspection. "On the way out, I just couldn't resist," Mattingly recalled. "I said, 'Admiral, do you remember me?' He looked me right in the eye and says, 'I sure do. You were the dumbest ensign I've ever known!'" The two men laughed and would eventually become the very closest of friends.

Following his flight training, Mattingly was assigned to a squadron located at Naval Air Station Jacksonville in Florida, where he found himself flying the same type of Douglas AD in which the Marine colonel had given him his first "real" flight. They were old, carrier-based airplanes, and certainly not a scratch on the newer jet fighters, but it was still flying, and Mattingly loved it.

He can still vividly remember the day a friend from a photo reconnaissance squadron told him that NASA was launching its first manned Gemini spacecraft from the Cape. His job was to fly down and take photographs of the launch, and he suggested that Mattingly grab himself an airplane and accompany him. "I had not been impressed with the space program at that point," Mattingly told interviewer Rebecca Wright. "I thought the pictures in the magazines of Mercury and Gemini weren't visually appealing. Airplanes are supposed to be smooth, and there's an elegance to them. . . . I can't imagine how anybody could be interested in that. It just had no appeal."

Managing to secure an airplane was a much easier thing back then, and Mattingly flew across to the Cape with his friend, listening in on all the radio frequencies he had been given. They circled the Banana River area, listening in to all the air-to-ground activities. Then the Titan lifted off with future *Apollo 16* crewmate John Young on board. "I heard the voice communications and when I saw some [McDonnell Douglas] F-4s that were trying to fly chase to take pictures of it and . . . this rocket walked away from them and just kept going," he recalls. Later that evening a friend asked what he had thought of the launch. "You know, I think that sounds like the most exciting thing anybody could ever do," he replied.

Mattingly was eager to get into test pilot school but knew that he had no chance as long as he remained with propeller aircraft. He also knew that if he was going to obtain a transfer to a jet squadron, he would have to make the move soon, while he was still young. After doing "all kinds of crazy things" and writing letters to "whoever gets this," he eventually moved to a twin-jet Douglas A-3 squadron, the newest and biggest carrier airplane in the navy. After five years of sea duty, Mattingly was advised to return to shore duty, as his navy career was stagnating and he was being left behind by many of his contemporaries. "I said, 'Well, I want to go to Pax River and the test pilot school.'"

Unfortunately, it was April, and Pax River had started a month earlier, so it was recommended that he attend postgraduate school to obtain a master's degree in aeronautical engineering and from there continue on to test pilot school. Mattingly said he did not want to do that, stating a preference for attending Harvard, but his idea received a less than enthusiastic response.

"They laughed at that," he recalled. "I've been on sea duty for five years. Right now I'm standing duty officer in a hangar. The squadron is back. We're just sitting here watching our airplanes get towed away while they bring in the new ones. I'm not going to any of the schools. I'm doing nothing. Why don't I detach early?"

Before Mattingly actually resigned from the navy, he received a call asking if he wanted to go to the Air Force Test Pilot School in California, which started later than Pax River. "I'd seen a little article that the Air Force school at Edwards was designated as the source for the Manned Orbiting Laboratory [MOL] program. It was going to be the military man-in-space program. According to this article, the source of those people would be the school at Edwards. So I thought maybe that's not all bad."

Mattingly drove out to California in the spring of 1965 and spent a year at Edwards, having "really just an absolutely fun year. There was nothing to do out there in the desert except fly, so it was really, really a lot of fun." It was at Edwards that Mattingly first met Charlie Duke, who was then serving as an instructor at the school.

Several months after his arrival, both NASA and MOL announced that they would be seeking additional astronauts in 1966. The selection process for military applicants involved applying to their parent service, which set its own criteria and convened a service selection board. This board then forwarded suitable applications to NASA and/or MOL for further consideration and ultimate selection. Navy applicants were required to designate their interest in only one program, while their air force counterparts were permitted to apply for both. "The Air Force guys all went and put their requests in, and you could check 'Do you want to go to NASA, Do you want to go to the Air Force program?' The Air Force guys could check both boxes. The Navy says, no, just pick one. So instructors, students, everybody was applying for both of these programs."

Mattingly and fellow naval classmate Ed Mitchell, who had just returned from the Massachusetts Institute of Technology (MIT), sat down and weighed their options. After long discussions they realized that they would not even have this opportunity if it had not been for the military. Therefore, they chose the military MOL program over the civilian NASA effort. "Besides, you look at all those press releases and stuff; we'll never get to NASA, so let's take the military program. So we checked [it] off."

All of the applicants for both programs were called to Brooks Air Force Base in San Antonio, Texas, for their physical examinations. Those who passed were forwarded to the MOL and NASA selection boards. After the initial screening the selection boards released their lists of applicants, who would be called for interviews. The MOL list came out first, and Mattingly and Mitchell were devastated to discover that neither had made it past the first round, although many of their air force classmates had made the cut.

"We were not selected by the Air Force screening board. A fellow who was in the school, his name was Lt. Col. John Prodan, said, 'You guys, I think you didn't get picked up because you had the wrong color blue in your uniform. Would you like to go to NASA?'" The two naval aviators responded enthusiastically but feared that as the application date had already passed and the navy did not allow them to apply for both programs, they would be out of luck.

Prodan said he would see what he could do. Mitchell and Mattingly were ranked first and second in their class at Edwards, just as fellow naval aviators Dick Truly and Bob Crippen had similarly dominated the previous class before moving onto MOL. Prodan suspected that there may have been a bit of inter-service rivalry at play and the air force was discriminating against its navy colleagues. The two would-be astronauts were interviewed by Prodan and then taken to Houston.

"That was a fascinating little experience, my first introduction to John Young and Mike Collins," Mattingly later recalled. "I was perplexed by John, couldn't figure him out. Mike was much more approachable and asked about flying and in particular wanted to know what I thought about the F-104 [Starfighter]. I told him it was about as much fun as anything I'd ever flown but wouldn't be worth much in combat. He furrowed his brow and I knew I had blown it."

The NASA applicants all returned to their regular duties to await the board results. The Aerospace Research Pilot School (ARPS) was divided into two six-month segments, aircraft and space, with the latter including a two-week aerospace physiology course at Brooks AFB. The course fascinated Mattingly, but he was also preoccupied with the thought that at any moment the astronaut selection would be announced.

Early one afternoon someone came to the classroom door and told Ed Mitchell he had a telephone call. The entire class watched him leave with great curiosity. When he came back, he did not say a word, but the smile that beamed from ear to ear made words unnecessary. The day ended with no further calls, and Mattingly's spirits began to sag.

After Mitchell's obviously triumphal return to class, Mattingly's classmates hung around and tried to keep up his spirits. They invited him to join them at a Mexican restaurant on San Antonio's River Walk, and he accepted. Just as they were about to leave, Mattingly got his call from Deke Slayton asking if he would like to become a NASA astronaut. "Obviously, yes!" he said.

Although he had been sworn to secrecy until the official announcement was made later in the week, the smile on Mattingly's face gave away the secret just as easily as Mitchell's had earlier. "So you cannot tell anybody, but you can't hide that from your friends. I mean, they can look and see. Everybody knew we couldn't say anything, but we celebrated anyhow."

The two secret astronauts and their classmates sat on a patio by the river that night drinking beers when a waiter came forward with a pen and napkin. "This lady over here would like your autograph," the waiter said, nodding to a woman in the corner.

"Somebody recognized me?" a startled Mattingly thought as his colleagues looked on in bemusement. "Wait a minute. How does she know? I said 'Who does she think I am?'"

The waiter replied, "Aren't you Dickie Smothers?" The whole gang collapsed in laughter at the suggestion that Mattingly resembled the younger of the musical-comedic Smothers Brothers duo. Mattingly signed his first autograph and handed the napkin back to the waiter. "I said 'No, but let me give you this and just hang onto it for a week!'"

The fifth group of NASA astronauts, in a self-deprecating nod to their Original Seven predecessors, dubbed themselves "The Original Nineteen" and reported to Houston in May 1966. Mattingly found a few familiar faces from the Aerospace Research Pilot School among his new colleagues, including Edwards instructor Charlie Duke.

Charles Moss Duke Jr. was one year older than Mattingly, and his first recollection of flight came from the early 1950s. He had been driving in a friend's convertible one afternoon and gazing up at the clear sky when he spied a contrail. "You didn't see many contrails back then," he later recalled. "And I said, you know, 'Gosh it'd be nice to make a contrail.' And I started dreaming about flying airplanes."

Duke went on to the United States Naval Academy in Annapolis, Maryland. "They gave me a couple of rides in an open-cockpit, bi-wing seaplane called the N3N Yellow Peril. And I was hooked from that moment on." He graduated from Annapolis in 1957 with a bachelor of science degree. At that time the air force did not have an academy of its own, and the Defense Department would allow one quarter of West Point and Annapolis graduates to volunteer for the air force. Duke chose to be among that group. "I'd fallen in love with airplanes at the Naval Academy rather than ships," Duke told Doug Ward in 1999. "I knew that's what I wanted to do. And maybe . . . another reason that I decided to go [was] because I really did get seasick!"

After pilot training Duke spent three years with the 526th Interceptor Squadron at Ramstein, Germany, flying F-86D Sabres, and it was there that the space bug first bit. "I was in flying school in 1957, and I'd just soloed in October when they launched the Sputnik. Four years later, of course, Gagarin and then Shepard went up, and I was still a lieutenant in Germany in a fighter squadron, and began to dream at that point about, 'Maybe this career I'm on, if I set the right goals, I could be an astronaut one day.'"

Back in the United States Duke knew that advanced education was a prerequisite for promotion in the modern air force, and he applied to attend MIT, graduating in 1964 with a master's degree in aeronautics and astronautics. "I got this engineering degree, but man, I really miss flying," he remembered. "And the next logical step was test pilot school. So I volunteered and went to school out there in '64." After graduating the following year, Duke stayed on at the ARPS as an instructor. "Two months later there was an ad in the paper that said 'NASA's looking for more astronauts, please apply.' And so a bunch of us from Edwards applied."

Unlike many of his air force colleagues, Duke tried only for NASA. "I went to my boss, a deputy commander of the test pilot school at Edwards. His name was Buck Buchanan [and I] said, 'Sir, you know I would like to apply; what do I do?'" Buchanan explained that Duke could apply for NASA and MOL, but Duke later recalled the stern admonition that followed: "If you apply for both, I guarantee you we are going to pick you for MOL and not let NASA have y'all."

"So I decided that I'd volunteer only for NASA, which I did," Duke explained, "and fortunately got selected."

In September 1965 there were thirty active astronauts for the remaining Gemini and early Apollo flights. At that time, however, NASA had ambitious plans for as many as forty manned Apollo missions including ten lunar landings and three orbiting research laboratories, all starting in late 1968. Such a program required a much larger astronaut corps to provide prime, backup, and support crews, and so NASA set out to select at least two new groups. Group 5 was the first of this new influx to arrive in Houston, and they were immediately immersed in training.

"Everybody did geology," Duke recalls. "Then we did spacecraft systems for four or five months. And then everybody got assigned to some sort of little engineering oversight job. I remember Stu Roosa and I got assigned to Frank Borman, who was sort of Head of the Propulsion System. These things were all sort of unofficial, as far as organization within the office. But that was where you concentrated your effort."

Mattingly and Duke soon found themselves fully in the swing of things. Mattingly served on the support crews for *Apollo 8*, *11*, and *12*, while Duke was on the *Apollo 10* support crew. At the specific request of Tom Stafford, Duke had served as CapCom for *Apollo 10*, communicating with the crew from Mission Control during the critical separation, descent, and subsequent linkup maneuvers they practiced ahead of an actual landing mission. Duke carried out his duties with such competence that it came as no surprise when Neil Armstrong requested his services as CapCom for the same landing phase of *Apollo 11*. Armstrong also recognized that Duke was highly experienced in LM activation procedures, which might prove a valuable asset during the first attempt to land humans on the moon. Simply put, Armstrong wanted the best available person on the job.

Shortly after *Apollo 11* returned to Earth, Slayton announced that the crew of the third lunar landing mission—*Apollo 13*—would be Jim Lovell, Fred Haise, and Ken Mattingly. Their backups would be John Young, Charlie Duke, and Jack Swigert. Mattingly and Duke were on their way to the moon: Mattingly scheduled to come within miles of its surface and be only the fourth person to orbit it alone and Duke—"if all went well, and I didn't

37. On 12 November 1971 Young (*behind wheel*) and Duke (*foreground, wearing glove*) observe the loading of LRV-2 into the Lunar Module *Orion* at the Kennedy Space Center. Courtesy NASA.

break a leg, and they didn't cancel the program"—to become one of the very first to walk on it.

There had been earlier problems for the crew on their way to the moon, but they had all been satisfactorily resolved. Almost forty-five thousand miles from Earth they had reported a mysterious stream of flaky, brownish-colored debris being shed from *Orion*. "I'm not normally a rabble-rouser," a concerned Young had radioed down to Mission Control, "but something funny is going on here." Duke described the flakes as being like "shredded wheat," and Mattingly added they were "like paint from an old barn." Young and Duke had then crawled into the LM to conduct comprehensive pressure and oxygen level checks, eventually describing everything as quite normal. The manufacturers of the spacecraft, the Grumman Aircraft Engineering Corporation, soon identified the source of the flaking material as thermal insulation paint, applied to specific areas of the LM to prevent it from overheating in the sun. They assured NASA it posed no threat to the mission or the astronauts.

Then potentially serious trouble had developed in *Casper*'s main guidance and navigation system some 139,000 miles from Earth. The problem involved a gyroscope-controlled, suspended navigation platform unit. This platform was meant to align itself to star fields and provide a reference against which movements of the spacecraft could be measured. Instead of swiveling freely, however, it seemed to have frozen in a single position (referred to as "gimbal lock"), rendering it useless for navigation. Mattingly had noticed a system warning light indicating gimbal lock while taking star sightings to align the spacecraft just before joining John Young and Charlie Duke in settling down for an eight-hour rest period. He punched off the alarm and straightaway reported it to ground control, but they had also noted a warning light on the Guidance Officer's console. Reassuringly, *Casper* was equipped with a backup guidance system that would allow the crew to return home safely, but there were strict rules in place, and Mattingly knew that if the primary guidance system was lost, so too was the primary goal of their mission. The moon landing would be canceled.

Fortunately, controllers located the source of the problem and radioed instructions up to Mattingly, who had also been working on it. After freeing the navigation platform unit, which immediately became inertial, he tried to realign it by taking star sightings. Some lingering paint flakes prevented him from doing this, however, so he took an initial sighting on the sun, which stabilized the platform. He would later fine-tune the alignment when he was able to take clear star sightings. Forty-five minutes after the alert began, the faulty system was back up and in service again. Surprisingly, the other two crew members had slept through everything, so Mattingly wrote out a quick note explaining to Young what had happened and taped it above his couch before settling down himself for a long-overdue sleep.

The flight of *Apollo 16* now continued smoothly, and on the second day their spacecraft slipped silently past the landmark 205,443 miles from Earth—the point at which the moon's gravitational pull finally exceeds that of Earth. As they closed in on the moon, the spacecraft's main engine was fired right on schedule, braking *Casper*'s velocity and enabling the crew to enter an initial orbit within some eighty-two miles of the surface in preparation for the fifth Apollo lunar landing—the longest and most ambitious yet attempted. Following the successful burn, mission commander John Young had jokingly remarked that he was renaming their craft "Sweet Sixteen."

Flight director Gerry Griffin, a veteran controller from earlier Apollo missions, was on duty for *Apollo 16* when the first of many problems struck. As Griffin would later recall, "things went progressively downhill" from the time he took over his shift and Young and Duke, aboard *Orion*, reported problems with their S-band steerable communications antenna, creating a radio link problem. The two astronauts had suited up and entered the LM on schedule and were conducting a final check of the craft's systems prior to undocking from *Casper.*

"For about forty-five minutes, things went well," Duke recorded in his autobiography. "Then when I began to power up our antenna for the primary communications link with Houston, our problems began. I couldn't get the antenna to rotate left to right in the yaw axis and point toward Earth. This resulted in a poor communications lock-on with the tracking network, and therefore they were not able to automatically uplink information to our computer." It would mean that Duke had to write down all the information and insert it into *Orion*'s computer manually. It was a time-consuming task when every minute was at a premium. The information also had to be downloaded before radio contact was lost as they flew behind the moon and out of range. They finally made it with one minute to spare, allowing the two spacecraft to separate and freeing *Orion* to begin the final descent to the lunar surface.

Duke had allowed himself to become excited after he and Young had completed their preparations, and then it was time for another milestone in their flight. "After making a small separation burn from the Command Module, we turned so that we could see *Casper* out our front windows. It was a beautiful sight. The spacecraft silhouetted against the lunar surface, rushing by beneath it."

Meanwhile, Duke was struggling with an ill-fitting lunar EVA suit. He had asked Houston for permission to loosen the laces to ease his discomfort, but they had said no, explaining that it would fit perfectly once it was fully pressurized in a vacuum. To make matters worse, just before separation he

38. Shortly after undocking from LM *Orion* on *Revolution 13*, *Apollo 16* CSM *Casper* was photographed above crater Valier's northwest rim on the moon's far side. Courtesy NASA.

found the instant feeder in his helmet had a faulty valve that allowed orange juice to keep squirting into his helmet, which was sticking to his face and hair. "NASA had designed a plastic drink bag to fit inside our spacesuits, since we were going to be working on the lunar surface for long periods of time. It was shaped like a hot water bottle and attached to our long underwear. A long plastic straw went from the bag, up through the neck ring of the helmet, right next to our mouths. To drink, we simply grabbed the straw between our teeth and sucked real hard." But now the juice was leaking into his helmet. "It seemed like every time I breathed, out would come one or two small drops of orange juice."

Duke complained to John Young about the sticky liquid smearing up his vision. His commander wholeheartedly agreed, griping that he too was sick and tired of the juice and the problems it was causing, including stomach cramps and gas, and peppered his complaints with some choice language. At this time Mission Control quickly interjected, reminding Young that they were transmitting to the world over an open microphone. An embarrassed commander quickly apologized, but the press corps later had a field day with his comments, leading the governor of Florida to issue a statement to the effect that it was not juice from Florida oranges causing the problems, as the liquid was actually an artificial substitute. Duke jokingly mentioned postflight that "Florida's native son, John, will never live that down."

With the communications problem fixed and the two craft separated, all systems seemed to indicate that the landing could proceed. Meanwhile, as Young and Duke moved farther away from *Casper* and went through their final preparations, Mattingly was getting ready to carry out a burn to change and circularize his orbit, placing him sixty miles above the moon and in the correct rendezvous path in the event Young and Duke had to abort their landing unexpectedly.

Then all hell broke loose. To the alarm of the two men aboard *Orion*, they heard Mattingly cry out: "There's something wrong with the secondary control system in the engine. When I turn it on, it feels as though it is shaking the spacecraft to pieces!" Young knew that the engine and control system provided their only means of getting back home safely. "Don't make the burn!" he instructed Mattingly. "We will delay that maneuver."

When *Orion* finally emerged from behind the moon and regained communication with Houston, Duke reported the problem. CapCom Jim Irwin told them to postpone everything until the following orbit so all could assess the situation. "We knew in our minds that it was very grim," Duke later stated. "It looked as if we had two chances to land—slim and none."

The problem now facing the crew of *Apollo 16* as they swept around the moon in their two spacecraft was caused by a recalcitrant steering system in *Casper*. The prime system itself was in good working order, but the fault lay in the backup system that would be utilized if the prime system failed. It was causing the CM's engine bell to oscillate sideways. Mattingly knew that if the prime system subsequently failed and could not be repaired, the crew would be marooned in lunar orbit, with no possible way of returning home.

While this drama was unfolding, ground controllers were also agonizing over whether to commit a possibly faulty Lunar Module to a hazardous touchdown in the Descartes area of the moon's central highlands. Wernher von Braun was in Mission Control that day, sitting in an emergency meeting with his eyes closed and his head in his hands, as he listened to possible solutions being thrown back and forth. Also in attendance was Apollo program manager, former *Gemini 4* and *Apollo 9* astronaut Jim McDivitt. Participating in his last mission before retiring from NASA, he later admitted that his first thought on hearing of the seriousness of the problem was very similar to Duke's: "We've lost the mission."

By that stage the three final Apollo moon-landing missions (*18*, *19*, and *20*) had already been canceled due to congressional budget cuts. Assuming the crew then orbiting the moon was able to return home safely, that would leave just one more planned lunar landing mission. They desperately needed *Apollo 16* to work. Alone in CM *Casper*, Mattingly knew the next few minutes would be among the most critical he had ever faced in his life.

As ground controllers wrestled with the problems aboard CM *Casper* in lunar orbit, the landing crew of Young and Duke aboard LM *Orion* were instructed to stay close to *Casper* until the situation could be resolved, in case it was necessary for them to re-dock with Mattingly in the CM and make a hurried return home. The crisis was certainly reminding many people of the aborted *Apollo 13* mission.

From Houston the flight directorate advised all three crew members that they were closely examining the steering problem along with engineers and technicians from North American Rockwell, who had designed and built the CM. They were further informed that a decision had been made to allow a maximum five additional revolutions of the moon, which would take around seven and a half hours. If the problem in the backup system had not been resolved by then, the landing would need to be scrubbed. *Orion* would then have to re-dock with *Casper*, allowing Young and Duke to transfer back into the CM for the sad journey back to Earth.

Mattingly was well aware of his responsibilities and the procedures to be followed at this critical situation and was able to work with the controllers to solve the problem. "At that time you're under pressure," he later reported, "and you are doing this from memory. Fortunately, I got bored a couple of days before flight, and I had gone through the flight plan and written down all the critical rules for everything that was out of communications link. And I flipped to it . . . yep, that's right!"

Eventually, the problem was identified as a defective yaw gimbal actuator that was designed to swing the engine bell left or right as directed by the pilot or the onboard computer. The defect, caused by an open circuit on a feedback system, was only apparent in the backup guidance system. It was finally determined that even with this fault, the backup system might still be used; any yaw oscillations could be overcome and would not affect the attitude of the spacecraft. This information was passed to the director of Flight Operations for Houston's Manned Spacecraft Center, Chris Kraft, and the flight director on duty at that time, Gerry Griffin. Based on these findings, Griffin made a historic determination that the landing could proceed on the next lunar orbit.

The go-ahead was relayed to a relieved *Apollo 16* crew as they emerged from behind the moon on their fifteenth revolution, and as they were now six hours behind schedule, it was decided they would have a sleep period after they landed on the moon and would undertake their first lunar expedition early the following morning, 21 April.

"Coming in like gangbusters . . . I can see the landing site . . . We're right in, John . . . Man, there it is—Gator, Lone Star . . . Perfect place over there, John . . . . A couple of big boulders . . . Contact . . . Stop. Well, old *Orion* is finally here, Houston. Fantastic! All we have to do is jump out of the hatch and we've got plenty of rocks . . . Man, it really looks nice out there . . . I'm like a little kid on Christmas Eve." Duke could not contain his excitement as the spindly legged *Orion* settled on the Descartes plateau. After spending much of the descent with their windows pointed out to space and with the lunar surface invisible beneath them, at seven thousand feet the guidance program maneuvered the vehicle to windows forward. Young and Duke had their first live glimpse of their home for the next several days.

The landing site had been chosen by the site selection committee of scientists and administrators, the decision based in part on detailed photographs of the area taken with a mapping camera by *Apollo 14* CMP Stuart Roosa. After selecting the Descartes highlands, a three-dimensional model of the proposed area was assembled on a large board with a camera suspended from a track above. Television monitors were placed in the windows of the LM simulator, and as the crew "flew" the simulator, so the view in the "windows" changed. "Thousands of times John and I came in for a landing and we'd pitch over and recognize one crater we called 'Gator'; one was called 'Lone Star,'" Duke recalled. "I could look out the right window of the simulator to the north and see North Ray Crater up there and John could look out his side and see the Stone Mountain down to the left. And when we did it for the first time, I mean for real, in flight . . . well, it looked exactly like the mock-up."

Craters and mountains and landing sites aside, Duke was looking for something else as they slowly descended to the lunar surface. "About three months before the mission," he related, "I'd had this dream about John and

I driving the rover up to the North Ray Crater and we came over one of the little ridges, and there's a set of tracks in front of us. And it's rover tracks! We started following these tracks. Well that dream was so real that one of the things, when I went to look north, was to see if I could see that set of tracks. Well of course, there wasn't any . . . but I did figure out that the surface wasn't as rough as we expected."

Orion settled on an almost perfectly level piece of lunar real estate. Young and Duke powered down the LM to its minimum levels to make up for the longer time spent in lunar orbit. This energy conservation was not without risk, however, as it also halted the mission timer that would provide them with their emergency liftoff times should they have to leave the surface in a hurry. If such a scenario also included the loss of communications from Earth, Young and Duke would have had to use their wristwatches to determine manually the exact moment for liftoff and rendezvous with Mattingly's *Casper*. Checklist completed and meal finished, the fifth lunar crew settled down for a rest period ahead of the first moon walk the following day.

Miles above them, Mattingly was only the sixth person to orbit the moon alone. After correcting *Casper*'s orbit to rendezvous with *Orion* when it returned from the surface, he had a full schedule of activities to keep him busy, including high-resolution lunar photography and orbital geology observations. Mattingly had sought a full schedule. "The Command Module pilot really wants to do something important," he remarked. His training had started some months earlier with a message to meet geologist Farouk El-Baz. "He had the patience and the enthusiasm to talk about what you could see from a distance," he recalled with evident respect. "He just made looking at the moon a meaningful thing. He spent untold hours with me trying to help me understand and look for things, and posed a series of questions, not to go look for the answer for this, but he was trying to teach me how to approach a problem with some curiosity that would lead me to see things that I would otherwise miss. He'd take me out in the field on a geology trip and say 'What do you see?'"

Along with fellow geologists Lee Silver and Bill Muehlberger, El-Baz taught Mattingly to identify and describe surface features from lunar orbit adequately. With their help and the use of photographs and field trips, he learned to see the big picture and understand not just what he was looking at but also what it could mean. It not only provided him with a sense of purpose but also with great satisfaction for the job at hand.

After instructing Mattingly on how to determine the slope of a hill by the amount of talus—or small rocks—gathered at its base, El-Baz presented him with a series of photographs taken by the lunar orbiter that showed some smaller hills in the mares area with talus formations that had the geologists stumped. "That's the kind of problem that would be interesting if somebody could explain," Mattingly recalled El-Baz having said during one training session. He later reported:

So on one of the passes, I came across the area where this phenomenon was observed. It was not even a question. One look out the window, and you said, "It's a bathtub ring." You could, in one view, instead of looking at this little piece but looking at a bigger piece, you could see that this . . . was solidified lava that at one point had been at this level, and it receded either because it gave off gases and collapsed or it was pushed up and then drained back down some. It was as clear as you could imagine. So there was a whole lot of stuff that I was able to do that . . . gave me a sense of this isn't all just for practice.

Like many space travelers before him and since, Mattingly took along his own music library. It kept him company in the darkness of the CM as he conducted his low-light photography experiments and provided the perfect accompaniment to the wondrous views outside his windows. "There's old Mother Earth," he called to Houston with snatches of music in the background. "Man, that's beauty, too. Never get tired of watching earthrise." Among the music he selected was Gustav Holst's *The Planets.* Completed in 1916, little could its English composer have imagined that within forty years of his death people would be sailing through space, gazing down at the moon from near orbit, listening to his orchestral suite, while their colleagues trod the surface below. "Putting music with these unbelievable sensations is just kind of the ultimate in exhilaration," Mattingly later reflected to spaceflight researcher Larry McGlynn.

Although he obviously would have liked to have joined Young and Duke on the surface, Mattingly was philosophical about his role and his experience. "I think, having talked to John, one of the few who had a chance to fly solo and . . . to go down and land, they are both unique experiences.

You really want to do both. I didn't quite get to do the second part—which I miss—but I had never trained [for] that configuration."

Eager to get their first EVA under way and having finished a breakfast of peaches, cold scrambled eggs, cinnamon toast cubes, and a lemon food bar, Young and Duke were ahead of schedule and chomping at the bit to get out of *Orion* and onto the lunar surface. As commander, Young was first to descend the ladder. "There you are, our mysterious and unknown Descartes highland plains," he declared to an eagerly waiting audience. "*Apollo 16* is gonna change your image!" He then added in a reference to himself, "I'm glad they got ol' Brer Rabbit, here, back in the briar patch where he belongs."

Charlie Duke quickly followed after Young, and together they set about unstowing the lunar rover and planting the flag. "Hey, John," Duke called as he composed a photograph. "This is perfect, with the LM and the Rover and you and Stone Mountain [and] the old Flag. Come on out here and give me a big navy salute." Capt. John W. Young, U.S. Navy, enthusiastically obliged the photographer by jumping three feet off the ground while snapping a smart salute.

It was actually while he was walking on the moon that some exciting news was relayed to John Young; Congress had, by a large margin, given the go-ahead to the Space Shuttle program. Young had no way of knowing that just nine years later he would be in command of the shuttle's maiden voyage into space.

Next on the list were the various experiments that had to be deployed around the LM. "I went around to retrieve the ALSEP," Duke reported. "There was a big crater about two meters behind us that we hadn't even seen. If we'd landed like three meters back to the east, we'd have . . . the back leg in that crater. It was amazing how things like that were sort of camouflaged. Without the right lighting conditions . . . you could miss some of these subtle features."

One of the key experiments on *Apollo 16* was to determine how much heat was flowing from the lunar interior. In order to do this, a ten-foot deep hole had to be drilled and a thermometer inserted. The experiment had originally been aboard *Apollo 13* but, along with the rest of the mission, had never made it to the lunar surface. *Apollo 15* had tried again but had been unable to deploy it to its ideal depth, and therefore it produced imperfect

39. In a photograph taken by Charlie Duke, CDR John Young leaps high and gives a "big navy salute" to the U.S. flag during the crew's first lunar EVA at the Descartes landing site. LRV-2 is parked in front of LM *Orion*. Courtesy NASA.

results. Duke had the benefit of an improved drill and within minutes had the thermometer inserted to its full depth much to the delight of the scientists watching intently from the Science Operations Room in Houston. Unfortunately, their elation was to be short-lived. Within moments Young's foot was entangled in one of the experiment's cables and, unable either to see it or feel it because of the bulky space suit, the cable was pulled loose as he shuffled away.

"Charlie?" Young called to his crewmate. "Something happened here . . . here's a line that pulled loose."

Duke could immediately tell what it was. "That's heat flow, you've pulled it off," he said despondently.

"I don't know how it happened. God almighty," Young replied. "I didn't even know it." He had accidentally disconnected the data tape that linked the heat flow electronics package to the main ALSEP transmitter. No sooner had it happened than Houston was already trying to figure out a way to repair the damage. Scientists, engineers, manufacturers, and astronauts worked through the night to find a solution, but it was not to be. Unfortunately, instead of third time lucky, the heat flow experiment had been foiled yet again.

Apollo 16 had a lot of ground to cover while on the moon, both literally and figuratively. Thanks to the lunar rover, Young and Duke could travel four miles in any direction instead of being limited to the immediate vicinity of the LM. The rover had originally been intended to debut on this mission, but with the cancellation of *Apollo 18* and *19*, it was added to *Apollo 15* and minor modifications made for its next performance.

"It was an incredible machine," Duke remembered years later. "It revolutionized lunar surface exploration." Their first road-trip was a short drive to two neighboring craters named Flag and Spook to collect a few specimens. Back at the LM, Young took the rover for a quick solo spin to test its handling and performance, while Duke filmed the first ever Descartes Rally.

"Is he on the ground at all?" a rather concerned CapCom called from Houston, upon seeing the live images of the rover bouncing over craters and leaving clouds of lunar dust in its wake.

"He's got about two wheels on the ground," Duke replied. "It's a big rooster tail out of all four wheels, and as he turns he skids. The back end breaks loose just like on snow. Man, I'll tell you, Indy's never seen a driver like this!" After more than seven hours outside, the tired, exhilarated, and dusty astronauts returned to *Orion* for their rest period.

With the mobility of the rover, Young and Duke covered a wide area, including traversing the 540-meter-high Stone Mountain. "We were three-and-a-half, four miles to the south," Duke later recalled. "We were several hundred feet above the valley floor. From this vantage you could look all the way across the Cayley Plains and the valley that we had landed in. You could see . . . [Smokey] Mountain and North Ray Crater. Right out in the middle there was our little Lunar Module that was . . . orange, and you could see that. And then looking off to the northwest over this way was—as far as the eye could see—was just the rolling terrain of lunar surface. It was really an impressive sight." At about this time Mattingly reported to CapCom Tony England that he had seen a bright glint from the rover as he overflew the area.

Sunday marked their final EVA and included a visit to a large rock visible from *Orion*. With little context to provide perspective, Young and Duke were undecided whether it was an average-sized rock nearby or a big rock in the distance. "We kept jogging and jogging, and the rock kept getting bigger and bigger and bigger," Duke remembered. "So we get down to this thing and we called it 'House Rock.' You know, it must've been ninety feet across and forty-five feet tall. It was humungous. And we walked around to the front side, which was in the sunlight, and . . . it was towering over us." The two astronauts used their hammers to remove a chunk of House Rock to bring home.

In all *Apollo 16* returned to Earth with 97.6 kilograms (215 pounds) of rocks and soil, by far the greatest haul of any mission to that point. Descartes had been expected to provide mainly volcanic samples, but despite the hard work and wide area covered by Young and Duke and the considerable quantity of samples recovered, it turned out to all be breccias or anorthosite—commonly found everywhere on the surface of the moon. Although not the answer they expected, geologists were able to determine that it was cosmic bombardment that had shaped much of the lunar surface, not the volcanic activity they had predicted. "Well," Mattingly quipped from lunar orbit, "back to the drawing boards, or wherever geologists go."

Before leaving the moon, the two moon walkers had planned to perform a "Lunar Olympics" for the television audience but had run out of time for the full demonstration. "I was going to bounce and set the high jump record," Duke announced to Houston, and started to bounce around by way of example. As he jumped for the last time, he began to topple over backward and disappeared behind the lunar rover. "That was a moment of panic," he later recalled. "I was in trouble. You could watch me scrambling . . . trying to get my balance. I ended up landing on my right side, and bouncing onto my back. My heart was just pounding." Duke was concerned that his fall had damaged the suit, possibly even punctured it. He was not the only one. Young came straight over and helped him to his feet. "Charlie, that ain't very smart," his commander reprimanded him. "Sorry about that," Duke agreed.

After more than seventy-one hours on the moon—twenty of them spent exploring the surface—*Orion* lifted off flawlessly and soon rendezvoused with *Casper*. The two spacecraft performed quick inspections of one another before docking. Young and Duke, covered in dust and abrasive lunar soil, using damp towels and every method available, valiantly attempted to clean up their spacecraft before opening the hatch. Mattingly greeted them with a vacuum cleaner, and together they transferred the samples and equipment from the LM to the CM. The following morning *Orion* was jettisoned, and because of a switch left in the wrong position, it immediately began to tumble uncontrollably. Mattingly hurriedly took evasive action to avoid a collision.

Once *Casper* was on its way back to Earth, Mattingly suited up for his deep spacewalk to retrieve film cassettes from the scientific instrument module (SIM) bay. The hatch was opened, and Mattingly worked his way down the outside of the spacecraft using hand- and footholds, all the while traveling at 5,000 mph and 180,000 miles from Earth. Duke popped out of the hatch to keep an eye on his crewmate and to expose a series of biological specimens to the cosmic rays of space.

"I looked at Charlie and said, 'That's really a neat picture,'" Mattingly reflected. "I looked up . . . and there wasn't a single star anywhere. There was this deep black picture. If you turned and looked, you could find the moon over there. I couldn't see the Earth anywhere. And that's all there was in this whole picture." Although he knew they were there, Mattingly was disconcerted by the absence of stars. Guessing they were blocked by his gold visor, he decided to take a quick peek, and instantly the heavens were again full.

"That's the only time that I had a sensation of being away," he remarked. "I don't know why, but around the moon, inside the spacecraft, I didn't ever have a sense being that far away until I looked out there when there were no stars and the entire world was within fifteen feet. And there was nothing else. That was a really powerful sensation. Never seen anything like it." After almost an hour outside Mattingly climbed back in, and *Apollo 16* continued on its way home.

The main parachutes deployed over the Pacific Ocean, and even before *Casper* hit the water, it was joined by a U.S. Navy helicopter. The spacecraft splashed down, rolled over, and was promptly collared by one of the rescue divers, who then stood on a raft alongside it waiting for the astronauts to climb out. Mattingly opened the hatch, only to have it promptly slam back in his face. He finally opened it again, and the three astronauts clambered free and into the raft. On the USS *Ticonderoga* a short while later,

Mattingly recognized the team of divers. “What happened out there with the hatch?” he asked.

“You don’t know?” the diver replied.

“No. What’s the matter?” Mattingly persisted.

The diver took him over to the spacecraft. “Let me open the hatch,” he said, reaching for the handle. The hatch was opened, and Mattingly was knocked backward. “The odor was unbelievable,” he confessed later. “Thank God our olfactory nerves disappear early in life. Living in a bathroom for ten days is not a desired position. *Life* magazine doesn’t tell you about those things. Boy, was that terrible!”

The end of the Apollo program marked the end of an era for NASA and saw many changes in the Astronaut Office. After serving as *Apollo 17* backup LMP, Charlie Duke became the first *Apollo 16* crew member to call it a day and retired from NASA in 1976. He went on to enjoy a successful business career before becoming a lay preacher.

John Young remained with NASA until 31 December 2004. In addition to his two Gemini and two Apollo flights, he also commanded the first flight of the space shuttle and the first international Spacelab mission, STS-9. He was the first space traveler to make six flights and the only person to fly Gemini, Apollo, and the shuttle. He eventually became NASA associate director (technical), but his outspoken criticism of shuttle safety policies in the wake of the *Challenger* disaster in 1986 very likely contributed to the space agency’s decision to retire Young from any further space missions. After nearly a quarter-century as a leading and record-breaking NASA astronaut, his wings had effectively been clipped.

Mattingly remained at NASA and was a key contributor to the development of the shuttle program. He flew in space twice more, as commander of the fourth shuttle flight aboard *Columbia* in 1982 and of *Discovery* on STS 51-C in 1985. Later that same year, after almost two decades as an astronaut, he retired from NASA. After returning to the navy, he was reunited with his old *Apollo 13* crewmate Fred Haise at Grumman’s Space Station division.

Ken Mattingly still has not had the measles.

10. One More Time

Melvin Croft

What more can we require? Nothing but time.

James Hutton (1726–97)

Those who witnessed the launch of *Apollo 17* at 12:33 a.m. on 7 December 1972 will always recall how night was spectacularly rendered into day through the explosive brightness of five gigantic, howling F1 engines, laboriously propelling three men and a three thousand–ton rocket upward toward a place called Taurus-Littrow.

There had been a last-minute hitch just thirty seconds before the scheduled liftoff, caused by a computer sequence malfunction. This had resulted in a two hour and forty–minute launch hold, but after calmly sitting out the delay on the launchpad, the crew was finally heading for the moon after a sensational liftoff that lit up the Florida east coast. As their Saturn V lunged ever higher into the night sky, twice-flown mission commander Gene Cernan and space rookies CMP Ron Evans and LMP Jack Schmitt were crewing not only the first and only Apollo night launch but what would become the sixth and final lunar landing of the Apollo program.

Schmitt, a geologist, had replaced the original LMP Joe Engle on the crew after pressure was exerted from Washington to include a scientist on one of the lunar missions. It was a tough decision to effect at the time but one that ultimately paid off handsomely for the scientific community. Intriguing discoveries would be made, which eventually did more to cast suspicion upon the origin of the moon than help to resolve its birth.

One significant contribution from the flight of *Apollo 17* was a welcome confirmation of NASA's strategy of cross-training Apollo astronauts with different backgrounds to ensure mission success—specifically, to land on the moon, gather data necessary for Earthbound scientists to build a coherent model of the lunar geology, and return safely. Elite pilots were trained as scientists, and top-notch scientists were taught to fly, but the risk potential was great, and NASA was initially too conservative to trust a scientist—even one trained as a jet pilot—to fly these hazardous missions. *Apollo 17* would be the ninth manned venture to the moon, however, and NASA had finally gained the confidence to trust a geologist as part of a lunar landing crew. A successful mission would prove that scientists, if chosen and trained carefully, were fully capable of supporting pilot astronauts to fly high-risk missions safely.

Like his mission commander Gene Cernan, Dr. Harrison H. Schmitt (more casually known as Jack) was not just ordinary at what he did. He was nurtured at an early age by his geologist father, had years of practical geological field experience, and was smart and pragmatic—a good combination for the first scientist to fly a lunar landing mission. Artist and *Apollo 12* LMP Alan Bean would later capture the essence of cooperation between scientist and pilot in his painting *Right Stuff Field Geologists*, in which he depicts Cernan handing Schmitt a sample bag containing freshly gathered lunar rocks, as "a team that included both operators who understood the scientific issues and scientists who understood the operational issues." Time will tell if NASA employs this best practice when space explorers eventually return to the moon and venture onward to Mars.

Riding a smoke-belching, gyrating monster into orbit undoubtedly excited the two rookie astronauts. Veteran flight director Gene Kranz writes in his memoir, *Failure Is Not an Option*, that even the ultra-quiet Evans joined in broadcasting an account of the night launch, reporting that "during each staging the fireball overtook us; then when the engine kicked in we once again flew out of the orange-red cloud into darkness."

Cernan relates in his autobiographical work *The Last Man on the Moon* that the conservative Schmitt also came to life during launch, with a running commentary on earthly weather patterns. No one should have expected the geologist to remain quiet as he drank in the whole Earth for the first time, after seeing it his entire life as one of the blindfolded men feeling his own part of the elephant. But according to Schmitt his weather reports were not entirely impromptu. "My father was an amateur meteorologist and he excited my interest when I was a boy," he would recall. "And so just as I suited up, one of my friends with that [meteorology] group brought in the latest satellite pictures . . . that gave me the whole southern hemisphere of the Earth. I had planned . . . to build on . . . what those satellite pictures showed . . . and try to experiment with how well I could forecast the weather."

With LM *Challenger* still snugly tucked away inside the Saturn V's third stage, between the rocket and CSM, the crew of *Apollo 17* abandoned the relatively safe harbor of the mother planet after several orbits, and all systems aboard CM *America* were dutifully checked out. Following procedures first demonstrated on *Apollo 10* and used on all subsequent lunar missions, *America* would then pull away from the Saturn's third stage, turn, dock nose to nose with the LM, and pull it free of the depleted rocket fairing. Then it would turn again and carry *Challenger* to the moon, mounted on the nose of the CM.

Nearing the end of the long coast to the moon, lunar gravity captured the spacecraft seventy-three hours and eighteen minutes after launch. The flight of *Apollo 17* proceeded with barely a hitch; the only real problem encountered was a faulty, shrill master alarm that had first disturbed the crew in Earth orbit and—until the cause was tracked down and resolved—had intermittently sounded on the outward journey to the annoyance of the tired crew. There was also a problem with some easily repaired faulty docking latches. One lesser problem came about when Ron Evans lost his surgical scissors, needed to open sealed food bags. Cernan and Schmitt each had a pair, but both were required while on the lunar surface and would be of little help to Evans orbiting the moon alone. Ground control eventually modified the flight plan for lunar surface activities so that they could leave a pair of scissors with Evans, but Cernan and Schmitt enjoyed ribbing Evans unmercifully about his oversight, saying that because of his clumsiness, he would have a lot of paperwork to complete over the lost utensil when they got back. With so few of the problems that had plagued previous lunar flights, the outbound flight was consequently filled with good humor and some voluminous pontificating by Schmitt.

The command of *Apollo 17* came easily and naturally to Gene Cernan. Today, more than thirty years after his retirement from NASA and (concur-

40. The crew of *Apollo 17* at a press conference in August 1971 following the announcement of their selection to the final lunar landing mission. *From left*: Harrison "Jack" Schmitt, CDR Gene Cernan, and CMP Ron Evans. NASA photo, courtesy Ed Hengeveld.

rently) the United States Navy, he still exudes confidence and control. In a decision that could easily have backfired for him, he had declined Deke Slayton's offer to be LMP on *Apollo 16* under the command of John Young in hopes of achieving a later command role for himself. Had he accepted that assignment, he would have gone down in history as the only pre-shuttle astronaut to fly three flights without a command. His bold gamble paid off when Slayton chose him as mission commander for *Apollo 17*.

For his part Cernan was quite content to focus on the overall mission and let his LMP, Schmitt, take the lead role in planning the surface activities while on the moon. He admitted that he liked the science, but he was a "big picture" guy who "looks at the mountains and [says], we need to go over there." Schmitt, he said, was "focused down here [in the details]. And between the two of us, I think we probably complemented each other as well or better than anyone else I could have flown with, including Joe."

Eugene Andrew Cernan was born into a hardworking blue-collar family in Chicago, Illinois, on 14 May 1934. His father taught him at an early age how machines worked and were put together. As a youth growing up during World War II, Cernan closely followed the fighting in the Pacific theater and was especially thrilled by movies of naval aviators landing on aircraft carriers. His knowledge of machines and fascination with airplanes was enough for him to set his sights on an engineering military career. He entered naval ROTC after graduating from high school on a partial scholarship at Purdue University, in Indiana, with the hope of garnering a commission in the naval reserves.

Promising job offers came Cernan's way after graduating from Purdue, but the lure of landing airplanes on aircraft carriers had not faded. In 1956 he was commissioned an ensign in the U.S. Naval Reserve but eventually applied for admission to the regular navy and was accepted. He received his wings on 22 November 1956, fulfilling a long-held dream. But he had yet to land on a carrier.

The young aviator rapidly progressed in his flying career, moving to Miramar Naval Air Station in San Diego, California, where he finally made that first carrier landing in 1958. While he was living his childhood dream, the Mercury Seven astronauts were blazing the trail into outer space. Their exploits excited him with the same passion he experienced as a child watching movies of wartime carrier landings, and he began to wonder how he might also become an astronaut.

Although Cernan believed the space race to the moon would end before he qualified, he nevertheless formed a new dream of riding rockets into space. As he later recalled, literally three weeks after Alan Shepard became the first American in space, President Kennedy challenged the country to land on the moon, even though Shepard "flew little more than a hundred-plus miles in space, and not more than fifteen or sixteen minutes." Alan Shepard then became Cernan's new hero. "The whole world was watching America's response to the Soviets then, and I remember after Al went up, someone said, 'Gene, how would you like to do that?' And I said, 'I'd love it—boy, just give me a chance, I'd love to do it.' But by the time I'd get ready . . . there won't be anything left to do."

He next enrolled in the Naval Postgraduate School in Monterey, California, seeking a master's degree in engineering. By the time John Glenn became the first American to orbit the earth in 1962, Cernan had developed serious thoughts about becoming an astronaut but still felt he needed much more experience to qualify. Fate intervened, when the navy contacted Cernan asking if he was interested in flying in Project Apollo. He could not believe his luck and was eventually selected as a NASA astronaut in October 1963. His astronaut career would eventually include a Gemini mission and two Apollo missions, and he had not even applied for the job.

Accounts from books written by Cernan, Gene Kranz, Deke Slayton, and others suggest that Jack Schmitt was a geologist's geologist—intelligent but practical; competent, brash, and aggressive; field savvy; ultimately a team player; a punster; one known to throw down a few drinks at parties; and much to the disdain of some pilots, a man who loved to talk about rocks. *Apollo 15* LMP Jim Irwin wrote in *To Rule the Night* that Schmitt "would lecture [about geology] to anyone who would listen." But Cernan also described Jack Schmitt as a person "tightly focused and spartan in his personal life." Seemingly, Schmitt's love of geology and the prospects of applying his geological skills on the moon brought out the other side of his personality known to many of his NASA colleagues. Like most good geologists, he enjoyed sharing his knowledge with others. One Apollo astronaut remembered that Cernan and Evans were somewhat tough on Schmitt during training because of his scientific background, although if so there seems to have been little if any negative impact on crew performance. There is no doubt he was passionate about his science; in November 2008 at the Kennedy Space Center (KSC) Alan Bean and Charlie Duke recalled Schmitt's enthusiasm but did not label him as being overzealous. Notwithstanding, good geologists like Schmitt are known to carry a metaphoric soapbox along with them wherever they go and are always willing to climb on it to preach their gospel.

Years later Schmitt ran a field trip to Space Center Houston for the 2006 American Association of Petroleum Geologists (AAPG) convention. Following a day of revisiting his old friend CM *America*, geologizing about lunar rocks, and attending expert presentations on lunar geology, Schmitt led the group on a personal tour of the Johnson Space Center (JSC). During the stop at historic Mission Control, from where all the lunar missions were directed, the young regular tour guide was almost speechless when asked if a moon walker could say a few words. Schmitt rose to the occasion as if he were back describing weather patterns on his flight to the moon, but after about ten minutes one could clearly see the tour guide fidgeting, eventually interrupting the twelfth man to walk on the moon to inform him that they had to leave, as the next group of tourists was becoming impatient with this unexpected delay. Surely there is a guide at JSC today telling his tour groups about the time he kicked "Dr. Rock" out of Mission Control!

Born into a family of explorers, Schmitt's selection into the astronaut corps and subsequently to the crew of *Apollo 17* should not have come as a surprise. His great-grandfathers on both sides were part of the American movement westward. His father, Harrison Ashley Schmitt, was a successful mining geologist in the southwestern United States. The younger Schmitt was born on 3 July 1935 in Santa Rita, New Mexico, a state he would later represent as a United States senator, following his tenure as an astronaut. Schmitt assimilated his father's "do anything and everything" attitude at an early age, which undoubtedly played a significant role in earning him a seat on *Apollo 17*. Growing up in the sparsely populated Southwest, he learned to hunt, fish, hike, and, most important, love geology.

Schmitt often accompanied his father on visits to the various mines where he worked. At eleven years of age he would meet his future professor Lee Silver, who, in *Our Man on the Moon*, recalled an evening at the home of the elder Schmitt: "I was a young graduate student in geology from the University of New Mexico at the time, working that summer for a Caltech geology alumnus, Vincent Kelly, and I went with him to the Schmitt home. I sat there in the front room listening to those two distinguished geologists and tried to absorb as much as I could. And there, sitting quietly in a corner of the room, was this young boy doing the same thing." That same young boy would one day convince Silver to help instruct the astronauts in field geology.

Schmitt was active in high school playing tennis and football and taking a lead role in the senior play as well as being president of the student council. He applied to only one university, California Institute of Technology (Caltech), forgoing an application to Princeton because of the lengthy form; fortunately, Caltech accepted him. Initially, he was interested in physics but soon made the transition to geology, which ultimately proved to be a life-changing decision. His time spent with his father paid off handsomely, giving him a considerable advantage over his peers; he consequently gravitated more to the graduate students than his fellow underclassmen.

During his senior year Schmitt applied for a Fulbright Scholarship as well as a National Science Foundation grant. He won both, but he chose the Fulbright and a stint at the University of Oslo conducting extensive geological fieldwork. The year in Norway taught him how to apply his knowledge acquired at Caltech, skills that undoubtedly served him well in his years working for NASA.

Following his return to the United States, Schmitt worked on his doctorate at Harvard and honed his geological field skills with the United States Geological Survey (USGS) in New Mexico, Montana, and Alaska. He was granted his PhD degree in 1964 and subsequently wrote Eugene Shoemaker at the USGS astrogeology group in Flagstaff requesting a job. Schmitt had met Shoemaker a few years earlier and was interested in his cutting-edge mapping of the moon. Shoemaker hired Schmitt and with uncanny foresight made him project chief for lunar field geological methods, an area in which Schmitt would make significant contributions working at NASA. Then the space agency selected Schmitt as a scientist-astronaut in June 1965 along with five other candidates.

Ron Evans was selected in the fifth group of NASA astronauts in April 1966. An accomplished naval aviator with over a hundred combat missions in Southeast Asia, Evans was quiet natured like Schmitt. Much to his surprise, he learned of his selection to the astronaut corps while on an aircraft carrier immediately after a bombing mission over North Vietnam. Most of the early astronauts were selected for the space program before their customary career path would have taken them to aerial combat in Vietnam; Evans was one of the few who did both. Born 10 November 1933 in St. Francis, Kansas, Evans graduated from Highland Park High School in Topeka and was subsequently awarded a degree in electrical engineering from the University of Kansas in 1956. He then achieved a graduate degree in aeronautical engineering from the Naval Postgraduate School in 1964.

While Evans would not get a chance to walk on the moon, his role in its exploration from lunar orbit would be no less significant than that of his two crewmates. Known as a smiling, often joking family man—his wife, Jan, told reporters before the flight that her husband did not have a worry in his body—he openly confessed that he would have loved to walk on the moon. "You're darned right I would," he said in a pre-launch press interview. "But I'm not crying sour grapes just because I can't go down to the moon, and I feel very fortunate that I even had the opportunity to get within 80,000 feet of the moon." Schmitt and Cernan dubbed him "Captain America" as a tribute to his patriotism and position as landlord of the CM.

Most astronauts on the Mercury, Gemini, and early Apollo missions strongly resisted science experiments on their flights. Each mission brought unique objectives and challenges necessary to develop the difficult and complicated techniques required to send men to the moon and return them safely, consistently pushing the edge of the envelope. These missions were risky, and unnecessary risks might have resulted in an accident that could have terminated the entire program or, worse yet, killed someone, or both. As Wally Schirra wrote in *Schirra's Space*: "When the original crew of the first Apollo was lost, I became deeply involved in deciding where we were headed. And when I realized we would try again with me in command, I resolved that the mission would not be jeopardized by the influence of special interests—scientific, political, whatever. I was annoyed by people who did not consider the total objective of the mission."

The "J" missions—Apollos *15*, *16*, and *17*—were, at least on paper, intended to focus on science, but shifting the mind-set would not be easy. Schmitt fully accepted that his chances of flying to the moon were slim at best, but he was willing to take that risk, and if he could not go there to ply the skills he had been amassing since childhood, then he would help prepare those who did make the cut. He would also assist in developing the program-critical Apollo Lunar Surface Experiments Package (ALSEP) but recognized that this experience was still not enough to place him on a crew. If he had a chance of reaching the moon, he needed to do more, so informally he took on additional tasks, such as working with Gene Shoemaker as coinvestigator to set up a lunar field geology program, training pilots in what to look for as they traversed the lunar surface, and collecting samples. "So my objective for much of that time," he explained, "was to make sure that these very bright, good observers who were pilots and test pilots became field geologists—to the degree that you can—without having ten, fifteen years experience of actually doing it."

Initially, the astronauts were taught basic geological concepts. There are two qualities required to be a competent, well-rounded field geologist: strong observational skills and the ability to take those observations and reconstruct the geological history of the field area, or—as geologists like to say—to tell the story. In reality it takes years to become a competent field geologist, and there simply was not enough time, so NASA focused on making the pilot astronauts good observers.

"Our early attempts were not as successful as we would have liked," Schmitt reflected. "Probably because we were running things like an elementary geology course, and we simply didn't have the time to turn the astronauts into geologists. So we had to figure out a way to turn them into reliable geological observers, people who could report accurately back to the geologists on Earth what they were seeing on the moon."

Geologists, probably more than other scientists, have to be creative. There is an old running joke about geologists: if you put five of them in a room and give them a puzzle to solve, you will get at least six answers. Unraveling complex geological terrain can be compared to reconstructing the events of an entire football game with only a handful of randomly selected plays and the final score all viewed with a fuzzy black-and-white television. Geological conundrums are rarely resolved quickly and usually only after numerous iterations of potential solutions. The theory of continental drift on Earth, for example, was first depicted by Antonio Snider in 1858 and was not fully resolved by new plate tectonic theory until the 1960s. While much would be learned about the moon during the Apollo missions, it would be foolish to believe anyone could unlock all of its hidden secrets during the short life of Apollo. The task of assimilating all the observations made on the moon would be done by the thousands of Earth-locked, backroom geologists when time was no longer a premium.

Schmitt also had the foresight to enlist some of his former professors, in particular the man he had first met in his childhood home, Lee Silver. With the help of numerous expert geologists, Schmitt now tailored geological training to the program needs. "And the way that we did that, ultimately, starting really with the *Apollo 15* crew, but some in the Apollo's *12* and *11* crew as well was to actually simulate missions out on geological sites. And it turned out it worked out very well. It kept the attention span up very high with the pilots, and they learned a lot of geology, as well as the procedures that they would have to use on the moon for collecting samples, taking photographs, making observations."

Some astronauts were more willing to endorse geology than others, while some had more natural ability for observational geology. Wally Schirra, in spite of his resistance to science experiments on his *Apollo 7* flight plan, was a good student. This was best exemplified at an autograph event in 2006, when he was handed an ordinary rock by pad leader Guenter Wendt as part of a humorous "gotcha." Later, during dinner, Schirra pulled the rock from his pocket, presented it to me (who was in on the joke), and, standing tall, correctly identified the rock. After all those years he still remembered his geology.

Schmitt had ample compliments for Neil Armstrong's geological skills. "Oh, Neil was a wonderful observer," he recalled. "We didn't have as much time with him, but he was a natural observer—probably the only one, really, of the pilots that you could say was a natural observer." Schmitt complimented Armstrong's geological skills during the 2006 AAPG Field Trip at Space Center Houston. He said that when Armstrong moved off camera during his lunar EVA, he collected a variety of rocks but then had the foresight to fill in the gaps between the rocks with lunar regolith, providing the best regolith sample from any mission and coincidentally an important additional data point to map helium-3 distribution on the moon. Schmitt believes that there are sufficient helium-3 deposits on the moon that it can be mined and transported to Earth, by which time promising research on nuclear fusion will provide an economical energy resource to power our expanding world.

Small pieces of ancient lunar crust, called anorthosite, were also found in Armstrong's sample, giving scientists confidence that large samples of primitive crust might be found elsewhere, eventually proven when Dave Scott and Jim Irwin discovered the so-called Genesis Rock during the *Apollo 15* mission. Remarkably, particles of orange soil were also found in Armstrong's regolith sample, portending a significant discovery that Schmitt would later make on *Apollo 17*. The value of cross-training pilots as scientists was already returning enormous dividends.

Schmitt also spent the better part of a year in 1965–66 learning to fly after being selected to the astronaut corps. His geological knowledge and field experience would be all for naught if he could not show NASA that he had at least a pinch of the Right Stuff. Like all highly motivated people, Schmitt tackled this task with determination and enthusiasm and learned to fly both jets and helicopters, highly impressive skills for an academician trained to sample terra firma rather than land on it. Schmitt says that he not only had to prove himself as a jet and helicopter pilot but show that he could also operate the spacecraft as well as any of the pilots. "I think that I did that," he recalls modestly. "And so when it came down to NASA headquarters basically telling Deke Slayton that he had to put me on the *Apollo 17* crew, Deke didn't have any problem with it."

There is, however, a touch of contrariety in Schmitt's statement. As the newly appointed director of the Manned Spacecraft Center in Houston at that time, Chris Kraft remembers Schmitt's inclusion in the *Apollo 17* crew a little differently in his autobiographical memoir, *Flight*. "Deke was livid," Kraft wrote. "He never trusted those scientists. They didn't have the test-pilot gene and he was blunt to their faces . . . about the chances that any of them would get a flight."

If Kraft's memory was correct, then Slayton must have softened his position many years later. In his own memoir Slayton states: "I didn't have anything against Schmitt. . . . And I had him ranked right where he should be—behind Joe Engle. . . . I thought it would be nice to send a geologist to the moon [but] I had at least one guy I thought was more qualified."

Meanwhile, Cernan continued preparations to get all this knowledge to the moon. Granted, the public had lost interest after the first or second landing, but the training and logistics of flying to the moon were staggering, and it was still dangerous. Cernan was also involved in the geology training that his LMP was spearheading, while at the same time Schmitt had to learn how to be a proficient LMP. "Even though Schmitt started with the advantage of scientific training and vocabulary," acknowledges Bill Muehlberger, *Apollo 17* principal investigator, "Gene Cernan has done a remarkable job of closing the gap."

There were some—Jim McDivitt, for example—who challenged Cernan's piloting skills and worthiness of a lunar mission. Cernan crashed a helicopter into the Indian River in 1971 while flying low and allegedly showing off, nearly killing himself and doing little to improve his chances of garnering a command position, much less another flight. Kraft quotes McDivitt in *Flight* as saying, "Cernan's not worthy of this assignment; he doesn't deserve it, he's not a very good pilot, he's liable to screw everything up, and I don't want him to fly." Fortunately for Cernan, Slayton had sufficient confidence to stick with him and assigned him as commander of *Apollo 17*; oth-

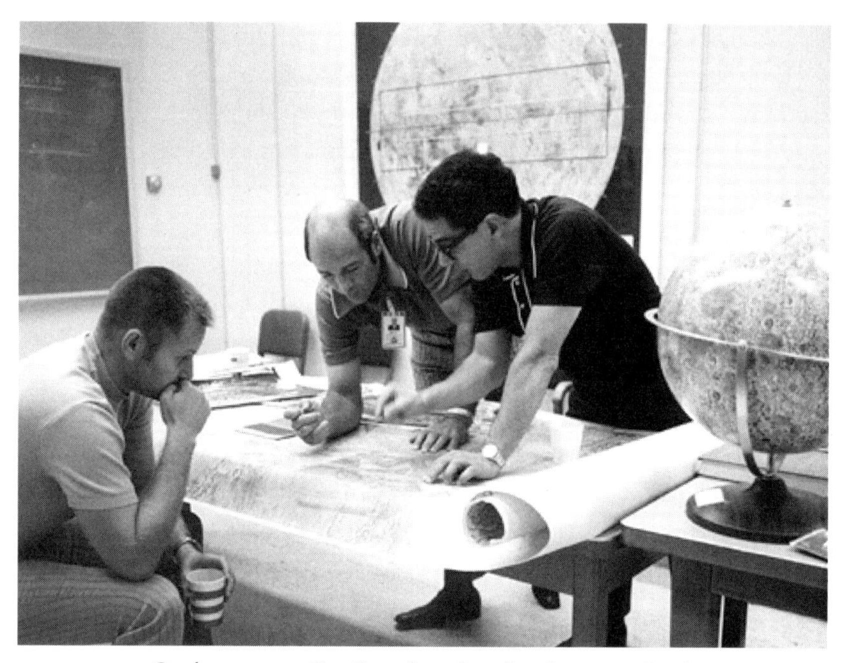

41. October 1972: CMP Ron Evans (*center*) studies a lunar mapping chart together with *Apollo 17* support crew member and mission CapCom, Bob Overmyer (*left*), and geologist Farouk El-Baz. Courtesy NASA.

erwise, Schmitt may have walked on the moon on that mission with Dick Gordon, once slated as commander of the canceled *Apollo 18*.

Evans spent considerable time separated from Cernan and Schmitt doing his own training for orbital activities, working with his own geological experts such as Farouk El-Baz, Dick Laidley, and Jeff Warner to ensure that he was well prepared to carry out his load of the science. El-Baz, principal investigator of the *Apollo 17* Visual Observations and Photography Experiment, recently stated that "Ron Evans was a keen observer. During his training cycle, he always spent extra time . . . constantly asking good questions . . . determined to do the job to the best of his ability. His training sessions were the longest among the CMPs [and] reflected his quiet but fiercely competitive nature."

This business of going to the moon was a full-time job and required long hours and full commitment, plus the ability to keep the whole shebang in perspective and not let it go to one's head. Cernan's daughter Tracy, then six years old, summed it up quite succinctly for him soon after he returned from his first trip to the moon, *Apollo 10*. One day she said: "Daddy? Now that you have been to the moon and back, are you going to take me camping like you promised?"

In addition to Schmitt's role in training the pilot astronauts, his geological background would be tremendously valuable on the lunar surface. But as Bill Muehlberger pointed out, he would also be a unique resource capable of assimilating what he and his fellow astronauts observed on the moon once he returned to Earth. "It's in the post-mission interpretive session that we hope Schmitt's long scientific experience will pay dividends," he stressed at the time. Indeed, Schmitt spent considerable time in the Lunar Sample Laboratory at JSC examining and pondering lunar samples, both before and after his mission.

A well-trained field geologist could also make other creative contributions. For instance, ground-based geologists wanted panoramic photographs of the sites sampled on the lunar surface; an experienced geologist could deliver even more. According to Muehlberger, it was Schmitt "who thought up the idea, instead of just standing at a point and pivoting around, [of] actually stepping sideways when you stepped [making] each pair of pictures that you took a stereo pair, so you could use that for near field as well as worrying about the distant things. Neat idea."

Schmitt was also involved in selecting the landing site for *Apollo 17*, improving the ALSEP instrument package, developing various tools such as shovels and scoops to be used by all the Apollo crews to simplify the collection of samples while wearing the bulky space suits, and ensuring these tools had storage room on the descent stage. It seems odd, but Schmitt seldom wielded the tool that nearly every geologist owns and covets, the geology rock hammer, even though he helped develop the one used on the moon. He explained this curious anomaly in the *Apollo Lunar Surface Journal*:

Although I had a lot to do with the design of the hammer, the handle was too big around for me. I couldn't grip it, and that was one of the reasons why Gene was the one who carried and usually used it. I had convinced myself that I would generally do better with the scoop. Also, Gene could put a lot more force into the hammer. He could grip it, and he also had longer arms. The thing it took Gene a long time to learn was that you don't hit rocks in the middle to break them. You hit them on the edges and pieces break off much more quickly. By the time we got to the Moon, he was pretty good at it. But I remember that, early on in training, he would go after a rock right in the middle of it.

Schmitt also put together a small group to evaluate the possibility of landing on the far side of the moon, a bold and risky proposal not particularly popular with management. Schmitt was determined not to give in easily and sold the recommendation hard, but according to Schmitt, Chris Kraft casually stopped him one day and said, "Jack, it's not going to happen." That settled the proposition.

The prospects for recent volcanic activity at Taurus-Littrow were initially spotted by Al Worden during *Apollo 15* and were attractive because of the potential to date when the moon had become geologically inactive. Schmitt liked the site because of the alluring scientific possibilities. Cernan liked it because of the challenges it presented for landing the LM in "an unexplored box canyon." He also knew he was up to that challenge. "I'm arrogant as hell, and I knew I could [land] better than anyone else. You had to have [that] arrogance, you had to know you could do better than [anyone else]." During post-mission comments Cernan admitted, "By the same token, I had a lot of personal pride that, if we had lost three-quarters of our automatic systems, I still could have landed the vehicle safely."

Schmitt, in spite of his aggressiveness and determination, was a consummate team player. He recognized the importance of Mission Control personnel and made a conscious effort to get to know them better, during and after work hours. Flight controller Sy Liebergot stated in his autobiography *Apollo EECOM: Journey of a Lifetime* that most of the astronauts at the Manned Spacecraft Center recognized the value of Mission Control personnel but did not see the need to become personally involved with them. Schmitt, according to Liebergot, "was the exception. He spent more time in the Control Center than any other astronaut."

For his part Schmitt recalled the need for an amicable spirit of cooperation. "So, I had a real good relationship with the Mission Control folks. We went out to the Singing Wheel together often . . . for barbeque and shuffleboard. And it was a really good crew. I really enjoyed my time with them." As Liebergot recalls, "The Wheel was *our* [flight controllers'] watering hole" located in Webster, Texas, near the space center. "It was a great place to eat . . . drink beer [and] have fun." And perhaps more important, a place to blow off steam and build bonds with fellow flight controllers.

Clearly, if a scientist was going to make a lunar landing, Schmitt had an advantage as the only geologist among the six original scientist-astronauts, and his relentless work in training all the astronauts in geology certainly did not dampen his chances. According to *Apollo 18* commander Dick Gordon, Schmitt's progress as a more than competent LMP did not hurt his chances either. As Cernan stated in his autobiography, "It had been no secret that Schmitt's star had been rising as the Apollo flights were trimmed back." He remembers when Schmitt was assigned to the *Apollo 17* crew: "Word came quickly to me from Dick Gordon that Schmitt could do the LM job, and I knew Deke would never have accepted him on the crew if Schmitt wasn't able to carry his share of the load."

The men whom NASA selected as astronauts were not only individuals with unique personalities but also strong team players dedicated to a primary objective—to accomplish the mission and return safely. Slayton was known to consider individual personalities when assembling his crews, but he also avowed that any astronaut was capable of performing any job. Most space enthusiasts remember the crew of *Apollo 12* as the most tightly knit, fun-loving team of all the lunar landings, whereas in his book *Carrying the Fire* Mike Collins described the *Apollo 11* crew as "amiable strangers," somewhat reserved and even cold at times. Yet in spite of their differences, both crews pulled off spectacularly successful missions. The *Apollo 17* crew would fly an equally impressive mission, but initially the two men chosen to land at Taurus-Littrow did not form an easy alliance.

The two moon walkers chosen to explore the valley of Taurus-Littrow had ample reason to be the least cohesive team of all the lunar landings. Before *Apollo 17* the crews of every NASA spaceflight were made up of men who had dedicated much of their lives to flying; Cernan and Schmitt were the first integrated crew, coming from vastly different backgrounds. Cernan hailed from the blue-collar Midwest, a highly talented naval aviator dedicated to the service and protection of his nation. Schmitt, on the other hand, was an academic—a student from Caltech and Harvard, the PhD who studied rocks. Then, in an unprecedented move, Schmitt had been thrust onto the *Apollo 17* crew at the expense of the amiable Joe Engle, which initially infuriated Cernan.

Cernan admits in his autobiography that Schmitt may have rightfully earned the privilege of exploring the moon, but a good commander protects his charges at all costs, and Cernan fought hard to keep Engle on his crew. But the die had been cast at higher levels, and Cernan knew he either had to accept Schmitt or the mission would be taken from him. Additionally, Cernan makes it clear that at first he and Schmitt just did not mesh. "On a first introduction, he usually came across as unlikable," Cernan wrote in his autobiography, "and his taciturn nature and brashness made it hard for people to get close to him. He didn't seem to care a whit. That was part of our problem, for Jack just wasn't my kind of guy."

There is an additional complication in crew makeup that bears scrutiny. Cernan regretted not having a navy command, and Apollo was his last chance to fill this gap. Before Engle was removed from his crew, Cernan had his wish, a command position and—perhaps even better—two rookies to nurture and protect. After Schmitt came on board the dynamics changed a bit; Cernan was still the commander with two rookies—until they landed on the moon, where Schmitt could not be considered a rookie. He would now be in his own element doing field geology, where he was usually in charge. What did this do to Cernan's confidence? Could he control this rock hound? Was he nothing more than a chauffeur to take the geologist to the moon to do real science at last? In *The Last Man on the Moon* Cernan recalls that some geologists believed he should become subservient to Schmitt once on the lunar surface. But good commanders never relinquish their authority, especially in such a perilous environment. "Not bloody likely," Cernan wrote. "I would give Schmitt great latitude, for I trusted him without question, but there was too much at stake, and my job was to see that everything was completed as successfully as possible."

In spite of these challenges, both men recognized each other's strengths and the opportunity to have *Apollo 17* remembered not only as the last Apollo flight to the moon but for its scientific discoveries. Cernan and Schmitt would eventually crystallize into an extraordinarily strong team.

Time has a way of softening the edges of the memory, and regardless of the rocky start to their bonding as a crew, Cernan makes it very clear today that Schmitt was a great LMP. Because of his scientific background, Schmitt brought talents to the mission that pilot-astronaut Engle simply did not possess and, as Cernan acknowledges, contributed in no small way to the success of the mission. "The bottom line is that Jack worked hard, studied hard. Was he a good pilot? No. But was he a good Lunar Module pilot? Yes. He was an outstanding Lunar Module pilot. He just knew what needed to be done."

Cernan and Schmitt set down in the gaping valley of Taurus-Littrow at 1:54 p.m. (Houston time) on 11 December 1972, just two hundred feet from their intended destination. Meanwhile, now flying solo in lunar orbit, Ron Evans transmitted a delighted message: "*Challenger*, this is *America*. I watched you all the way. Looks great!" He then began to initiate the orbital science investigations assigned to him. Landing was a busy time; at one critical stage Cernan had to tell Schmitt to be quiet, although in Schmitt's defense he was just doing what he had been trained to do. "All I had to do was keep Jack Schmitt from talking," Cernan recalled with a laugh. "I said, 'Jack, I don't even want to hear it from you!'"

Schmitt was busy performing LMP duties immediately after touchdown, checking pressures, circuit breakers, and batteries before taking a quick peek outside at the new field site. "Oh, man! Look at that rock out there," he exclaimed. Cernan was already basking in the beauty of the new sights. "Absolutely incredible," he added. Cernan landed with 117 seconds of fuel remaining, prompting Schmitt to suggest, "We should have hovered around a little bit; gone and looked at the Scarp." Cernan replied "No, thank you. I like it right where we are." Schmitt responded with a smile and a laugh.

Based on crew feedback from earlier flights, it was felt that a rest period immediately after landing was not necessary, and so, as planned, the first lunar EVA began shortly after touchdown. Following his commander, Schmitt descended the LM ladder, ready to take those first steps onto his geological paradise. Before he had his moon legs, however, he fell and dropped their only pair of scissors, which promptly "vanished into the lunar soil." Fortunately, with the help of his expertly trained geological eyes, he located them a few minutes later. It seemed almost as if *Apollo 17* was cursed with a penchant for losing scissors.

Originally, Cernan and Schmitt had been assigned ten stations at which to ply their geological skills during three lunar EVAs, each of them exceeding seven hours, but several changes had to be made due to time lost setting up the ALSEP experiment package during the first EVA. For three Earth days boots, tools, Rover wheels, and occasionally the space suit of a toppled astronaut accosted the surface regolith of Taurus-Littrow, leaving behind a randomly disturbed landscape in exchange for more than 250 pounds of rock and soil, in excess of two thousand photographs, hours of fascinating video, and countless geological secrets. Their three traverses included craters Steno, Nansen, Lara, Shorty, Camelot, Van Serg, and the North Massif and Sculptured Hills.

Like the launch and coast to the moon, they encountered few problems on the surface, the exception being when Cernan accidentally snagged one of the Rover's fenders with the handle of his rock hammer during EVA 1 and ripped it off. Initially, he repaired the fender using good old duct tape, but with everything covered in greasy moondust, it did not hold for long. Later Cernan received instructions from Houston to use four sturdy geological maps to repair the fender. Was science being sacrificed? Hardly, according to Bill Muehlberger. Smiling, he said he and the crew had pored over the maps so many times at the Cape that "they knew it all by memory. Fact is, the main use for the geologic maps that they had on the moon was to replace the fenders they seemed to tear off!"

Listening to the two explorers go about their duties one would not immediately suspect that one was an expert geologist and the other a flyboy; they were both speaking a combination of NASA and geology jargon. Coincidentally, their voices, including accents, are similar enough that it takes a bit of practice to distinguish one from the other when listening to mission audiotapes—both voices rushing with excitement at each new surprise, such as the discovery of orange soil at Shorty. From the mission transcripts:

Schmitt: *Wait a minute—*

Cernan: *What?*

Schmitt: *Where are the reflections? I've been fooled once. There is orange soil!*

Cernan: *Well, don't move it until I see it.*

Schmitt: (Very excited) *It's all over! Orange!*

Cernan: *Don't move it until I see it.*

Schmitt: *I stirred it up with my feet.*

Cernan: (Excited too) *Hey, it is! I can see it from here!*

Schmitt: *It's orange!*

Cernan: *Wait a minute, let me put my visor up. It's still orange!*

Schmitt: *Sure it is! Crazy!*

Cernan: *Orange!*

Schmitt: *I've got to dig a trench, Houston.*

Who would have believed that a Right Stuff jet pilot such as Cernan could get so excited about orange dirt? It sounded like two kids on Christmas morning. "Geology was part of my job in this period of time in my life," he later mused. "And I think that means I was pretty well trained to be a geologic observer. I think that was true in the case of most everybody who went; but I had the advantage of having a professional geologist with me."

Meanwhile, Evans was consumed with his own exploration efforts aboard CM *America*. Housed inside a bay in the Service Module were instruments designed to take measurements of the thin lunar atmosphere, construct a high-resolution temperature map of the lunar surface, and collect data that would be used to make a geological map of the shallow lunar subsurface. Captain America would also snap over five thousand frames of fascinating high-tech photographs for the backroom scientists stranded on Earth. In addition, he made visual observations of the lunar surface, complementing Cernan and Schmitt's work. Following Schmitt's discovery of orange soil at Shorty, Evans was able to see large areas of orange tinted soil from orbit. Geologist Farouk El-Baz takes up the story:

At my location in the Mission Control Center, I began to think of ways to check if the discovered locality was unique. Based on my knowledge of Ron's sharp eye and keen observation abilities, I planned a scheme to ask him to (1): focus on Shorty Crater and locate the "orange soil" along its rim; and (2): check if there were similar sites elsewhere on the Moon along his visual track. The idea was if that soil occurred only at that site, it would strengthen the possibility of relatively recent volcanic origin. If it were omnipresent on the Moon, then we have to think of some other explanation.

The flight director resisted this plan due to mission time constraints and doubts that Evans could discern such a small area, but El-Baz insisted, confident of Evans's observational abilities. El-Baz, nicknamed "The King," won the battle and years later boasted of his prodigy: "Ron swiftly conveyed that he identified the orange spot on Shorty's rim . . . amazingly he went on to describe numerous sites of a similar color giving us exact locations. Some of these were on the opposite side of *Mare Serenitatis*, some 600 kilometers away from the *Apollo 17* site. This was the first and most significant hint that what Jack and Gene were sampling at the Taurus-Littrow site was not unique, but rather common on the Moon." Evans acknowledged one surprising fact. "You know, to me the Moon's got a lot more color than I'd been led to believe. I kind of had the impression that everything was the same color. That's far from being true."

Data from previous missions had already shown that the interior of the moon at one time consisted of molten rock but had since cooled and was no longer geologically active as the Earth is today. The orange soil held significance in that it suggested recent volcanic activity, which would help resolve when the moon had cooled. Evans's perceptive eyes, plus later analysis back on Earth, disproved the young volcanic origin of the orange soil. As El-Baz explains, "The color was due to the prevalence of zircon-rich glass beads as old as the [four-billion-year-old] host basaltic rock that formed the plain at the landing site," exposed when Shorty was excavated by an impact. Evidence of the recent volcanic activity that geologists had predicted prior to the first landing and still clung to when planning the *Apollo 17* mission continued to elude them.

It had been a magnificent team effort. Great team players in sports seem to be able to assess the situation quickly and decide when to take the shot or pass off to a teammate; this ability applies to teams of all disciplines. Cross-training may have been necessary and desirable during the Apollo program, but sometimes it simply made more sense for the expert to do the job, as Cernan explains: "You can't specifically train for every moment of the mission [so] you train for knowledge of each other and you train for teamwork. And, when you come up with something new and different, you just defer to each other's capabilities. In this particular case, Jack had pretty-much surrounded the orange-soil project . . . and had in his mind what we needed to get. So we went ahead and did it."

Schmitt also acknowledged these unwritten ground rules. "I think there have been others, but this [discovery of the orange soil] is a particularly good example of where Gene deferred to me in organizing how we would approach a geological problem, just as I would defer to him on the Rover stuff and things like that. It worked out pretty well; that had all been sorted out, almost without us knowing it, during the training cycle."

In fact, Schmitt played a central role in determining on the fly how each station would be explored. Bill Muehlberger was responsible for revising, if necessary, the established plan for each station and passing it on so the CapCom could ensure that each objective was carried out on schedule. But the value of having a savvy geologist on the surface is that he could quickly survey the situation at each station and describe his observations so that Muehlberger in the back room could also change the plans and pass them on via the CapCom. Muehlberger explains: "What we ended up doing was, on the moon Schmitt got off and ran around telling us what he was seeing, which meant that we changed [our plan]. In effect, he was running the mission from the moon. I was the official one, but what the heck? I can't see that stuff like he can. Besides that, he knows it better in the first place." Schmitt writes in *A Field Trip to the Moon* that "while Gene dusted off the equipment on the Rover, I would look over the sampling area, giving a general description of the geology and what we would try to do."

The two explorers made a great team, but nevertheless they had their foibles. Leave it to the geologist to be different; Schmitt devised an unorthodox method similar to cross-country snow skiing to navigate his way across the lunar surface, whereas Cernan used the now-traditional "bunny hop." Schmitt assessed his ability to negotiate the hummocky lunar terrain by explaining that, having spent time in Norway and learning cross-country skiing, he felt that it was better to run with a stride: "You could go six, ten kilometers an hour quite easily. If I had had a couple of ski poles, I could probably have got up to about fifteen and held that for some time. I think I could have outrun the Rover!"

Schmitt's negotiation across the barren landscape eventually earned him the name "Twinkle Toes" because of his frequent falls, most notably at Lara Crater. In Schmitt's defense his arms were fatigued and sore having had to grasp a faulty camera bracket tightly en route to Nansen so that it would not fall off. In addition he was working alone with a defective bracket intended to hold sample bags, and solo sampling was a learned exercise that he had not yet mastered on EVA 2. Inevitably, he stumbled several times and then fell on his chest while transferring samples from the scoop into sample bags. CapCom Bob Parker jested, "Be advised that the switchboard here at MCS has been lit up by callers from the Houston Ballet Foundation requesting your services for next season." Schmitt laughingly responded, "I should hope so," as he playfully tightroped on one leg for a bit before he fell again into a push-up position, eventually rebounding upright. Schmitt admits this was not his only tumble. "The first thing that I remember is that when I stepped down from the [LM] ladder—I think it was my left foot first—it got on the side of a rock with these little . . . beads of glass on it, and slipped. I can remember hanging onto the ladder while my foot was slipping off to the side."

Cernan, however, was never really concerned over his LMP's circus-like acrobatics. "This is not meant as a criticism, but I think that Jack tended to fall more than the rest of us did. And it's maybe because he became more aggressive. . . . You developed a personal technique—the way you skipped or hopped, the way you got on and off the Rover, the way you got up when you fell." Likewise, Schmitt never felt like he was at risk. "It had to do with just a long experience in using the suits, dealing with them, of knowing the people who made them, and how they were made. . . . I felt certainly that I could do almost anything . . . and everything would be alright. . . . I never even thought about it."

Upon returning to *Challenger* following EVA 3, both explorers had chores to accomplish before climbing back into their field vehicle and only ticket home, including work at the ALSEP, taking another gravity measurement, collecting a sample to test for descent engine effluents, as well as transferring rock samples for the flight home. Schmitt climbed back inside *Challenger* first and began dusting himself off, while Cernan eloquently spoke of the end of an era from the surface of the moon. "As we leave the moon and Taurus," he declared, "we leave as we came, and God willing, as we shall return, with peace and hope for all mankind." Then, as he prepared to lift his foot from the lunar soil for the final time, he added: "As I take these last steps from the surface for some time to come, I'd just like to record that America's challenge of today has forged man's destiny of tomorrow. Godspeed the crew of *Apollo 17*." He then ascended the ladder and the final rung of the Apollo lunar surface program.

Their time on the surface had set numerous records, including an astonishing twenty-two hours and five minutes of surface EVAS. Following a

42. This image of LMP Jack Schmitt is believed to be the last photograph taken of an Apollo astronaut on the moon. Taken by Gene Cernan at the conclusion of *Apollo 17*'s third and final lunar EVA. Courtesy NASA.

much-needed rest period, they began preparations to return to lunar orbit and dock with *America*, during which Schmitt claims that Cernan playfully suggested he climb out and photograph the launch. Just prior to liftoff Schmitt was busy fussing with a camera, and after realizing that the camera would not run unless he held it, Cernan told him with a degree of agitation: "Okay. Now, let's get off; forget the camera." The last words spoken from the lunar surface were "Three, two, one—ignition!" uttered by Schmitt as *Challenger* lifted off and accelerated upward on its second launch of the mission, in search of Evans and *America*. Instantly at ignition, all communication between *Challenger* and Houston evaporated into a loud static requiring all messages to be relayed via the CM for the four-minute interlude. Cernan repeatedly directed Schmitt to reestablish communications, but there was nothing that could be done; the glitch was later determined to be caused by interference from *Challenger*'s engine plume.

Although the mission was now in its final phase, important work remained. To accomplish it the crew stayed in lunar orbit for two additional days in order to complete the orbital observation program. Following the Trans-Earth Injection burn and eventual capture by Earth's gravity, Evans got his time in the spotlight during a heart-stirring sixty-six-minute EVA to retrieve three film canisters from the Service Module scientific instrument module (SIM) bay on the way home. Cernan was concerned for Evans's safety because of the nightmare of complete exhaustion he had encountered on his own *Gemini 9* EVA. Cernan cautioned, "Okay, babe, when you get out there, just take it nice and slow and easy. You got all day long." Schmitt added: "Nice day for an EVA, Ron. Go out and have a good time."

After exiting through the hatch of *America*, a tethered Evans made his way, hand over hand, along the blistered side of *America*. He first mounted a TV camera on a pole and then proceeded to traverse to the SIM bay to recover the three film cassettes. He seemed almost intoxicated with glee as he climbed back and forth between the SIM bay and *America*'s hatch to deliver the cassettes to Schmitt, skillfully harvesting the fruits of his labors. Chuckling no less than twenty-four times, humming "dum-de-dum" a dozen times, waving to his family on camera, and delightfully launching stray fragments of thermal insulation as he went about his business, Evans was exultant. "Hey, this is great," he beamed. "Talk about being a spaceman!"

In San Antonio in 2006, in a conversation I had with Alan Bean, he compared his own lunar EVA and his later Skylab spacewalk. "The lunar EVA was rehearsed so many times that although it was spectacular, it seemed normal," he explained to me. He also invoked the world of science fiction in describing his orbital EVA on Skylab: "There wasn't anything under me when I stepped out." Evans obviously shared Bean's elation, so when CapCom Bob Parker instructed him to terminate the spacewalk, he cleverly made an end run to avoid the inevitable by videoing the moon and Earth. He gained some extra EVA time, but after several minutes Parker again instructed Evans to get back inside. Then all that was left was the long drift home and then reentry. It was later noted that the spacewalk had taken place on the sixty-ninth anniversary of the Wright brothers' first powered flight.

Schmitt eventually found Evans's scissors on the way home but cheekily kept the discovery secret, delighting in his friend's forgetfulness. He had been working on something else at the time: "I was looking up behind a soft bag in the lower equipment bay, and what should I find in there but Ron's scissors floating up, hidden behind in there. Well, I threw a bunch of hand signals to let Gene know that I had found them. We didn't let Ron know . . . saying that the IRS were going to be very interested in what happened to those scissors when we got back! And it wasn't until a few months later at a flight controller party that we finally presented Ron with his scissors."

The mission ended with a gentle, pinpoint splashdown in the South Pacific, four hundred miles southeast of the island of Samoa, and was signaled by Evans saying, "Oh, we've got a tin can [aircraft carrier] with us," closing out the final chapter of the last manned mission to the moon in the twentieth century. "This is America and the crew is doing fine," Cernan reported as helicopters from the recovery carrier USS *Ticonderoga* rapidly headed their way. The crew was bringing back a record 258 pounds of priceless lunar rock and soil samples, which scientists would promptly hail as the richest bounty yielded by the moon.

The splashdown may have marked the climax to the final Apollo lunar mission, but Cernan says all three quickly became annoyed at *Apollo 17* being called "the end." "We got tired of hearing that," he asserted, "and I said hey, we are not the end—we are just the end of the beginning. Apollo is the beginning of a whole new era in the history of mankind. Not only are we going to go back to the moon, but we are going to be on our way to Mars by the turn of the century."

Unfortunately, Cernan's predictions would prove overly optimistic, but the synergies of two highly skilled aviators and an experienced field geologist provided an outstanding closing to the United States's Apollo program.

In *Return to the Moon* Schmitt summarizes the value of the Apollo missions: "A great beneficiary of Apollo has been and continues to be the science of the Earth, the planets, and the solar system." The lunar landings greatly enhanced our understanding of the moon; robotic missions and telescopes had brought much knowledge, but there were still many unknowns before Apollo, such as whether the moon was still geologically active, prone to earthquakes and volcanic activity, and, most important, how it was formed. Schmitt continues: "there came a first-order understanding of the origin and history of the moon. Debates related to specific questions about lunar origin and history continue." Although the origin of the moon remains an enigma, the opportunity for the Apollo astronauts to observe up close the geological relationships between the lunar highlands, dark Maria, breccias, impact melts, and heavily cratered landscape has allowed scientists to answer many of the questions regarding the moon's geological history.

Scientists commonly held three theories to explain the origin of Earth's companion: it formed from debris generated when our planet took a large impact by a Mars-sized object; it was an existing body captured by Earth's gravity; or it formed at the same time as the Earth. The impact theory currently garners the most support based on the age of the moon rocks and their composition; in general the moon and the Earth are the same age, and the rocks are made of the same minerals found on Earth, although small but significant differences exist. The orange soil that Schmitt discovered at Shorty Crater, however, challenges the impact theory.

As Schmitt explains in volume 1 of *Apollo 17: The NASA Mission Reports*, "The primary importance of pyroclastic glasses [orange soil] relative to lunar origin lies in their deep origins and in the composition of the absorbed volatiles on the surfaces of the small glass beads and devitrified glass beads . . . complicating the case for the moon being the result of a giant impact on Earth." Simply translated, this means that easily evaporable compounds, such as water and sulfur associated with the volcanic glasses, are chemically inconsistent with the impact theory; they would most likely have been vaporized from the heat of the impact. "The consensus of the scientists is still trying to make that giant impact work," Schmitt adds, "but I frankly don't think it's going to work over the long haul. I think we're going to have to have a different theory than that. The orange soil is right in the middle of that debate."

The oldest lunar rocks returned to Earth, about 4.6 billion years old, came from the lunar highlands. The presence of anorthosite in the highlands is solid evidence that the moon, similar to the Earth, was once molten and subsequently formed a crust, mantle, and possibly a core. Anorthosite, a rock composed primarily of the mineral plagioclase feldspar, forms when a mass of molten rock cools slowly. The darker and denser minerals solidify first and sink to the bottom of the liquid, leaving the lighter-colored and less dense minerals such as plagioclase on top, where they eventually crystallize to form a crust. Because the molten rock cools very slowly deep beneath the surface, where high temperatures exist, the crystals have a long time to grow large and are visible to the naked eye, in contrast to the dark, fine-grained basalts of the lowlands.

Following the formation of the crust about 4.4 billion years ago, the moon was subsequently pounded relentlessly by celestial bodies for the next 700 million years, an ongoing barrage known as the "Great Bombardment." The largest collisions violently thrust large masses of anorthositic crust outward and upward, forming the lunar highlands. Cernan and Schmitt collected samples of ancient lunar crust from the highlands of the North and South Massifs, accomplishing one of the key mission scientific objectives and adding tangible evidence to the growing database necessary to decipher the moon's early history.

Over the next billion years molten rock sourced from the remaining magma was forced upward from depth up through the crust to the surface through volcanoes and fissures filling in shallow depressions in the crust excavated by the large impacts. The molten rock quickly crystallized into fine-grained, dark-colored basalt; when subjected to lower surface temperatures, the crystals formed quickly, without adequate time to grow large, and consequently formed fine-grained rocks with crystals that cannot be seen with the naked eye. These basalts formed the dark Maria, leaving lighter-colored naked highlands protruding and beckoning for lunar explorers searching for ancient crust. Such basaltic rocks were gathered from the floor of the valley of Taurus-Littrow by Cernan and Schmitt and are thought to make up the bulk of its substrate, or base.

The volcanic activity finally ceased when the internal heat engine of the moon slowly died out, about three billion years ago. The higher-density basalts never reached equilibrium with the crust by sinking as they would have on Earth, most likely because the moon's crust had already cooled and resisted downward settling of the heavier basalts. These differences in density are responsible for the lumpy gravity anomalies known as mascons, which affect the track of spacecraft orbiting the moon.

Early lunar theories, based on maps made from satellite photographs and telescopic observations, held that many of the craters on the moon represented recent volcanism, suggesting that the moon had not completely cooled until late in its history. In fact one of the key objectives of Apollos *14* through *17* was to seek out evidence of recent volcanic activity, and the failure to find such evidence certainly implies that the moon has been geologically dead for billions of years.

Meteorites then pelted the surface to create more craters, breccias, impact melts, ejecta rays, and the stark landscape visited by a dozen twentieth-century sojourners. To this day Cernan vividly recalls this dark landscape. "[Taurus-Littrow] was a valley surrounded by mountains higher than the Grand Canyon. . . . It was the most magnificent desolation, as one of my colleagues used to describe [the lunar terrain]."

There are very few rocks older than several billion years on Earth due to plate tectonics and erosion, which continually recycle rocks, destroying old rocks and turning their remains into new ones. Therefore, geologists know very little about the Earth during the first 2 to 3 billion years of its existence. Plate tectonics and erosion do not exist on the moon; therefore, rocks 2 to 4 billion years old are common, providing a glimpse of this missing part of Earth's history. About 3.8 billion years ago the rate of bombardment of the moon by meteorites subsided to about that of today. Schmitt theorizes that there could be a connection between the termination of heavy bombardment and the beginning of life on our planet; "what might be most interesting to note," he proposes, "is the match between the appearance of isotopic evidence of life on the water-rich Earth with the end of the great bombardment, both occurring 3.8 billion years ago may have finally permitted simple, replicating life to form at the surface of the Earth."

The lunar material brought back to Earth, data collected by scientific experiments deployed on the moon, and the orbital observations will continue to be studied and pondered for years. A mere six landings produced 842 pounds of rock from 2,200 separate lunar samples available to study, the bulk of it stored in the Lunar Sample Laboratory at JSC, much of it still waiting to be analyzed. Literally, the surface has just been scratched.

More than forty years after humans took the first baby steps beyond our remarkable blue marble, what do the early space pioneers think of the decision to focus on the space shuttle and International Space Station (ISS) in lieu of moving on to a permanent colony on the moon and visiting Mars? Certainly many, including astronauts, believed that humans were destined to land on Mars by the end of the twentieth century. Were the space shuttles and ISS the appropriate building blocks for the next step beyond, or have

43. On 25 April 2000 at the National Air and Space Museum in Washington DC, Harrison Schmitt delivered a talk concerning the latest theories on the moon's origins and evolution. He continues to tour the United States and the world promoting a return to the moon and citing potential benefits associated with that initiative. Courtesy John Youskauskas.

they steered us away from the opportunity that Apollo presented? Schmitt obviously expected more progress: "I don't think very many people who were really active in the program, even with the cancellation of *Apollo 18*, *19*, and *20*, thought it would be over twenty-five years before we go to the moon again or go into deep space again. . . . I mean, it'd be ridiculous to have done all of this and then not do something else."

Schmitt adds in *Return to the Moon*: "It could be said, in light of subsequent history that for many, such hope was misplaced. Indeed the world and United States did not build on the promise of Apollo." He is actively doing his part to keep the promise alive and has ambitious plans to build a commercial enterprise to return to the moon to mine helium-3.

Cernan also laments the best space technology of today. "The problem is [the ISS] doesn't go anywhere," he acknowledges. "As valuable as the station is today, quite frankly it is not interesting and it is unexciting. The guys and the gals up there are doing one heck of a job. They are head and shoulders above us in terms of their capabilities—what they know and what they can do—if they are given a chance to do it. . . . We opened the door to the future, to knowledge, to the universe. And that's as far as it got—it just got cracked open."

Unfortunately, we cannot ask Ron Evans for his thoughts on NASA's current direction. He remained with NASA and served as backup CMP for the ASTP mission, later becoming involved in the development of NASA's space shuttle program, in which he was responsible for the launch and ascent phases of the shuttle flight program. He finally retired from NASA in March 1977 in order to become a coal industry executive. Tragically, he died of a heart attack in Scottsdale, Arizona, on 6 December 1990, at the relatively young age of fifty-six, and is survived by his loving wife, Jan, and two children.

The legacy of *Apollo 17* should not be that of the last lunar landing of the Apollo era but, rather, one of science, teamwork, and confirmation that scientists should be part of the first crews to return to the moon and onward to Mars. *Apollo 17* and subsequent Skylab flights proved that properly chosen and trained scientists are fully capable of performing duties required to carry out complex missions safely into outer space. The first flights back to the moon should carry geologists, and the first flight to Mars should also include additional specialists such as geochemists and biochemists to search for signs of hidden life, both fossil and living. Unfortunately, the current group of active astronauts who are likely to walk on the moon when NASA returns does not include field experienced geo-astronauts like Schmitt, although one, Mike Fincke, has a geological degree. Hopefully, NASA will seek field geologists like Schmitt in future groups.

Average Americans whose tax dollars paid to make John F. Kennedy's vision a reality and continue to support the shuttle and ISS do not really care much about the science or how NASA did the nearly impossible; they mainly want to know what it felt like to go to the moon. Only twenty-four men have made the fairytale journey to the moon; three made it twice, including the *Apollo 17* commander Gene Cernan.

Cernan and all the astronauts have most likely been asked the same questions over and over ad nauseam, and understandably many have tired of answering them. But with Cernan, even after nearly four decades of answering stock questions, it seems as if he is still trying to invent the right words. You can hear the excitement in his voice and see it in his face as he digs deeply into three days of distant memory to relive the quest of a lifetime.

Here we are, thirty years later, and the questions are how does it feel, what does it look like? Were you scared, did you feel any closer to God? They want to have the answer to the question of how does it feel . . . to stand on the moon, and look back a quarter of a million miles at the overwhelming beauty and majesty of our planet. It's an incredible, incredible experience. You know, you can climb the highest mountain, walk on the depths of the deepest ocean on this planet of ours, and you still are on planet Earth, you still are touching this star of ours in the heavens. But when you land on the surface of the moon, you get out of that spacecraft, all of a sudden I realize . . . I am no longer in motion in space as we did all the way to the moon. I am standing on another planet in this universe that is not Earth. I try to describe it in one sentence . . . to know what it is like to stand on the surface of the moon, let your imagination wander and put you on God's front porch, looking back home. That's what it was like to me.

Jack Schmitt addresses the question more pragmatically, simply comparing it to a trampoline. "Well, if you imagine what an infinite trampoline is like, then that is a little bit what it is like to walk on the moon." Schmitt focused more on the science and practical aspects of the mission, little of this "what did it feel like?" stuff for him. Even so, it seems unthinkable that a geologist would cast away his rock hammer prior to liftoff from the moon instead of bringing it home as a souvenir. After all, artist and LMP Alan Bean brought his home and uses it to add texture to his canvasses before he begins painting. Bean did not consciously bring his hammer home. It was hanging on his suit and not on the checklist of items to be discarded on the moon, so he simply forgot about the hammer during the hectic hours of liftoff preparation. During the AAPG-sponsored field trip that Schmitt led in 2006 the question was posed: "Jack, why did you throw away your rock hammer?" Instantly, without a hint of hesitation and with a glint in his eye, he fired back, ever mindful of weight constraints, "Bring back more rocks!" Alan Bean suggested that by the time of *Apollo 17* it is possible that the hammer had been added to the checklist of items to discard. Regardless, when Cernan prepared to throw the hammer away, Schmitt pleaded for the honor. Schmitt's response exemplifies the enthusiasm for exploration and science that he brought to the program. NASA's decision to include a geologist on a landing mission was undeniably an overwhelming success.

Almost eleven years after Kennedy challenged the nation to do the impossible, *Apollo 17* returned home, closing out what will undoubtedly be remembered as one of humankind's greatest achievements of the twentieth century. Time will tell if history judges the twentieth century for making those first steps or for detouring away from the obvious next steps.

11. Beyond the Moon

Colin Burgess

It is not man's evolution but his attainment that is the greatest lesson of the past and the highest theme of history.

George Macauley Trevelyan (1876–1962)

Twelve Americans make up one of the most elite groups in the history of our planet, as the only human beings ever to have walked on another world. Fully four decades after Neil Armstrong and Edwin "Buzz" Aldrin set foot on the pulverized surface of the Sea of Tranquility, declaring their epic journey a peaceful accomplishment "for all mankind," a whole new generation of people gaze up in wonder at the sight of a full moon and ponder when humans will once again be transported to the magical but inhospitable place Aldrin described that momentous day as "magnificent desolation."

In December 1972 the crew of *Apollo 17* lifted off from the Valley of Taurus-Littrow aboard LM *Challenger*, bringing to an end the Apollo lunar landing program. The frantic race to be the first to reach the moon may have concluded, but it had also sparked the beginning of human migration from our blue planet, once described by Russian rocket pioneer and visionary Konstantin Tsiolkovsky as our "cradle of humanity."

Three more missions to the moon should have taken place after *Apollo 17*, but savage congressional cuts in NASA's budgets for 1969 and 1970 had forced the space agency into a situation in which *Apollo 18*, *19*, and *20* had to be dropped, albeit with great reluctance, from its flight manifest. History will recall that while the resultant monetary savings were relatively small, the science lost as a result of this penny-pinching exercise was priceless. Over three additional lunar expeditions and using modified Lunar Rovers, a total of six men would have ventured out of their landing craft and traveled far afield on geological journeys of increasing complexity that would undoubtedly have been rich in scientific interest and value.

Apollo 18's Lunar Module was scheduled to land in Schroter's Valley, the site of intriguing transient lunar surface phenomena and possibly even volcanic activity. The two-man landing crew of *Apollo 19* would then have explored the collapsed lava tubes of Hyginus Rille. The most hazardous but ultimately beneficial mission of all would have been *Apollo 20*. The Lunar Module was to have been upgraded to allow the two-man crew a stay of up to six days on the lunar surface, enabling them to visit and excavate the floor of the immense Copernicus Crater.

But as the moon-landing program continued beyond *Apollo 11*, as more and more astronauts left crisp, lingering footprints in the lunar soil, so the American public was growing increasingly apathetic and began to query NASA's mandate. Many resorted to asking their congressional representatives why further millions were being spent in pursuing what they saw as a few miserable moon rocks, when more pressing earthbound problems were being starved of finances. By now people had become inured to the televised sight of playful astronauts bounding across the lunar terrain; the once-indomitable spaceflight fervor was clearly at an ebb, and the writing was clearly on the wall for Project Apollo. The United States had beaten Russia to the moon, the euphoria of conquest had abated, and the budgetary shears could now be unsheathed.

The end result was almost inevitable; NASA's last three planned lunar missions would be unceremoniously expelled from history, along with the space agency's ambitious but unfulfilled plans for other far-sighted and enterprising journeys beyond the glory that had been Apollo.

Joe Engle had led life in the fast lane as one of the nation's most respected and skillful pilots. Born into a farming community in Dickinson County near Abilene, Kansas, on 26 August 1932, by early 1971 Engle had just about done it all. He had flown some of the hottest planes in the sky and on no less than three occasions had piloted the mighty rocket-powered X-15 research aircraft to an altitude greater than fifty miles, thus qualifying him as a U.S. Air Force astronaut. As the backup Lunar Module pilot (LMP) for *Apollo 14*,

44. As the backup crew for *Apollo 14*, this was the crew that should have rotated onto the final lunar landing mission, *Apollo 17*. *From left*: LMP Joe Engle, CDR Gene Cernan, and CMP Ron Evans. Courtesy NASA.

he would normally have progressed under the rotation system then in place to the prime crew for *Apollo 17* and walk on the moon. Or so he thought.

At the age of twenty-three Joe (not Joseph) Engle had earned a commission in the U.S. Air Force through their ROTC program at the University of Kansas, where in 1955 he received a bachelor of science degree in aeronautical engineering. In October the following year he married the former Mary Catherine Lawrence; they would eventually have two children, Laurie Jo (born in 1959) and Jon Lawrence (born in 1962).

Engle entered flight training in 1957. Over the next four years he served as a fighter pilot operating with the 474th Fighter Day Squadron and the 309th Tactical Fighter Squadron out of George Air Force Base in California and from bases in Spain, Italy, and Denmark. Having received a personal recommendation from legendary test pilot Chuck Yeager, Engle was next assigned to the air force's Experimental Flight Test Pilot School (EFTPS) at Edwards Air Force Base, graduating as a member of Class 61-C in April 1962. That October, again with the personal recommendation of Yeager, he progressed to the advanced seven-month course at the air force's Aerospace Research Pilot School (ARPS), also located at Edwards. Here he would undertake formal and comprehensive military astronaut training as a member of Class 3, eventually graduating as a U.S. Air Force astronaut designee in May 1963 and ready to fly into space on any future military space research program.

The following month Engle was assigned as a test pilot with the Fighter Test Group at Edwards, and that same month he joined the X-15 research program. He would make a total of sixteen flights aboard X-15-1 and X-15-3 between October 1963 and October 1965 and on his final three flights would exceed the fifty-mile limit designated by the U.S. Air Force as the boundary of space flight, thus earning his astronaut wings. He achieved his highest altitude of 280,600 feet (53.14 miles) on 29 June 1965. At the age of thirty-two Joe Engle had achieved the distinction of becoming the youngest American astronaut to fly into space.

Fellow X-15 pilot Milt Thompson, who had been assigned to the research program at the same time as Engle, recalled him as being "an excellent pilot" and hard worker:

He really took the job as a test pilot seriously except occasionally when his exuberance overcame him. That happened on his first X-15 flight. After he had completed the familiarization maneuvers, he slow rolled the X-15. That maneuver really shocked the engineers in the control room. They did not immediately recognize it as a slow roll. They assumed the worst and thought Joe had a control problem. In a research program, the pilot simply did not add an extemporaneous maneuver to the flight plan. Joe was thoroughly chastised by [senior X-15 pilot and operations officer] Bob Rushworth after the flight.

Joe went on to become a straight arrow after that incident. In fact he became one of the best X-15 chase pilots in addition to being one of the better X-15 pilots.

In April 1966 NASA made life even better for Captain Engle when he was selected in the space agency's latest astronaut group. He had subsequently paid his dues in the astronaut corps; he'd worked hard and was eventually looking forward to what would undoubtedly be the pinnacle of his career and his life—he was going to walk on the surface of the moon. But fate still held a few sad ironies in store for the affable young test pilot from Kansas.

Two years earlier, in July 1969, Neil Armstrong and Buzz Aldrin had become the first men to walk on the moon, and NASA had accomplished humanity's most sublime, intricate, and ambitious engineering and scientific enterprise. While it was hardly "the greatest week in the history of the world since the Creation" President Richard Nixon had alluded to when greeting the crew on their return to Earth, the race to the moon had nevertheless drawn Americans together like very few events of the twentieth century. But the president, for all his public enthusiasm and slick hyperbole, actually cared little for spaceflight activity.

A Republican Party conservative, Richard Nixon did not believe his government should be spending large sums of money on the space program, or any social programs for that matter. Well before the first lunar landing some Republicans were calling for Project Apollo to be scrapped, but Nixon was shrewd and ignored their appeals. He had an ulterior motive; should the first lunar landing fail, the blame would be placed squarely on the previous Democratic administrations of Kennedy and Johnson. To that end NASA's administrator Thomas Paine (appointed to that position by President Lyndon Johnson) had not been replaced by a Republican appointee, as was the custom.

In June 1969 Nixon appointed Vice President Spiro Agnew to head a task force that would determine what directions the nation's space program should take during the 1970s and 1980s. Following the success of *Apollo 11*, the president began to cut the NASA budget quite savagely, while a public once addicted to Apollo grew ever more disenchanted and apathetic with each successive lunar mission.

Part of the problem, surprisingly, was NASA itself. In their public statements and literature they had portrayed *Apollo 11* as the culmination of a grand effort to get to the moon, rather than the beginning of a new era of exploration and scientific discovery. Other reasons for the public disquiet were the increasing pressures of poverty, social change, and the Vietnam War—financial and moral issues that were dividing the country. Both the Congress and White House noted the shift in public mood and kept it in mind when considering future space funding. As always, NASA presented an appealing target for budget cuts.

This anti–space exploration attitude was adequately summed up at the time by New York congressman Ed Koch, who declared, "I just for the life of me can't see voting for monies to find out whether or not there is some microbe on Mars, when in fact I know there are rats in the Harlem apartments."

Originally, ten lunar missions were planned—*Apollo 11* through *20*, which would have placed twenty men on the surface of the moon. In September 1968 Spiro Agnew's task force presented its report to President Nixon. Instead of taking any immediate action, the president let it sit on his desk for four months without making any comment. The report contained a number of interesting possible directions for the U.S. space program. The first option was a rapid program similar to Apollo, involving a crewed journey to Mars in 1983, which would require funding of nine billion dollars per year from 1980 onward. The final decision would have to be made in 1974. The second choice was a Mars expedition to leave Earth in 1986. This would cost eight billion dollars per year from 1980 onward and would require a commitment by 1977. The final option would defer a Mars flight until the year 2000, with a decision to proceed needing to be made by 1990.

One of the main elements in each of these plans was the development of a reusable space vehicle that would make routine flights to an Earth-orbiting space station—the genesis of the space shuttle. Another requirement for the Mars flight was a rocket powerful enough to send humans and their supplies on the journey. A team headed by Wernher von Braun had drawn up plans for a nuclear-powered rocket engine called *Nerva*, which would provide sufficient thrust for both the shuttle and a Mars vehicle.

In October 1969 a revised launch schedule set the *Apollo 19* landing back to November 1972 and *Apollo 20* to May 1973. Between the *Apollo 18* mission (then planned for February 1972) and *Apollo 19* three orbital flights were also tentatively on the manifest. They would be training flights preparatory to the establishment of the first U.S. space station in 1975. This program went by the inauspicious title of Apollo Applications (later changed to Skylab), and it provided for a short-term, Earth-orbiting space station, fundamentally assembled using leftover hardware from the lunar landing program. At the time, Flight Operations director Chris Kraft from Houston's Manned Spaceflight Center (later renamed the Johnson Space Center [JSC]) expressed some concern at the busy schedule. "It's going to be difficult," he declared, "to handle both Apollo and Apollo Applications from an operational point of view as well as a people point of view in 1972."

A further reason for delaying the later Apollo landings was a possibility they would be made in remote lunar areas such as the large bright-rayed crater Copernicus, just south of Mare Imbrium, and the rim of Tycho, also a relatively young crater probably of impact origin in the southern lunar highlands. The Tycho Crater was rated by scientists as a top-priority landing site.

Meanwhile, the flight schedule continued. *Apollo 12*, under the command of Charles "Pete" Conrad, virtually mirrored the success of the first lunar landing flight, with a touchdown in the Ocean of Storms on 14 November 1969. Conrad and his LMP Alan Bean spent thirty-one hours on the moon's surface before blasting off to rendezvous with Dick Gordon in lunar orbit and returning home with around sixty-three pounds of moon rock and soil.

By 1970, however, NASA was struggling. The Nixon administration was pouring billions of dollars into the yawning maw of the Vietnam conflict, and "soft" programs such as Apollo were rapidly plummeting down the preferred funding list. It was a quandary for NASA administrator Tom Paine, who told a press conference about the need to align space operations with the 1971 fiscal year budget. "We recognize the many important needs and urgent problems we face here on Earth," Paine began.

America's space achievements in the 1960s have rightly raised hopes that this country and all mankind can do more to overcome pressing problems of society. The space program should inspire bolder solutions and suggest new approaches. It has already provided many direct and indirect benefits and is creating new wealth and capabilities.

However we recognize that under current fiscal restraints NASA must find new ways to stretch out current programs and reduce our present operational base. [NASA will] press forward in 1971 at a reduced level, but in the right direction with the basic ingredients we need for major achievements in the 1970s and beyond . . . we will not dissipate the strong teams that sent men to explore the moon and automated spacecraft to observe the planets.

In April that year the life-or-death flight of *Apollo 13* dramatically emphasized the colossal risks involved in manned space exploration. But with the crew safely back on Earth, public apathy set in once again.

Following the first three Apollo lunar missions, it had become abundantly clear to everyone close to the civilian agency that NASA was going to have to pare *Apollo 20* from the flight schedule due to budgetary woes. President Kennedy had been right when he prodded Congress to fund the moon race back in 1961; landing men on another world was a *very* costly exercise. Enough Saturn V launch vehicles had already been budgeted to see out the lunar landing program with *Apollo 20*, but Apollo Applications also needed a Saturn V to haul the massive station into orbit. Shortly after *Apollo 11* had lifted off and before the first historic landing had taken place, Paine signed off on the decision to switch Apollo Applications to a "dry lab" concept that would require a Saturn V launch vehicle. His announcement did not automatically mean the cancellation of *Apollo 20*, as hopes prevailed that there might yet be another Saturn V production batch.

But it soon became increasingly clear to Paine that Saturn V production would not recommence, and production of the smaller Saturn IB would also cease. Ultimately, he had no other option; *Apollo 20* was officially canceled on 4 January 1970. The Saturn V rocket, now freed up, would be allocated instead to the Earth-orbiting program. Even *Apollo 18* and *19* now began to look a little doubtful.

Had *Apollo 20* not been canceled, the prime crew would have initially included Pete Conrad (CDR), Jack Lousma (LMP), and Paul Weitz (CMP). Conrad, Lousma, and Weitz were now able to concentrate on full-time training for the space station. When asked about it for this book, Jack Lousma explained that, though the crew had not officially been named to the lunar flight, "it was understood informally."

When Alan Shepard's crew was assigned to *Apollo 14*, Gene Cernan, Ron Evans, and Joe Engle were also named as their backups. Using the backup/skip-two formula, Engle now realized the three of them were next in line for *Apollo 17*. Shortly thereafter, the prime crew for *Apollo 15* was announced as Dave Scott (CDR), Jim Irwin (LMP), and Al Worden (CMP).

Pete Conrad, who had already walked on the moon on *Apollo 12*, might have been the only man to set foot on the lunar soil twice had *Apollo 20* gone ahead. But even before the flight was scrubbed, Conrad had decided it was a long shot anyway and instead threw himself into training for the space station. Informed speculation is that his place as mission commander on *Apollo 20* could have been filled by *Apollo 14* CMP Stu Roosa. Conrad desperately wanted his *Apollo 12* CMP and good friend Dick Gordon to join him in the space station program, but Gordon had decided to tough it out in the hope that *Apollo 18* would still fly. He wanted to walk on the moon this time, not just fly around it. He and his crew had been assigned as backups for *Apollo 15*, which placed them squarely on the agenda for *Apollo 18*.

Then, in the summer of 1970, the budgetary ax dropped yet again. On 2 September 1970 *Apollo 15* and *19* were officially canceled. The flight designated *Apollo 16* retained its original lunar destination of Descartes Plain but was renamed and became the "new" *Apollo 15*, together with the original *Apollo 15* crew. *Apollo 17* now became *Apollo 16*, while *Apollo 18* was renamed *Apollo 17*—and was announced as the final manned lunar flight.

At the same time, the recommendations of Agnew's task force were shelved. After the enforced gutting of the space agency's manned and unmanned programs, directly attributable to the White House, Paine resigned from NASA. Another result of the wind-down in space activities was the human cost. From a peak nationwide employment of 400,000 people in 1965 to 180,000 five years later, the decimation of the space program had a devastating effect in the states that had built up large aerospace industries around the Apollo program.

The states most affected in this latest crisis were, however, electorally important to President Nixon for the upcoming 1972 election. California, Florida, Texas, and other southern states were being targeted by the White House in order to return Nixon to office. As these states were being hit hard by the space recession, it was decided to unveil another large space project. Thus, in January 1972 Nixon announced that he had given approval to NASA to begin development of a space shuttle that was expected to fly in the mid to late 1970s, providing a means for the country's continued human exploration of space until the end of the century.

A deep gloom still prevailed at NASA over the savage cutbacks to the lunar program. But had *Apollo 18* and *19* flown as originally scheduled, the crews would have been, respectively, Richard Gordon (CDR), Harrison Schmitt (LMP), Vance Brand (CMP) and Fred Haise (CDR), Jerry Carr (LMP), and Bill Pogue (CMP).

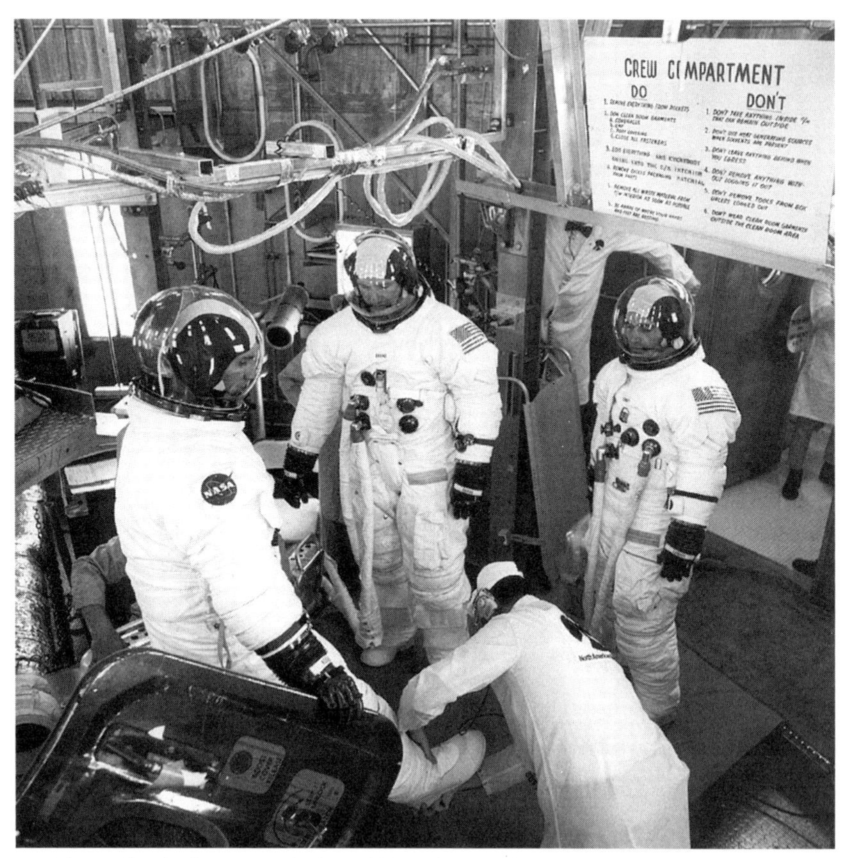

45. The *Apollo 15* backup crew, which would have flown to the moon had the *Apollo 18* mission not been canceled. *From left*: Richard Gordon, Vance Brand, and Harrison Schmitt. Courtesy NASA.

Once again, the crews were never officially announced, but when Haise, who would have commanded *Apollo 19*, was asked to confirm his role, he replied:

When I was assigned after Apollo 13 *as the backup commander to John Young [*Apollo 16*], it was the start of my training cycle to eventually fly* Apollo 19. *I had Bill Pogue assigned as the* CMP *and Jerry Carr as the* LMP. *However, in September of 1970 the last two missions were canceled which was another disappointment for me [after* Apollo 13*]. At that time, Stuart Roosa and Ed Mitchell were assigned to join me in fulfilling the "deadhead" backup crew assignment. That is a long way of saying that I, nor anyone [else], actually started the specific training on the* Apollo 18 *or* 19 *missions.*

It was further decided that a Saturn V would launch the space station core module as originally scheduled, and the station would be revisited by Saturn IBs and CSMs up until June 1973. Beyond that, according to an official statement by associate administrator for Manned Space Flight, Dale Myers, spaceflight activities would remain "at a standstill as far as orbital operations are concerned until the shuttle comes on line, hopefully in the 1976–77 time period."

Unfortunately for the astronauts, there were just too few missions for too many candidates, and a lot of them had missed out on being assigned. The scientific community was also up in arms. Why appoint a group of scientist-astronauts as NASA had done back in 1965, they argued, when not even one of them would fly to the moon?

Three lunar scientists expressed their concern about the cancellation of *Apollo 15* and *19* in a September article in the *New York Times*. Thomas Gold of Cornell University said, "It's like buying a Rolls Royce and then not using it because you claim you can't afford the gas." Nobel Prize winner Harold Urey declared, "I think the American people and Congress should realize that the moon is an extremely old object . . . this gives scientists a way of studying an object that goes back to the very beginning of the solar system." He added that saving forty million dollars by canceling the two flights was "chicken feed" in light of the twenty-five billion already spent on the Apollo program. Finally, former scientist-astronaut Brian O'Leary weighed in, saying: "The scientific community has become disenchanted with NASA. The present decision seems ridiculous."

The *Times* then offered its own solution:

Throughout the last decade this newspaper opposed the top priority then accorded Project Apollo on the ground that too much money was being diverted from urgent social needs. But now that these huge sums have been spent the need is to obtain the maximum yield, scientifically and otherwise, from that investment. Surely NASA, which has been able to reach the moon, can find a better solution than the one now afforded for adjusting to austerity in space research. One desirable alternative would be to enlist foreign resources in the exploitation of Apollo technology, perhaps by offering to send teams of British, French or Soviet astronauts on the journey pioneered by Apollo 11.

The heat was truly on for NASA, and the agency began to weigh the possibility of flying geologist Harrison "Jack" Schmitt on what was now the final lunar mission, *Apollo 17*. At the time, Schmitt was assigned to back up the *Apollo 15* crew as LMP, along with Dick Gordon and Vance Brand, but with the cancellation of *Apollo 18* that mission was now essentially a dead-end job.

When Dick Gordon got wind that NASA was considering sending Schmitt to the moon, he began jockeying to have his entire crew replace that to be headed by Gene Cernan. There were no holds barred in this last-gasp effort to secure the final lunar flight, and it was left to Chief Astronaut Deke Slayton to make a decision, albeit with extreme pressure being exerted by NASA Headquarters to get a scientist on the flight.

Meanwhile, both crews trained hard, seeking to impress on everyone that they were the best available for the job. Cernan certainly did not help matters for his crew when he crashed a helicopter into the Banana River near the Patrick Air Force Base in Florida during training. Cernan, who was on a solo training flight, managed to swim away from the craft just seconds before it burst into flames and sank in eight feet of water. He suffered minor cuts and singed eyebrows and eyelids but did not require any medical attention apart from a precautionary checkup, so the intense rivalry continued anew.

By August 1971 it was time to announce the chosen crew for *Apollo 17*, and Cernan's crew got the nod—with one exception. Joe Engle had been dropped to make way for Jack Schmitt. Schmitt felt that Gordon and Brand should also have made the flight as the three of them were already a crew in training, and he jeopardized his own selection by making a personal appeal to Slayton. But the decision had been made; Deke Slayton would not change his mind, and Dick Gordon had to accept the call. It later came to light that Slayton had originally submitted the Cernan-Evans-Engle crew to HQ as his recommendation, but it had been rejected. He was told in no uncertain terms to get Schmitt on the moon, and his hands were officially tied, but he still chose Cernan and Evans over Gordon and Brand.

Joe Engle was visibly upset for some time and not a little bitter about the decision, but he finally came to accept the fact that he had been dropped from *Apollo 17* for political reasons. When asked by a reporter how he felt about missing out on the flight, he said the toughest thing of all was telling his kids that their daddy was not going to the moon.

NASA was still not out of the financial woods, and the embattled space agency received another jolt in October 1971, when one of Nixon's deputy directors, Caspar Weinberger, told new NASA administrator James Fletcher that the president was giving serious thought to scrubbing *Apollo 16* and *17* from the program. On 3 November Fletcher sent Weinberger a comprehensive ten-page report on why the missions should go ahead, explaining the impact such a move would have on national goals, science programs, and such things as employment in NASA and the aerospace industry. In his summation Fletcher wrote:

From a scientific standpoint these final two missions are extremely important, especially Apollo 17 *which will be the only flight carrying some of the most advanced experiments originally planned for* Apollo 18 *and* 19, *cancelled last year. With what we have learned from* Apollo 15 *and previous missions, we seem to be on the verge of discovering what the entire moon is like: its structure, its composition, its resources, and perhaps even its origin. If* Apollo 16 *and* 17 *lead to these discoveries, the Apollo program will go down in history not only as man's greatest adventure, but also his greatest scientific achievement. Recognizing the great scientific potential and the relatively small saving ($133 million) compared to the investment already made in Apollo ($24 billion), I must as Administrator of NASA strongly recommend that the program be carried to completion as now planned.*

Nixon obviously relented, and the last two Apollo missions went ahead. Jack Schmitt got to walk on the moon on *Apollo 17*, and having his trained eye at the landing site of Taurus-Littrow was beneficial, engendering many valuable discoveries and gathering some spectacular samples for return to Earth. During his lunar expeditions Schmitt described the landing area as "a geologist's paradise if I ever saw one."

On 14 December 1972 Gene Cernan became the last person to leave footprints in the lunar soil. As he took a final, lingering look around before climbing up into the Lunar Module, he remarked, "As we leave the moon at Taurus-Littrow, we leave as we came, and, God willing, as we shall return, with peace and hope for all mankind." The Apollo lunar landing program was for all practical purposes at an end.

A final comment on the canceled Apollo flights came from an astronaut who would have walked on the moon had *Apollo 19* taken place. Jerry Carr, originally assigned to the mission as LMP, was circumspect as he looked back on his lost opportunity. "The Apollo era was, indeed, a dramatic and exciting time for all of us [and] it was a black day for me when the Apollo program was cancelled. *Apollo 19* was to have been my mission, but when it was cancelled I was assigned as commander of the third Skylab mission. As it turned out I feel very fortunate that I had the opportunity to participate in the effort to expand human knowledge in the area of long duration space flight."

A full twenty-eight years before the turn of the century the final Apollo lunar landing mission splashed down safely, bringing to an end the Apollo lunar program. Today Apollo remains a distant memory of a time when an ebullient NASA, backed by an equally enthusiastic Congress and White House, was able to turn the dreams of science fiction into the reality of twelve Americans walking on the surface of the moon.

Sadly, the landing sites for those lost three Apollo missions remain unexplored but still beckon—worthy targets for future generations of astronauts perhaps but not until well into the twenty-first century. The irony of where the United States could have been now was eloquently stated way back in 1985, as a former Apollo astronaut was watching the launch of the first military space shuttle. "To think," he wistfully remarked, "we could have been going to Mars today."

Before the advent of the first space stations, the popular concept of these orbiting habitats was one of giant, wheel-like structures, no doubt inspired by Walt Disney's futuristic programs on space exploration, featuring the visionary predictions of Wernher von Braun, and by the epic Stanley Kubrick film *2001: A Space Odyssey*. The reality is that our first Earth-orbiting space stations, and those that will undoubtedly follow, were ungainly looking constructions that had far more to do with functionality as space laboratories than a display of elegance in orbit. Literally crammed from floor to ceiling with a packed labyrinth of research modules, apparatus, and experiments, the first orbiting outposts greatly advanced our knowledge of science, medicine, weather patterns, and the universe.

In its nine months of operation the country's first space station, Skylab, provided many complex engineering problems, some of which threatened the survival of the program. Conversely, Skylab gained renown as an orbiting laboratory of immeasurable scientific value. During the station's 3,896 revolutions of our planet, three visiting crews confirmed that the resources of space offer new and unique approaches to science and technology and the way we look at the Earth and our universe.

In 1963 the U.S. Air Force conceived plans for a small orbiting station to be known as the Manned Orbiting Laboratory, or MOL. This military station had its origins in NASA's Gemini program but was principally planned for use in covert reconnaissance operations. Seventeen military candidates were selected, and these exemplary pilots underwent initial training to occupy the station two at a time. The station concept was canceled in 1969, however, before any operational flights could be carried out.

NASA created the Apollo Applications Program (AAP) in 1965. The civilian space agency had been tasked with developing long-term uses for Apollo program hardware in order to maintain a U.S. presence in space beyond the manned lunar landing missions. Eventually, NASA decided to convert the leftover S-IVB second stage of a Saturn IB booster, once intended for use on an Apollo Earth orbital mission, to an orbital workshop configuration. The entire Skylab station, prefabricated by McDonnell Douglas, could be sent up in one launch so that it was virtually ready to receive the first of three planned science crews in their Apollo spacecraft.

It was a massive object to contemplate hurling into space, weighing in at around one hundred tons. A number of launch options for placing the cylindrical station into orbit were considered, but the cancellation of the final three manned Apollo lunar flights had freed up a Saturn V booster. The fully fitted AAP station (later renamed Skylab) could now be placed into orbit in its entirety.

The unmanned prototype station, designated Skylab 1 (SL-1), was launched from the Kennedy Space Center on 14 May 1973 atop a two-stage version of the Saturn V launch vehicle. The launch appeared flawless, with the Saturn V disappearing into heavy clouds, and soon after the station was successfully inserted into a near-circular orbit 269 to 275 miles above the Earth, inclined at fifty degrees to the equator.

Once it had been established that a stable orbit had been achieved, ground controllers waited for confirmation that the workshop's solar arrays had deployed. But they never received the signal. Instead, controllers in Houston

46. The SL-1 launch, carrying America's Skylab space station into orbit. Courtesy NASA.

received very low electrical power readings as well as indications that the station's internal temperature was climbing.

Later analyses of launch data indicated that excessive launch vibrations, unforeseen in simulations, had shaken loose a protective micrometeoroid and thermal shield, which had then been torn apart in the supersonic airstream. Debris impact had also caused the loss of the station's port solar array, crucial to providing electrical power. Of even greater concern, the second solar panel had become jammed in an unfurled position by an aluminum strap and failed to deploy as programmed. An integral part of the station was the Apollo Telescope Mount (ATM), and its batteries were also straining to supply the massive workshop with power.

NASA and contractor personnel now had to salvage the $2.5 billion Skylab mission in the face of these critical problems. Apart from a substantial reduction in power, the loss of the meteoroid shield not only meant a loss of thermal protection for the following crew members, but solar heating would soon raise the crippled station's internal temperature to an unsustainable 52°C.

The commander of the first Skylab mission, SL-2, was the likable Pete Conrad, a veteran of three previous missions who, as one of the *Apollo 12* crew, had walked on the moon. Following that mission he had tried to convince his two crewmates, Alan Bean and Dick Gordon, to follow him into the Skylab program, but only Bean heeded his suggestion. "By the time that I got to go to the moon, I knew that the program was in deep yogurt and that I would never go back again," Conrad told fellow navy aviator Don Pealer during an interview for *Quest* magazine. "I knew that I had to go over to Skylab if I wanted to fly again. That was it. . . . I tried to convince Dick to transfer if he ever wanted to fly again. But Dick wanted to go the last sixty miles." To Gordon's dismay those tentative plans fell through with the cancellation of the final three Apollo missions.

For the first of the three planned Skylab crews, Conrad had been assigned two rookie crew members: Paul Weitz and scientist-astronaut officer Joe Kerwin. Like Conrad, both had been aviation officers in the U.S. Navy.

Conrad's SL-2 crew had been scheduled for launch the day after the core Skylab module had been put into orbit, but their mission was postponed to allow repair equipment to be developed and tested as well as for the crew to be trained to use them. One of the crew's first options was to conduct a fly-around of the crippled station in order to survey and assess the damage for the possible later launch of a replacement Skylab B station that NASA had in storage. Meanwhile, the temperature inside the orbiting station had reached 90°C, but Mission Control was rapidly learning how to orient the station at optimum angles to avoid excessive overheating. Doing so, however, was

47. The first crew scheduled to occupy the Skylab station. *From left*: Joe Kerwin, CMD Pete Conrad, and Paul Weitz. Courtesy NASA.

consuming irreplaceable nitrogen attitude control gas in Skylab's tanks, so repairs would have to be expedited if the station was to be saved.

Following eleven days of intense activity and contingency training in a huge water tank at the Marshall Spaceflight Center in Huntsville, Alabama, the SL-2 crew was finally launched aboard an Apollo spacecraft, *CSM 116*, on 25 May 1973. A Saturn IB rocket would hoist them to an orbital linkup with the overheated space station.

Once the rendezvous and docking had been skillfully achieved and preparations made for the first repair effort, mission commander Conrad would undock the Apollo spacecraft and watchfully fly it over to the unfurled array so the fully space-suited crew could assess the situation. "We found the exact cause of the jammed boom since we could clearly see it," Conrad told Pealer. "We did the best we could by sending television images back to Earth and gave our verbal descriptions of the situation to the ground. I knew that we were going to work on it sooner or later."

Eventually, Conrad had maneuvered right up to the station and got the Apollo craft's hatch opened. While Joe Kerwin clung tightly to his colleague's ankles to prevent him from floating away, pilot Paul Weitz stood up through the open hatch and tried manually to drag the solar array panel free using a hooked pole. When this crude method failed, he attempted to cut the panel straps using a modified branch lopper. Seventy-five minutes later, with both craft moving into darkness, a frustrated Conrad called off the hazardous attempt and redocked with the crippled station before assessing some alternate methods with his crew.

Once inside Skylab the following day, the three astronauts were at least able to remedy the problem of the missing meteoroid shield through the novel method of pushing a specially created twenty-five-foot parasol device through an eight-inch aperture in Skylab that had been designed to expose scientific samples to the space environment. Once the golden Mylar fabric sheet was fully through, the apparatus was unfurled like an umbrella opening and brought down to shroud and protect much of the exterior. The deployment operation was deemed a success, although the three crew members had to take frequent rest breaks due to the oppressive heat.

By the fourth day temperatures within the station had been reduced to a comfortable but limited working level. Three days later Conrad and Kerwin carried out a dangerous four-hour extravehicular activity (EVA), during which they finally managed to free the remaining solar array using wire cutters. It then unfurled and deployed fully, but dangerously fast. As the beam swung out, it caught the startled astronauts by surprise, knocking them away from the station. Fortunately, they were tethered to the vehicle, so they were safe and even exultant as they made their way back to the airlock. As power began flowing into the workshop's batteries, the station quickly powered up sufficiently, and the Skylab project had been saved through a combination of ingenuity, planning, persistence, and sheer courage.

On 22 June, having successfully completed their twenty-eight-day science mission, the crew strapped themselves into their Apollo spacecraft ready to return to Earth. Having to repair the station meant that they had only completed about 80 percent of their assigned scientific tasks, but through their extraordinary efforts they had saved not only a multimillion dollar spacecraft but an entire research program. They had brought the imperiled station back to life and ensured that it would remain safe for the next two crews to occupy while also proving that humans can live and work in space for nearly a month with only minimal adverse physical effects.

48. The second Skylab crew: Jack Lousma, Owen Garriott, and Alan Bean. Courtesy NASA.

The second Skylab crew (SL-3) was launched to the unoccupied space station on 28 July and docked with the station the same day. Mission commander Alan Bean, together with pilot Jack Lousma and scientist-astronaut Owen Garriott, then occupied Skylab for a record-breaking fifty-nine days. This crew performed (and even exceeded) their assigned tasks with such efficiency that they also made up for any experiment shortfalls from the first crew's dramatic tenancy.

Following their departure, Skylab remained unoccupied for fifty-two more days before the SL-4 crew of mission commander Jerry Carr, pilot Bill Pogue, and scientist-astronaut Ed Gibson was launched on 16 November to continue the work on what would prove to be the longest, and consequently most productive, of all three of the Skylab crews. Among vital experiments they would conduct over the next eighty-four days was one devoted to tracing the development of spots on the sun, which can cause radio interference and have other influences on life back on Earth.

In addition to completing numerous solar and Earth observations, two of the SL-4 crew members also performed a lengthy spacewalk, during which they photographed the comet Kohoutek, then looping around the sun before retreating into the far reaches of the solar system. This crew would also

49. The third and final Skylab crew (SL-4): Jerry Carr, Bill Pogue, and Ed Gibson. Courtesy NASA.

be called on to make a number of minor repairs to station equipment, but when faced with an impossibly demanding experiment schedule, they were forced to discuss the excessive workload with ground controllers, following which a more realistic schedule was agreed upon and put in place. It was an important lesson for future space station planners.

On 8 February the final Skylab crew abandoned the station, leaving it parked in a drifting Earth orbit. They had traveled some 34.3 million miles, with the three crews chalking up an overall total of about 71.5 million miles in Skylab's 171 days of human occupancy.

In all, the nine Skylab crew members performed ten EVAS, totaling more than forty-two hours in order to carry out necessary external repairs and maintenance work, retrieve experiment packages and film cartridges from telescope cameras, and photograph celestial objects. They had conducted roughly two thousand hours of vital experiments covering a vast and varied array of medical, scientific, and astronomical disciplines, in the meantime proving that human beings could safely adapt to living and working for protracted periods in the microgravity of space.

With the development of the space shuttle now under way, it was hoped that an early shuttle flight might dock with Skylab and help boost it into a higher orbit, allowing for the retrieval of several exterior-mounted experiments and a possible rehabitation by future crews. But sadly for the Skylab program, the shuttle schedule slipped, causing these and other possible rescue plans to be abandoned as orbital decay slowly and ineluctably drew Skylab closer to a fiery end. On 11 July 1979 Skylab slipped out of orbit and was incinerated in the colossal heat of reentry over an area covering the Indian Ocean and some sparsely occupied areas of Western Australia, where scattered debris were later recovered.

A full account of the Skylab program can be found in the Outward Odyssey series book *Homesteading Space: The Skylab Story*, by David Hitt, Owen Garriott, and Joe Kerwin.

12. The Last Apollo

Geoffrey Bowman

> First, inevitably, the idea, the fantasy, the fairy tale. Then, scientific calculation. Ultimately, fulfillment crowns the dream.
>
> Konstantin Tsiolkovsky (1857–1935)

I was born in Bangor, Northern Ireland, in 1954. As a young teenage space enthusiast, I followed the exploration of the moon on television and in the newspapers, always longing to see the launch of a Saturn rocket with my own eyes. As a law student at Queen's University, Belfast, I took up gliding and made my first solo flight in June 1975, but my enthusiasm for flight beyond the atmosphere never diminished. The following month I crossed the Atlantic to see something I simply refused to miss: the last launch of a Saturn rocket, carrying the last Apollo spacecraft into orbit. This account, based on diary notes I kept of the journey, is for me incredibly evocative of that extraordinary era. Interspersed into the narrative are several reflections by the three American astronauts and two Soviet cosmonauts who took part in the historic Apollo-Soyuz Test Project (ASTP) mission, which came to be universally known for its historic "handshake in space."

Between November 1973 and July 1975 no astronauts departed from the Kennedy Space Center by rocket, but the exploration of space continued. At the end of March 1974 *Mariner 10* flew past the innermost planet Mercury and returned clear photographs showing almost half of the little world's surface.

In that same period no less than eight Soyuz two-man capsules were launched from the remote Soviet launch site in Kazakhstan, although only six missions actually succeeded. On 27 September 1973, more than two years after the loss of the *Soyuz 11* crew, a redesigned *Soyuz 12* lifted off from the Baikonur Cosmodrome carrying Vasily Lazarev and Oleg Makarov on a two-day mission to test the craft's maneuverability. Among the most prominent of the craft's improved safety features was the removal of the third couch, providing room for the two cosmonauts to suit up, thus preventing a tragedy similar to that suffered by the crew of *Soyuz 11*. It was followed just three months later by the week-long flight of *Soyuz 13*, carrying Pyotr Klimuk and Valentin Lebedev.

In July 1974 *Soyuz 14* linked up with the new *Salyut 3* space station, allowing veteran cosmonaut Pavel Popovich to spend fifteen days on board along with his flight engineer Yuri Artyukhin. A further docking attempt in August by cosmonauts Gennadi Sarafanov and Lev Demin (at forty-eight the first grandfather to fly into space) failed, and *Soyuz 15* had to make an emergency landing at night. *Soyuz 16*, in December 1974, was simply a dress rehearsal for the forthcoming Soviet-American joint mission. During the six-day mission Anatoli Filipchenko and Nikolai Rukavishnikov used an imitation docking ring attached to the front of their Soyuz spacecraft to simulate a number of dockings. They also raised the cabin pressure and oxygen content to match that of the Apollo spacecraft, which would allow crew transfers to take place during the ASTP mission. The *Soyuz 16* flight, although launched in secret, demonstrated to concerned sections of the American space industry that the Soyuz spacecraft and its systems were up to the task.

Soyuz 17 (Alexei Gubarev and Georgi Grechko) and *Soyuz 18B* (Pyotr Klimuk and Vitali Sevastyanov) both involved Skylab-style occupations of the newly orbiting *Salyut 4* space station. These missions lasted twenty-nine and sixty-two days, respectively, but sandwiched between them was a space "first" of a most unwelcome kind, which sent shivers down the spine of anyone with an interest in the Apollo-Soyuz Test Project, including one Belfast law student.

On 5 April 1975 cosmonauts Lazarev and Makarov, who had flown previously on *Soyuz 12*, blasted off from the Baikonur pad with the intention of visiting *Salyut 4* for a planned two-month habitation. Initially, the launch went smoothly, but when the second stage of the booster rocket ignited, the spent first stage failed to separate properly. Dragging the spent stage behind it, the rocket veered off course. The Soyuz spacecraft's guidance system automatically aborted the flight, and the capsule blasted free of the wayward booster, subjecting the cosmonauts to enormous g-forces during what was effectively a suborbital reentry. But their chief concern was political: they definitely did not want to land in China, where a different brand of communism meant that the welcome they received would have been decidedly chilly.

In the event the Soyuz capsule parachuted safely into the mountains of Soviet Siberia, two hundred miles from the Chinese border. Lazarev and Makarov had to wait a day to be rescued but were found to be in good health, having unintentionally made the shortest flight since Gus Grissom's Mercury flight in 1961. Their aborted flight was subsequently designated *Soyuz 18A*.

The successful *Soyuz 18B* mission was launched on 24 May 1975. One knock-on effect of the earlier abort would be that in July 1975 two separate Soviet spacecraft were in orbit at the same time, performing entirely different tasks. The second craft was the Soviet half of ASTP. For the Americans that joint flight was truly the end of an era: Apollo's last hurrah, indeed the last American "expendable" spacecraft before the dawn of the reusable space shuttle, still several years down the line. For me Apollo-Soyuz was a unique opportunity to experience the whole flavor and atmosphere of a remarkable endeavor that would soon come to an end.

I already knew that the Apollo spacecraft would be entrusted to Tom Stafford, who had commanded *Apollo 10* in May 1969 and later rose to the rank of brigadier general in the U.S. Air Force. The Command Module pilot would be Vance Brand, who might have gone to the moon if *Apollo 18* had not been canceled. The third crew member was the venerable Donald "Deke" Slayton, aged fifty-one at launch, who had been robbed of an early Mercury orbital flight when a minor heart irregularity had grounded him in 1962.

Alexei Leonov, veteran of the 1965 *Voskhod 2* flight, and Valery Kubasov, who had flown aboard *Soyuz 6* in October 1969, would be the cosmonauts to form the Soviet half of the orbital linkup. Leonov, the first man to "walk" in space, was one of the Soviet Union's most respected and genial cosmonauts, a man who counted many American astronauts among his friends. But how had this international venture begun?

Vance Brand: *There were early contacts by officials from NASA and the Academy of Sciences in the Soviet Union that got it started, but it took quite a few years in the Cold War to reach a point where I guess both sides thought it was worth trying.*

Tom Stafford: *What came about was that Academician [Mstislav] Keldysh on the Soviet side and also Dr. [Robert] Seamans from NASA met at some international conference and said, you know, the United States have been to the moon, but the Soviets have done a lot in space, so perhaps we ought to do something coöperative together, so we can work together. And so from that point we started working on it, working with the astronauts, and of course the Russian Academy of Science, and that's how it came about.*

Alexei Leonov: *The discussions between Russia and America began in 1972, [between] President [Richard] Nixon and Premier [Alexei] Kosygin, and President of the Russian Academy of Sciences, Keldysh. The decision of these very clever people was, let's do it, with our space programs. Everyone on the Earth should know about a new spirit between our countries. This was very important. . . . At that time, we believed it to be a very dangerous situation . . . between our countries. I was a military man at that time; it was a very crazy time for people. In 1959, Tom Stafford was flying in West Germany. I was flying in East Germany. Our countries' airfields were only twenty-five kilometers apart. Every time Tom Stafford took off, I took off. Every time I took off, Tom Stafford took off. We flew, not far away from each other. We could see each other. But that was 1959. . . . It was truly another time, another people. I believe many people on the Earth came to understand each other much better than they did before.*

Deke Slayton [from his autobiography *Deke!*, with Michael Cassutt]: *Oddly enough, one of the things that moved the joint flight closer to reality was a fictional movie called* Marooned, *based on a novel by Martin Caidin. In the original version, published in 1964, a Mercury-Atlas 10 astronaut is rescued by a Soviet cosmonaut. . . . The movie never made much money in the United States, but it apparently impressed the Soviets that Americans were ready to consider international flights—especially to demonstrate the concept of space rescue.*

Tom Stafford: *So finally in April of 1972, President Kosygin and President Nixon signed a document.*

In 1971 I had joined the British Interplanetary Society, a highly respected organization formed in 1933—a time when anyone who talked about a trip to the moon was truly considered a lunatic. In the early 1970s the society's *Spaceflight* magazine carried a series of advertisements from a firm named Transolar Travel, which offered to take enthusiasts to Florida to see the later Apollo launches. I was intensely frustrated to know that I could have gone to see the launch of *Apollo 16* or *17*, or Skylab, but at the time it hardly seemed within my means. Crossing the Atlantic was not such a commonplace holiday experience in those days, particularly for a lone teenager. We are not all Jack Londons.

Apollo's lunar program and Skylab passed into history. I never saw a Saturn V launch, and to this day it grieves me to the core. Then, one day in 1974 I opened the August edition of *Spaceflight*, and on the inside front cover was Transolar Travel's latest lure—a fifteen-day visit to Florida departing on 6 July 1975 to witness the launch of the Apollo half of ASTP. As soon as I saw the advertisement, I knew I had to go. The price of £198 was daunting, but I had savings from part-time jobs, and I was not about to let mere money stand in my way.

My trip to Florida was slightly delayed to 8 July to accommodate changes in the itinerary, which pushed up the price by a dizzying £20. In high spirits, having survived the very real ordeal of Second Year law exams and having made that first deeply satisfying solo glider flight, I flew to London and met up with the rest of the ASTP tour party, after spending the night at a Heathrow hotel. My mother had been kind enough to donate the cost of the Belfast-London flight and the hotel.

On Wednesday, 9 July, I boarded a BOAC "Jumbo Jet" for the nine-hour flight to Miami. It was a novel experience for me, crossing an ocean inside the belly of this impossibly huge metal beast, and the hours flew past, punctuated by glorious views of Bermuda and the Bahamas. Arriving in Miami at 5:30 p.m. local time, we transferred to the very comfortable Sans Souci Hotel on the famous Miami Beach. I slept like a log that night and rose early enough the next morning to make a short tour of the neighborhood. I remember being most impressed by the idiosyncratic Art Deco architecture along the beachfront, but a close examination of some of the hotel facades revealed that Miami Beach had seen better days and needed a facelift, or at least a lick of paint.

Then we were off in the private air-conditioned bus that would stay with us throughout the trip. The journey north along the Florida coast was interminable, partly because there was little to see except mile after mile of hotels, restaurants, and beaches but mainly because I was brimming over with anticipation. Our destination was Cocoa Beach, another few miles of hotels, motels, bars, and restaurants stretched along a narrow sand spit that formed a southerly extension of Cape Canaveral.

Cape Canaveral—a mass of brackish lagoons, swamps, islands, and a sandy point jutting out into the Atlantic. Cape Canaveral—home of the Kennedy Space Center, the Holy of Holies. The Home of Apollo. As we drove along the freeway, I caught a glimpse of a huge white rectangle on the horizon, contrasting starkly with the clear blue sky. I realized, with a sharp intake of breath, that I was looking at the Vehicle Assembly Building (VAB), the nursery of Saturn rockets. From that immense structure the *Apollo 11* launch vehicle had emerged on the first stage of its journey to the moon. I kept the VAB in view for several miles before losing it behind low scrub and bushes, but it was a potent reminder that I had almost reached my destination.

We checked in at the Atlantis Beach Lodge, a comfortable low-rise motel set on the Atlantic shore. The heat of the Florida afternoon was fiercely oppressive after the air-conditioned bus, and it was a relief to enter the spacious coolness of a well-furnished motel room. I was sharing my lodgings with an affable, rotund space enthusiast in his thirties named Mike Dunsford, who was on his third visit to Cocoa Beach, having previously witnessed the launches of *Apollo 16* and Skylab.

Shortly after our arrival the heavens opened, and rain the like of which I had never seen before bounced off the palm-fringed central courtyard of the motel and beat a tattoo on the roof overhead. Just before dinnertime the rain stopped as suddenly as it had begun, and the sky cleared. Groups of guests wandered down to the beach, and there in the distance I could see the sand curving round in a great arc toward Cape Canaveral. Along the horizon crouched rows of dark rocket gantries, mostly disused but clearly identifying the southeastern perimeter of the Space Center.

There were several people already on the beach talking in American accents. Conversations were struck up, and I soon found myself chatting with a technician identifying himself as Ed Procrasky, who was working on the

Titan-Centaur rockets, which were due to launch the twin Viking probes to Mars the following month. Here was a man who shared responsibility for ensuring that the giant launch vehicles would successfully deliver, to a precise course through space, the two most sophisticated space probes ever constructed, probes that would actually search for life on another world. I also reflected on the fact that each Titan-Centaur packed roughly twice the punch of the Saturn IB, whose launch I had crossed an ocean to see.

The following day, Friday, 11 July, I discovered a NASA Visitors' Center in one of the central buildings of the motel complex. It was a veritable Aladdin's cave of Apollo and Skylab memorabilia, and I was able to buy several ASTP mission badges. The motel conference center was doubling as a press room, and I managed to scrounge a complete ASTP press kit, which I still have today.

During the first four days of my visit to Florida the weather was a constant source of concern. To me it was oppressively, unbelievably hot, although I heard locals say it was "cool for the time of year." The humidity was almost unbearable, and stepping out of an air-conditioned room into a solid wall of heat was an experience. Although the mornings and evenings were usually dry and sunny, clouds regularly built up in the middle of the day and dumped heavy deluges of rain and even hail, interspersed with violent thunderstorms around the very time the launch was due, at 3:50 p.m. The possibility of a postponement was an ever-present nightmare. We could have coped with a short delay of a few days, but I was also concerned that the launch might proceed in marginal conditions with low overcast and a brief tantalizing glimpse of the rocket before it disappeared into the clouds. To have come all this way and to be cheated of my goal did not bear thinking about.

Careful study of the local newspapers did nothing to ease my concerns. On 11 July the *Today* newspaper headline reported, "Apollo Weather May Be Rainy." On the same day the *Orlando Sentinel Star* noted that the Saturn-Apollo launch vehicle had been struck twice by lightning, without mishap. On 14 July the *Miami Herald* announced, "Stafford Worried about Weather." If Tom Stafford was worried, I had every right to follow suit. I read that local weathermen were predicting a 23 percent chance of lightning clouds over the launch site at ignition time, and there was a particularly discomforting note that this was "the peak of Florida's lightning season." Ap-

50. The author at the Kennedy Space Center in 1975. In the distant background, on Launch Pad 39B, is the Saturn IB rocket. Courtesy Geoffrey Bowman.

parently, there were even plans to dump millions of tiny aluminized strips of nylon into the storm clouds from high-flying aircraft to "short-circuit" the electrical buildup and discharge the lightning. After *Apollo 12*, when the ascending rocket was hit twice by lightning, I knew NASA would not risk a launch if there was the merest hint of lightning activity in the area.

Thankfully, on 14 July the *Today* newspaper was more upbeat and headlined news that "Storm Activity on Launch Day May Decrease." The forecast was for partly cloudy skies and temperatures in the mid-80s F.

Everywhere I looked, everywhere I went in the days leading up to the launch, there was no doubting that Apollo-Soyuz was the biggest show in town. Every Cocoa Beach motel and restaurant along North Atlantic Avenue seemed to have a huge neon sign screaming out, "Go with God, Apollo crew, Soyuz crew" or "Good Luck, Astronauts" or "Go Apollo, Go Soyuz." The Cold War had apparently been forgotten or at least set aside for an interlude of cozy international relations courtesy of five spacemen.

Valery Kubasov: *There were lots of differences between American and Russian management. But when it came to the astronauts, they worked the same way as the cosmonauts in Russia; it felt like the same kind of system.*

Vance Brand: *Our technology, which was essentially the Apollo technology—Com-*

mand and Service Module, Saturn boosters, things like that—was being used for the last time. We were at that point developing the space shuttle. So maybe the technology, had they used it, wasn't so valuable at that point. And besides, had they wanted to duplicate our modules, which they never did, it would have taken several years to do that. So that didn't concern us very much.

Tom Stafford: *There were issues to overcome. I don't think there were many misgivings. The main thing is we had the capability . . . both sides signed up to do the effort. The biggest things were learning the Russian, and the timelines of the mission. Learning the Russian was the hardest. I'd been to the moon, flew* Gemini 6, *did* Gemini 9, *so the PR was no problem either.*

Valery Kubasov: *Oh, it is very difficult to understand Tom Stafford's language, because it's not the English language—it's Oklahoma language! I believe that it was more difficult for us. That's why we swapped Russian and English language for the mission. Americans had to speak in Russian, and Russian cosmonauts had to speak in English.*

The mission filled page after page of all the newspapers circulating in Florida and most had printed special souvenir editions on launch day. News of the flight, viewed from every conceivable angle and commented upon by every columnist, was interspersed with advertisements adopting an Apollo-Soyuz theme. A full-page ad for Winn Dixie Supermarkets ("The Beef People") reminded customers of the chain's unswerving support for the space program and added that "All Systems Are GO for Apollo-Soyuz!" On a more spiritual note the Peace Lutheran Church of Palm Bay prayed "not only for those who go down to the sea in ships afloat in blue waters but also for those who ascend above and beyond the 'wild blue yonder.'" Christ's Universal Associate Church offered the blander message: "A Safe and Successful Journey."

I was fascinated by some of the more colorful stories associated with the forthcoming launch. Who would not have felt sympathy for Walter Kapryan, the Apollo launch director, who had to oversee the countdown and launch while hobbling around on crutches? The *Miami Sentinel Star* revealed that on 11 July he had been walking on a Florida beach and had stepped on a large beached catfish. One of its protruding spines pierced his foot, cut a tendon, and lodged in a bone.

The prelaunch atmosphere around Cocoa Beach was electric. It was difficult for me, on my first visit, to determine whether it was always this way or whether this was a final fling before the long empty years. In motels and bars tourists mingled with NASA officials and Cape employees preparing for their last send-off. Most of the larger establishments had temporary NASA press offices, and I used increasingly ingenious excuses to obtain press kits and photographs. Everywhere people were reminiscing about "the good old days," and there were many laments about "the end of an era." Along with the boisterous sense of anticipation for what was about to happen, there was certainly something of a fin-de-siècle mood. For many local workers this would probably be the end of the road. With no more manned missions for at least four (it turned out to be six) years, there would be fewer rocket engineers and support personnel, therefore fewer waiters, waitresses, and bar staff.

In the heady buildup to the launch, however, it was easy for most people to cast aside negative thoughts, at least for the time being. At this moment, July 1975, the bars and restaurants were doing a roaring trade, and the motels were full. According to the local press, all large motels in the Cape area were booked solid, with most reservations having been made up to twelve months in advance. Local commentators were saying that anywhere between a half-million and a million people would crowd into Brevard County for the launch, and some seventy thousand were expected within Kennedy Space Center itself. The numbers would eventually prove to be significantly smaller, but that was of little comfort to anyone caught in a huge prelaunch traffic jam.

On Monday, 14 July, the day before the launch, I boarded our chartered bus around nine o'clock, and we set off on a forty-minute journey to the space capital of the world, the home of Apollo. Driving from Cocoa Beach to the Kennedy Space Center was a longer and more complicated business than a quick glance at the map might have suggested. A network of interlinked lagoons and swamps turned a distance of about twelve miles (as the crow flies) into a long journey of exploration across bridges spanning the Banana River, Merritt Island, and Indian River, through the town of Cocoa, northward along the freeway, then back across the Indian River before entering the Space Center. At the Visitors' Information Center we exchanged

51. The prime U.S. and Soviet ASTP crews. *From left*: Vance Brand, Tom Stafford, Alexei Leonov, Valery Kubasov, and Deke Slayton pose beneath a representation of their docked spacecraft at the Kennedy Space Center. Courtesy NASA.

our tour bus for one operated by NASA, complete with camera-challenging tinted windows and a cheerful, knowledgeable courier.

We then drove for miles through industrial areas and past a few abandoned launch complexes, including the site of Alan Shepard's Mercury spaceflight in 1961, the achievement commemorated by a full-scale replica Mercury-Redstone rocket and a plaque with a bas-relief of Shepard's face. To borrow a phrase, everyone took pictures of everyone taking pictures. We also stopped at a building—I can't remember its purpose or function—which contained a remarkable and unique artifact. The *Apollo 15* mission had originally been planned as the last of the "first wave" of lunar flights, essentially identical to *Apollo 14*. Then it was upgraded to a J mission, the "second wave" of sophisticated long-duration landings supported by lunar rovers. The additional weight required heavier and more sophisticated Lunar Modules (LMs), so the original flight-rated LM for *Apollo 15* was replaced and became surplus to requirements. It remained at the Kennedy Space Center (and can still be seen there today, inside the Saturn V Visitors' Center). There I was, peering down into a large glass-walled room at a squat silver, gray, black, and gold contraption built to land two men on another world. I stared long and hard and felt privileged to have seen this symbol of the modern age. There is a remarkable life-sized replica of an LM in the London Science Museum, but each time I see it I remember that I have seen the Real McCoy.

Our tour in the bus with tinted windows took us from one spectacle to another. Eventually, we disembarked into the oppressive damp heat, and I will always vividly recall stopping in my tracks and staring in awe at the concrete bulk of Launch Complex 39A, crowned by the dark red metallic finger of the launch gantry that had provided support for *Apollo 11* and all of the rest of the moon-landing flights. Now abandoned, its work done, the immense tower of tubular steel loomed over the pad. Even stripped of the huge crane that had topped it off in better times, the gantry was over four hundred feet tall. There would be opportunities for photographs and film, but for a few moments I just gazed up at the enormity of that gantry and marveled at the titanic forces that had seared its paintwork with their hot breath. This was the focal point of Project Apollo in its heyday and a place I will always regard as hallowed ground, the place where human beings first began a monumental journey to another world.

I dragged myself back to the present day and turned my gaze slightly to the left. In the distance, perhaps a mile away, was the abandoned gantry's twin brother resting on a huge Mobile Launch Platform on Pad 39B. This gantry was still surmounted by a crane, and above it rose a thin white pillar, the lightning conductor that had already been put to good use that week. Embracing the rust-red gantry was the even more massive silver-gray Mobile Service Structure, a maze of steel tubing and girders that cocooned and protected the slim white structure barely discernible in its depths. Peering through my binoculars, I could just make out the object of so much human activity in this isolated subtropical swampland: the Saturn IB rocket with its gleaming conical tip. Only twenty-seven hours before the launch, I was looking at the Last Apollo.

There may have been dozens, even hundreds, of workers milling around like ants within the Mobile Service Structure, but I couldn't see them. It was almost as if the rocket and its support tower had been abandoned for the day. Or perhaps even rocket engineers take a lunch break. We returned to our bus and were driven along a service road toward the remarkable Vehicle Assembly Building. Parallel to the road lay the 130-foot wide Crawlerway, down which the Saturn rockets were transported, crunching along at a snail's pace to the distant launchpads. At this point a few statistics are unavoidable:

The underlying foundations of the Crawlerway had been rolled, vibrated, and compacted to an almost rocklike density. The whole thing was topped with a layer of loose Alabama river stones between six and eight inches deep.

The Crawlerway was designed to bear the weight of a Saturn V rocket and its Mobile Launch Platform (MLP) riding together on top of a gargantuan diesel-powered Crawler transporter. Each of NASA's two "Crawlers" weighs 6 million pounds (or 2,678 tons) and is 114 feet wide, 131 feet long, and moves on four pairs of caterpillar tracks, one pair at each corner. Each link of each track weighs one ton.

Resting on top of the Crawler during the six-hour, three-and-a-half-mile journey to the pad, the MLP itself weighed 12 million pounds (5,357 tons). From its base, resting on the Crawler, to the top of its crane measured 445 feet.

The Saturn V rocket, the raison d'être for all this heavyweight ironmongery, was 363 feet from top to bottom and had a comparatively modest unfueled weight of 860,000 pounds (about 380 tons).

The total weight crushing down on the Crawlerway when a Saturn V was transferred to the pad at just over a half-mile per hour was about 8,400 tons, and occasionally the Alabama river stones had to be raked smooth after its passage. I could never establish why Alabama was selected to provide the stones.

As the system depended on the two identical Crawler transporters, it is gratifying to record that they not only served faithfully throughout Apollo, Skylab, and ASTP but remain in service to this day transporting space shuttles (and the Ares I-X test vehicle) safely to the launchpads.

Our bus then traveled the three miles from our viewing point near Pad 39A, and we saw the VAB looming in the distance, gradually expanding to fill the view out of the front window. As we drew nearer this enormous hatchery for moon rockets, everyone on the bus was distracted by yet another marvel. We stopped at the end of the Crawlerway beside the enormous first stage of a Saturn V, resting horizontally on three support stands, its five immense engines staring blindly across the flat swampland. Reminding me of nothing as much as a beached whale, this relic of a dying era was already revealing flaking paint and creeping corrosion, insidious products of the damp, salty air. In time this first stage would be joined by the rest of the great rocket, lying on its side close to the VAB as a tourist attraction. More recently, it has been lovingly restored, repainted, and displayed inside a weatherproof custom-made exhibition hall. But however spectacular an attraction it makes, viewing it in 2007 in all its restored glory did not alter my feelings of deep regret that it was never given the opportunity to fulfill the purpose for which it was created.

As the tour group stood slowly melting onto the asphalt in the ferocious midday sun, our guide shepherded us together and led us a short distance to experience something that is simply not available to visitors today. Security considerations in the shuttle age make the VAB strictly off-limits to tourists, but in 1975 the Kennedy Space Center was passing from the triumphant Apollo era into a period of limbo awaiting the shuttle. And so it was that I was able to enter what was once the world's largest building, volume wise.

The VAB is truly breathtaking, enormous, stupendous. Superlatives trip off the tongue and echo around the vastness of its interior. Virtually cubical, the main section of the building is 518 feet wide and 526 feet high. Such dimensions are hard to judge at a distance as there are no other tall buildings in the vicinity, but standing just outside the VAB, I had followed the lines of wall panels and rivets up and up and up . . . until my neck protested and I almost fell over backward. Now, standing inside and slowly revolving on one heel to take it all in, I felt like a mouse in a cathedral. In every direction a latticework of steel support beams crisscrossed and soared heavenward. The ceiling was visibly foreshortened as I put further strain on my neck to record some movie sequences that came nowhere close to doing justice to the place. Built to house up to four Saturn Vs simultaneously, the building seemed all the vaster for its emptiness. The hectic days of preparing enormous rockets would come again, but on this day in 1975 the VAB seemed to be pausing for breath, and only the echoes of astonished tourist voices filled the void.

It was noticeably cooler inside, and not just because we were spared the wrath of the sun. The VAB requires powerful air-conditioning, without which—so the legend says—clouds would condense at ceiling height and the world's only indoor rain would fall. In place of Saturn rockets the VAB housed two exhibits associated with the Apollo-Soyuz joint mission. There was what seemed to be a mock-up of a complete Apollo spacecraft resting vertically on the tapered fairing that protected the spindly LM during launches to the moon. Beside it lay a very impressive representation of the Apollo and Soyuz spacecraft as they would appear docked in orbit. The Soyuz was only a full-scale replica, but the Apollo was the genuine article: one of the leftovers that would never make it into space. Its silvery surfaces gleamed in the harsh floodlights, and I couldn't help but wonder whether it had once been intended to sit on top of a Saturn V, with the moon as its destination.

I was the last member of our group to leave the building. For a lingering moment I simply gazed around its vastness, fixing in my memory what cameras could never properly record.

For our final port of call we returned to the Visitors' Information Center to spend a couple of hours wallowing in every conceivable aspect of American space history. It really was a case of treasures of ever-ascending value. After examining a one-tenth scale cutaway model of a Saturn V, I passed on to a genuine Mercury capsule that had flown an unmanned suborbital trajectory in 1960 as a test of the system that would eventually hurl Alan Shepard into space. Next to the Mercury was the *Gemini 9* capsule flown by Tom Stafford and Gene Cernan in 1966. The same Tom Stafford was at that very moment in the astronauts' quarters a few miles away, making final preparations to return to orbit.

Then I saw something that stopped me in my tracks—a truncated cone eleven feet five inches from top to bottom, with a base diameter of twelve feet ten inches. Its surface was a dull brown color, streaked, burned, and charred in places. The visible part of its broad, blunt base had clearly been ravaged by enormous heat and stress. Its cramped interior had provided safe shelter for three men journeying home from the edge of disaster amid worldwide celebrations and sighs of relief. It was the only surviving remnant of *Apollo 13*: the spaceship named *Odyssey*.

I have no idea how long I spent circling this symbol of triumph out of calamity, peering at every window, thruster exhaust port, rivet, and bolt. Incredibly, the craft was guarded only by a low, clear Plexiglass surround, which dipped to allow a clear view of the interior. I was actually able to stretch my head, shoulders, and upper torso inside that celebrated spacecraft, the center of so much worldwide concern during those remarkable days in 1970. Although much of the interior equipment had been stripped out, it was still not a place for the claustrophobic. Having once peered inside an Apollo capsule, I could not fail to be even more mightily impressed by the focused discipline and courage it must have taken to span the distance between Earth and moon in such a cramped environment. But this was no Victorian steam-driven artifact. An almost identical version of this space vehicle rested on top of its Saturn launcher nine miles from where I stood, and as I stared at *Odyssey*, my thoughts flitted from the recent past to the events that would unfold in the next few days, and particularly the very next day.

As I write these words, I am looking at a photograph, framed on the wall of my study, of a younger, slimmer version of myself standing rather solemnly in front of the *Apollo 13* spacecraft. I have no idea who took the picture for me—probably some passing tourist like me—but it reaches out to me from the past that was my twenty-first year. The singed, dinged object behind me stands out in stark contrast to the shiny unused Apollo spacecraft and LM, and the slightly weather-beaten Saturn V first stage, which I had seen earlier that day. They had all been built to play their part in a journey to the moon that had never taken place. *Odyssey*, the *Apollo 13* Command Module, had actually made such a journey; it had served its purpose by preserving its human cargo in the most challenging of circumstances and was now resting in a place of honor representing what I might wholly inadequately describe as the true spirit of Apollo. What a day. But as they say, the best was yet to come . . .

It was an early start for me on the morning of Tuesday, 15 July: launch day. Two separate countdowns had been progressing through the night, with the numbers approaching zero at the Soviet Baikonur Cosmodrome, where the Soyuz rocket ready to carry Alexei Leonov and Valery Kubasov was due to blast off at 8:20 a.m. Florida time. For most of the tour group at the Atlantis

Beach Lodge the plan was to rise early, watch the Soyuz launch live on TV, have breakfast, then prepare for the bus trip to KSC for the main event. All these years later it is easy to talk about watching a live Soyuz launch, but in 1975 it was unprecedented for the precise launch time to be announced in advance and for live TV to beam the spectacle around the world.

To complicate matters further, the Russians had prepared a second, backup Soyuz vehicle, which sat on another launchpad twelve miles to the north. Both rockets were counted down simultaneously to the T-5 hour mark, and if the prime Soyuz and its crew had been unable for any reason to launch on schedule, the reserve crew was ready to step into their shoes. This was, in fact, the crew of Filipchenko and Rukavishnikov, who had flown the *Soyuz 16* "dress rehearsal" flight the previous December. The Russians were absolutely determined not to be the weak link in the chain.

Discussing the mission with me in October 2008, Vance Brand recalled the aborted launch of *Soyuz 18A*, which caused a degree of consternation within NASA at the time:

We were at Star City and through our sources we found out there had been an aborted launch. I may be telling stories out of school here, but Tom Stafford asked General Shatalov [commander of cosmonaut training] if he had anything he could say about it. The general was kind of surprised we knew about it! He was the only Russian at Star City who was aware of it right at that moment. The Russians weren't real quick to give out information even to their own people [but] I don't think it impacted our mission at all. Whatever the problem was, they took care of it. It never changed anything about our training.

As the final minutes of the Soyuz countdown unfolded, I sat on the end of my bed watching live color TV from the middle of the vast, empty steppes of Central Asia. Mike Dunsford and four or five others were clustered around the set, and we exchanged hushed comments about the novelty of the situation. In the past Soyuz launches were only announced *after* the cosmonauts were safely in orbit. Delays were not announced in advance. It began to dawn on us that if anything were to go seriously wrong, we might not be making our eagerly awaited bus trip later that morning. It was all very well having a backup Soyuz waiting in the wings, but if Leonov's rocket blew up on the launchpad, nobody seriously expected the understudies to be launched as if nothing had happened. We weren't saying it out loud, but we desperately wanted the Soyuz launch to succeed because we wanted to see our Apollo blasting off on time.

Thirty-three years later, I asked Brand whether he had any nagging doubts about the Soyuz launch. "I don't think we were very concerned about it. The Russians, despite what we just talked about on the abort, had a pretty good track record for launches. And, you know, any little nail in the hoof of the horse could have stopped the show, like we could have had an abort, they could have had an abort, or something else."

It was 8:20 a.m. The final seconds ticked away, and the Soyuz booster filled the screen. An umbilical arm separated from the rocket, a dull red glow appeared at the bottom of the picture, and suddenly the Soyuz was leaping skyward much more rapidly than a Saturn. I felt momentary concern as the live picture broke up in a blaze of colored bars and static, but then a fresh view of the ascending rocket appeared on the screen, and all seemed well. It was soon a brilliant blob of light, trailing a fiery tail into the sunny afternoon sky. Nine minutes later *Soyuz 19* was in orbit, and the Apollo-Soyuz Test Project was under way.

The Apollo crew members were still fast asleep as the Soyuz rocket roared into space. They were woken up an hour later and shown a video replay while consuming the traditional astronaut breakfast of orange juice, steak, eggs, and coffee. To Brand it "looked like a good launch."

Back in Cocoa Beach, the next four and a half hours were a blur of preparations. For some time before traveling to Florida, I had been considering how best to observe and record the Apollo launch. I knew that the tour organizers had arranged for us to watch from a viewing area adjacent to the NASA Parkway, some three and a half miles from Launchpad 39B. This was one of the best observation sites and was actually closer to the rocket than even the official press site, which was optimized for Pad 39A. I had also worked out that at 3:50 p.m. the midafternoon sun would be more or less behind us, perfectly illuminating the white-painted rocket. From the press site the Saturn would stand partially in its own shadow. I gave much thought to the equipment I would need: a pair of binoculars seemed an obvious choice, but I could hardly view the launch through binoculars and operate the cine camera at the same time. Or could I? I eventually decided to strap the binoculars to the top of the cine camera with heavy rubber bands, hoping that I could watch and film simultaneously. Of course this meant I couldn't take ordinary photographs, but such were the limitations of having only one pair of hands.

I had also decided months earlier to bring my portable tape recorder with me: the faithful unit that had recorded televised highlights of every mission from *Apollo 12* to Skylab would now hear the roar of a Saturn for real. This, of course, was before the era of camcorders for all. Back then ordinary tourist cine cameras recorded pictures but not sound. I had borrowed a tripod from a friend and found it a surprisingly bulky item, even when folded up. With the cine camera mounted on the tripod and the binoculars on top of the cine, I would have a cumbersome but—I hoped—quite stable and functional viewing platform. The scary thing, looking back, is that I never tested the rig to make sure it wouldn't fall apart in use. I simply had faith in my ingenuity. With all this equipment packed into a carrier bag, I climbed aboard the chartered bus. We left the motel a little later than planned, at 1:15 p.m., but we were all confident we would reach our viewing point in good time. The morning news was reporting smaller crowds than anticipated.

As I checked my recording equipment in the motel, the Apollo crew was heading for Launchpad 39B. Brand recalled the moment when it dawned on him that he and his fellow astronauts were the focus of attention for so many people in and around the Space Center: "We went up in the elevator and stepped out onto the catwalk that goes to the 'White Room' next to the Command Module. I was the last one to get in: they strapped Tom and Deke in ahead of me, so I had a little time just to stand there and look around, and I was amazed—I heard later there were over a million people around the area at the time of the launch. I was looking at people down the beaches, on the causeway, everywhere. It was quite a sight."

Shortly after returning home from Florida, I sat down and wrote a detailed account of the launch, starting at T-2 hours 35 minutes, the time we left the motel. I really don't need to refer to it to recall the main events, but it is useful to confirm some of the timing of events on the outward journey.

We followed the same route as the day before, but what a difference. Almost immediately we encountered heavy traffic, although it was nothing compared to what awaited us on the main freeway. After a full hour on this major traffic artery we were moving fifty feet, stopping for several minutes, moving another fifty feet—we could have made the journey faster on foot.

I could see in the expressions of others on the bus that they shared my concern we would have very little time to prepare our equipment at the viewing site. In fact, after one particularly long halt, tendrils of real doubt began to creep into the back of my mind. I tried to banish any thought of watching the launch from the roadside, and it did not greatly encourage us when the bus driver, in a lame attempt to relieve the tension, began to point out interesting plants and flowers at the side of the road.

Finally, we found ourselves on the approach to the main gates of Kennedy Space Center. We had been on the road for an hour and a half, and it was now T-1 hour 15 minutes. There were still a number of buses and cars ahead of us, each being checked with what seemed to be unnecessary zeal by the security guards. We crept painfully slowly toward the gates, and I was calculating how long it would take to inch all the way to the viewing site and how long we would have to prepare. But suddenly we were through the entrance and inside KSC, where, as if by magic, the line of traffic melted away. Our driver was able to accelerate for the first time since leaving the motel, and within ten minutes we had pulled to a halt in a large parking area. It was T-1 hour 5 minutes. All was well—or so I thought.

The weather was behaving itself, thankfully. The glorious blue skies of the previous day had returned with only a modest buildup of midafternoon cloud, mostly out to sea. The direst predictions by the weather forecasters a few days earlier were clearly not coming to pass. I ruefully recalled telling someone I hoped we would fry on launch day, and it was becoming clear that we very well might. The heat was fierce, and the humidity sapped the will to move any faster than a slow walk, but I wasn't complaining. I was at the viewing point three and a half miles from the launchpad—as close as any members of the public were allowed to go—and I had an hour to take stock of the situation and prepare for the climax of my worship at the altar of the great god Apollo.

I was standing on a flat expanse of grass surrounded by groups of onlookers of all ages but primarily younger people. Cars were parked here and there, and teenagers and twenty-somethings wandered around in flared jeans, tee-shirts, and baseball caps. Some heavily tanned hippie types lounged around in frayed shorts, their torsos exposed to the heat of the sun. Officials were obviously anticipating the possibility of sunstroke or heat prostration: there was a small first-aid tent alongside which stood a massive, strangely hearse-like white ambulance with its tailgate wide open. The ambulance officers were leaning against the side of the vehicle, probably praying for good health to bless the crowd, for at least the next hour or so.

I gazed toward the east, and there it was: the focal point of hundreds of thousands of visitors to Florida. A brilliant white needle pointing skyward, poised on top of a short, stubby metal tower that rose from the deck of the huge Mobile Launch Platform. When I had last seen the Saturn IB launch vehicle and its Apollo spacecraft, they were all but hidden behind the Mobile Service Structure, but now they stood free, ready for the signal to leap into space. Slightly to the left of the Saturn-Apollo vehicle was the rust-red gantry topped by its lightning rod. Five metal arms reached across from the gantry and seemed to embrace the rocket. These were the swing-arms along which supplies of liquid hydrogen and oxygen and electrical power coursed during the countdown. In the intense heat of the Florida afternoon the liquefied gases that would power the Saturn into orbit tended to boil off through pressure relief valves, creating a visible white plume close to the top of the first stage. Additional supplies of these highly volatile propellants were fed into the fuel tanks to make good the losses until the final moments when the swing-arms released their grip and folded back against the gantry. Only the relatively stable kerosene in the first stage was immune to the heat of the day.

The uppermost swing-arm allowed crew access to the Apollo capsule and also served as an escape route in the event of an emergency. It was still pressed up against the capsule as I gazed through my binoculars but would shortly pull back slightly before retracting fully a few minutes before the launch. This would leave the crew only one escape route if a fire or explosion occurred in the final moments of the countdown or during the early stages of the launch: the slim white escape tower mounted above the Command Module would blast the capsule high into the sky with potent acceleration, allowing the main parachutes to float the astronauts back to a relatively smooth landing.

Although I had much work to do, I could barely tear my eyes away from the binocular view of the rocket. It was awesome. Although similar to scenes I had witnessed countless times on television, it had the obvious enhancement of depth, of three dimensions. It was the real thing, crouching and ready to go, shimmering very slightly in the hot afternoon air.

To the left of the launchpad, from my perspective, stood a small wooded area of what seemed to be conifers. I could not move too far to my left, or the trees would obscure the rocket. Between my position and Pad 39B was an area of scrubland with ragged gorse-like bushes partially obscuring the base of the mobile launcher. If I moved farther back, more of the pad would become visible, but there would be more people in front of me blocking my view. If I moved to the front of the friendly, good-natured crowd, the bushes would obscure the entire launch platform. I looked for the best compromise. To complicate matters further, individuals in the crowd kept doing what I was doing. As I shifted to find the best position, so did they. At one point a man about my age with a bushy Afro hairstyle planted himself twenty yards in front of me, his head entirely blocking my view of the rocket. It was pointless reacting to every movement in the crowd, so I concentrated on preparing my photographic gear.

By T-45 minutes the blazing sun had heated my cameras and tape recorder to such an extent that I could barely touch their black casings. Not for the first time I wondered why on earth such precision instruments always had black cases or bodies, or both. Did manufacturers not know about the sun? At first I tried not to think about what the heat was doing to the film emulsion and the plastic audiotape, and then I gave all of the equipment a brief respite by placing it in the shade of the first-aid tent.

By T-30 I had set up the borrowed tripod at what seemed like the best compromise location. After attaching the cine camera, I rested the binoculars on top and stretched at least a half-dozen strong rubber bands around both items to hold them rigidly together. The whole assembly was wobbly and misaligned at first, but a few wads of paper stuffed between them did the trick. In the presence of cutting-edge technology, Heath Robinson still had a role to play. It was a simple matter to align the fields of view so that anything visible through the 10 x 50 binoculars would appear (more or less) in the cine camera's field of view. Precise alignment was unnecessary as the cine zoom lens had a maximum magnification of four.

At various locations around the viewing site, the KSC had conveniently placed loudspeaker towers, which relayed the countdown and the launch commentary to the crowd. At T-20 I set up the tape recorder, aiming the microphone at a compromise position between the distant rocket and one of the loudspeaker towers. More by luck than judgment, I had positioned it perfectly: the sound recording is an enthralling evocation of that afternoon's events, still fresh and clear some thirty-five years later.

By now I was regularly having to wipe the perspiration from my forehead and hands. My cameras and tape recorder were almost too hot to touch. Tension was mounting among the crowd, and young children kept asking distracted parents when the rocket was "going up." I continued to adjust the cine camera and binocular assembly, shooting off some footage of the liquid oxygen vapor venting from the Saturn rocket. In this age of advanced video and CD technology, it is worth remembering that a Super 8 film cartridge only ran for three minutes twenty seconds, which left very little room for artistic experimentation or test shots. I had put a new film cassette in the camera that morning, and I wanted to keep a minimum of three minutes of film for the launch itself. I also took several still photographs through a telephoto lens showing the rocket on the pad and found time to ask other people to take a few pictures for posterity as I proudly posed beside my makeshift viewing kit, with the Saturn-Apollo in the distance behind me, a barely discernible speck without the aid of a long lens.

It was T-5 minutes. Almost a year had passed since seeing that advertisement. After all that time and all those preparations, after a journey of 4,500 miles to another continent, there were only five minutes to go. And I had a crisis. Groups of people were anxiously shifting places, finding a better view, moving to avoid taller people who had positioned themselves right in their line of sight. I was suddenly confronted by the unpleasant possibility that I had brought all this photographic equipment across the Atlantic Ocean just to record a selection of American hairstyles.

My heart sank as the rocket disappeared from view, and I considered trying to move the tripod again. No, there just wasn't enough time to make all the adjustments. If I couldn't move the cameras, perhaps I could move the people. Frustration and adrenaline lent me boldness, and I made a series of thirty-yard forays to the worst offenders directly in front of me. I kept telling myself they had just as much right as I had to see the launch—in fact, many of them were taxpayers who had actually contributed to the cost. As affably and tactfully as possible, I gently persuaded several groups to move slightly to one side or the other. As I look back on it, I cringe with embarrassment, but at the time I was *driven*, and it was the only way. Perhaps my ill-concealed anxiety struck a few chords. Perhaps it was empathy. Whatever the reason, I suddenly found the lens of my cine camera peering down a narrow corridor devoid of heads or backs.

Meanwhile, the launch commentary boomed out across the crowds, counting off the final moments. The early part of the countdown had been punctuated by built-in "holds" to allow any final problems to be resolved without delaying the launch, but now everything was ready, as the announcements indicated:

This is Apollo-Saturn Launch Control. We're at T-3 minutes 52 seconds and continuing the count. Launch Operations Manager Paul Donnelly has addressed the crew: "Tom, Deke and Vance, the launch team want you to know they saved the best till last. Good luck and Godspeed."

A head strayed in front of my view of the rocket. "I'm sorry to trouble you," I said again, "but could you just move a little bit to one side?" Bless him, he moved.

T-2 minutes 45 seconds and counting. Everything continuing to move well . . . the second stage LOX tanks now pressurized, and the first stage tanks have been pressurized. We'll have pressurization on all fuel tanks by the thirty-second mark in the countdown.

Barely a minute left. Everything was going according to plan now. I realized how fortunate I was with the weather and the lighting conditions and the clear view of the rocket—clear so long as the shifting groups of people would just stay put. I was quite prepared to grovel if it would keep the camera's view clear.

T-55 seconds. We'll have a "go" to switch to internal power shortly. All of the tanks now pressurized. We're switching to internal power . . . first stage, second stage, and the Instrument Unit now on internal power, approaching the thirty-second mark in our countdown . . .

Only a half-minute to go. Had I forgotten anything? Was there really a film cartridge in the camera? Why hadn't I installed fresh batteries to be on the safe side? Had the heat damaged the camera or the film? It was like going on holiday and turning back to check that the lights are off and the doors locked—except I didn't have any time.

52. The Saturn IB lifts off to begin the great ASTP "chase in space" and eventual rendezvous and docking with *Soyuz 19*. Courtesy NASA.

Everything proceeding smoothly. We'll get a guidance release at the seventeen-second mark. The engines will actually start, ignition sequence starts, at 3.1 seconds in the countdown. We'll hold down till thrust builds up . . .

I knew that the first stage engines were due to ignite several seconds before the count reached zero: liquid-fueled rockets need time to build up to maximum power. I wanted to be sure the cine camera was running well before the engines lit up, and now I was fumbling for the trigger switch and the lock that would keep the camera running even without pressure from my finger. Panic—I couldn't find the lock. Relief—there it was. Locked and running.

Engine ready light on. 10, 9, 8, 7, 6 . . . . . .

For better or for worse the cine camera was on its own. I was staring through the binoculars, held remarkably steady on their makeshift mounting. The gleaming white rocket stood motionless and serene on its pedestal, half-filling my field of view, shimmering very slightly in the bright sunlight. Even though I knew what was about to happen, it was hard to believe that it was real. Was I really there? Was this all in my imagination? I continued to stare, unwilling to blink for fear of missing the moment of engine ignition. It *was* real. This was really it.

. . . 5, 4, 3, 2, 1, Zero, launch commit. We have a liftoff. All engines building up thrust. Moving out, clear the tower.

In those final seconds my senses almost seemed to be overloaded. I was vaguely aware of many voices joining in the countdown, chanting out the mantra of the reducing numbers. Everything happened so quickly, events tumbling over one another. I was staring intently at the Saturn. I heard the count reach 4, and then just before 3 a bright, turbulent orange pillar of fire stabbed downward from the base of the rocket as the fuel and oxidizer ignited in what from a distance was total silence. A billowing cloud of smoke and flame expanded on either side of the launch pedestal, and as thrust built up to its full potential of 1.6 million pounds, the orange brightened to an intense sunshine yellow almost too painful to look at. That instant of ignition briefly burned into my retinas and is permanently etched on my memory. The savage brilliance of the exhaust flame is so much brighter and more impressive than any film or photograph can ever portray. It is the distinction between three dimensions and two. It was utterly breathtaking, as much for the suddenness of its arrival as for the searing intensity of the fiery light.

In the logical recesses of my mind I knew that sound waves were rolling toward me across the flat marshland at almost 750 mph, but in that instant the silence of the engine ignition was almost as remarkable as the color and the brilliance. The absence of sound added to a momentary air of unreality, a feeling of detachment from the unfolding events that gripped me before reality dawned. When the countdown reached zero, the swing-arms pulled back, and the Saturn rocket rose smoothly and majestically into the air, balanced on a dazzling shaft of fire. How often had I seen such a spectacle on television? But this time it was for real.

Clouds of smoke, tinted orange by the light of the rocket's glare, expanded out in all directions over the launchpad and out of the binoculars' field of view. I heard myself hissing "Go! Go! Go!" under my breath. Around me the crowd was clapping and cheering wildly. I heard a loud "Woohooh!" from close by. The Saturn cleared the top of the launch gantry and surged straight upward into the pale blue sky, passing across a single horizontal band of distant thin cloud. Then, after thirteen seconds of unearthly silent flight, the racing sound waves enveloped me. What I actually heard was an enormous BOOM like an explosion, followed by a deep grumbling, growling, throaty, crackling roar that increased in intensity, undulating and reverberating in great surging ripples of low-frequency noise. It was very loud, though not deafening. It penetrated the body and the mind. It was exhilarating and intensely physical. The crowd seemed stunned into silence, or perhaps I could no longer hear them over the clamor of the Saturn's birth roar.

The rocket began to accelerate, climbing much more steeply than I had anticipated. As I followed through the binoculars, which remained at a fixed height mounted on the tripod assembly, I was forced to bend my knees and sink lower and lower, until my thigh and calf muscles screamed in protest. If only I could have sat on my heels, but at that point the rocket began to arc out over the Atlantic, and I had to remain crouching in a position entirely unsuited to a world ruled by gravity. Somehow I avoided all but a momentary trace of camera-shake.

As the Saturn surged rapidly away, I was pretty much mesmerized, staring wide-eyed almost along its length, straight into the engine exhaust, which was still so intensely bright as to cause dazzling after-images. There was momentary relief as it passed through a thin patch of hazy cloud that glazed the intensity of the flame. Then it emerged again into clear sky, although at the edge of the binoculars' field of view I could see an encroaching bank of cumulus cloud moving from right to left, threatening to block the Saturn from view.

Suddenly, as the brilliant spear of light skirted the edge of the cloud, it threw off a thick white contrail as if the rocket engines were extruding shaving foam. By now the Saturn had reached the zenith of its arc. Although still accelerating upward at supersonic speed, to the watcher on the ground it seemed to curve over and descend, dragging the contrail behind it and flying under the intervening patch of cloud. The formation of the contrail, a beautiful white curving scimitar of water vapor standing out against the blue sky, was the cue for wild clapping and cheering from the crowd. As if a spell had been broken, they released the pent-up emotions that had briefly surfaced at liftoff but had been swallowed up by the engulfing roar of the engines.

A minute and a half had elapsed since the launch. The Saturn, with its Apollo crew, was now a brilliant, elongated blob of light fading into the distance, accelerating toward orbit. Then, as the crackling rumble of the rocket exhaust faded, reality intervened in the form of a loud PA announcement: "Please stay in your cars. Please do *not* move your cars yet."

Dragged back to earth, I switched off the cine camera. To the naked eye the rocket was now just a bright dot in the sky, approaching the point when the first stage would burn out and fall away, leaving the second stage to push Apollo onward and upward into orbit. I heard helicopters in the distance, possibly helping with crowd control, and anticipated traffic chaos. Now the announcements were encouraging us to leave the area, but few spectators were taking notice. We had come to see the launch, and we had absolutely no intention of moving away while the rocket was still visible. I switched the cine camera on again in an effort to capture the separation of the two stages at about two minutes twenty seconds after launch but was slightly too late to catch the brief flash of the retro-rockets. I watched as the tiny white arrow split in two, the empty tail lagging behind while the head pulled away, at the apex of a faint vapor trail that gradually faded away. The astronauts were now over forty miles out over the Atlantic, thirty-six miles high, and almost out of sight.

I was surrounded by a hubbub of voices, shouts, whoops, and animated conversations. I left the tape recorder running to capture crowd noises interspersed with NASA PA announcements and occasional bursts of conversation between the astronauts and Houston, from where the flight was now being controlled. Almost ten minutes after liftoff the spacecraft entered Earth's orbit and the crowds began to disperse. I couldn't help noticing that virtually every face shared a common expression of excitement, amazement, even joy. I was surrounded by a sea of dopey grins, and I didn't need a mirror to realize how well I was blending in. Someone kept saying, "Wow!" and "Incredible!" It was me. I had seen a launch: the last Apollo. It had been beautiful, spectacular, moving . . . everything I had hoped for, and more. I was elated.

Inside the spacecraft, strapped into the center couch, Vance Brand had an entirely different perspective on the launch:

You're lying there, it fires up; the hold-downs are released, you get a bump in the back a little bit. Not objectionable, but you know you're on your way. There was a lot of shake, rattle, and roll the first two-and-a-half minutes before staging. Of course we had helmets on, we were suited, talking on intercom; it was noisy enough—you had to raise your voice a little bit to talk on the intercom. I don't know if I heard it or felt it, but there was a "bang" at staging and then it seemed like on the upper stage it was smoother, and when we were approaching orbit the vehicle was getting lighter, so the g's were increasing. It was a little more than on the shuttle, because the shuttle throttles back at 3 g. It was a little more, but not as much as the ride people got on the Gemini. Gemini was a real sports car!

Before packing away my equipment, I took a final few photographs then ran the cine camera until a whine told me the film had run out. I focused the binoculars once again on Pad 39B: it was silent and empty. The Saturn had risen. I now faced a long and unpleasant walk back to the bus, which had been parked about a mile from where the tour group had been deposited. Carrying a heavy load of equipment through fierce heat and high humidity was by no means easy, and it had been almost four hours since I had drunk any fluids. I had developed the raging thirst of dehydration. A refreshment stall on the way to the bus parking lot had been stripped as if by a plague of locusts, but for an extortionate price I was able to buy the last can of lemonade, which was not only unchilled but noticeably warm to the touch. Fanta has probably never tasted better.

Arriving back at the bus, I wallowed in blissful air-conditioned coolness. On our return journey nobody seemed to mind that it took us longer to travel a few miles than Apollo took to circle the Earth. We still wore our fixed grins and spent the time swapping stories and impressions and playing and replaying launch recordings. Apollo was on its second orbit, somewhere over Europe, when we arrived back at the Atlantis Beach Lodge tired, sunburned, and sweat-stained but ecstatic. This had been one of the most memorable days of my life, and I savored the sweet taste of success. The launch had been perfect and would be a precious memory I will take with me to the grave.

It would be almost four weeks before I had my first look at the cine film of the launch. It surpassed all my hopes. Exposure and camera aiming were perfect, and when I played the launch recording alongside the film, it transported me back to the noise, heat, and spectacle of that incredible afternoon. The film, showing a wider-angle view than I had seen through the binoculars, also revealed something almost beyond belief. In the final seconds of the countdown, just before engine ignition, all heads were facing toward the Saturn rocket—except one. Clearly visible is a man in his mid-forties with short dark hair, a white shirt, and a cheap camera slung round his neck. He is walking toward me, his back to the launchpad. As the engines ignite and the count reaches T-1, he raises the camera to his face, turns on his heel, and takes a photograph. Who was he? What was he thinking of? Why would anyone interested enough to come to a privileged viewing point within Kennedy Space Center turn their back on the center of attention at the crucial moment and miss the awesomely beautiful unleashing of the power of the Saturn?

It was the morning after the day before. Ironically, the successful start of the Apollo-Soyuz Test Project represented the climax of my visit to Florida, and the mission itself could only be something of an anticlimax, however successful or historic. In my diary, which I had brought with me to Florida, I described the launch as a "sound-and-light spectacle . . . etched on my memory for all time." On Wednesday, 16 July, I wrote: "Although the ASTP mission has only begun, there is an end-of-term feeling around the Cape. The almost tangible pre-launch atmosphere has gone along with the Apollo. 'That's all until the shuttle' is the general feeling."

I called at several of the NASA press offices in various motels and bars along North Atlantic Avenue and managed to scrounge a few photographs of the launch and several press releases. I also bought up copies of every newspaper I could find, which, needless to say, gave blanket coverage of the launch and the mission. The headline in *Today* was "Apollo's Spacemen Begin Friendly Pursuit of Soyuz." The *Sentinel Star* trumpeted, "Space Drama Starts," and the *Miami Herald* shouted, "Right Behind You, Soyuz." It was all good, dramatic stuff, although *Today* managed a thoughtful and sober editorial headed, "End of the Beginning," which read: "In a very real sense the launch . . . of an Apollo spacecraft for a destined rendezvous with a Soviet Soyuz ship in Earth orbit marks the end of the beginning in the American manned space program. It is somewhat ironic that this concluding scene in the first act of the space drama is a co-starring role with the Soviets—the people who stimulated the competitive advance of our space program."

The newspaper also reported that between 600,000 and 750,000 people had crowded into the Cape area to see the launch. Although the numbers fell short of some earlier predictions, representatives of the KSC reported the largest crowd ever to gather within the precincts of the center: 85,705 eyewitnesses to history.

On Thursday, 17 July 1975, the groundbreaking rendezvous and docking of Apollo and Soyuz was due to take place just after midday, Florida time. A special TV link had been set up in a conference area of the Atlantis Beach Lodge to allow our tour group to watch the docking together. After breakfast many of us gathered in front of the set, watching eagerly as live pictures were beamed down via the ATS-6 communications satellite, showing a clear view of Soyuz from Apollo. Houston gave the go-ahead in an announcement that was both lighthearted and laden with symbolism: "Moscow is go for docking. Houston is go for docking. It's up to you guys: have fun!"

In this first international space docking the Soyuz spacecraft would be passive, leaving the docking maneuvers to the larger and more versatile Apollo. By prior agreement each crew would speak the other's language, which made for a certain stiffness in the dialogue.

Tom Stafford gradually nudged Apollo toward its target. In both control centers astronauts and cosmonauts mingled with government officials and guests, all crowding around TV monitors, watching and listening. In Houston, Jacques Cousteau and the Soviet ambassador joined moon walkers David Scott and John Young. In Florida I watched the same TV picture of the gleaming Soyuz growing bigger and bigger. Stafford had considerable experience of rendezvous and docking, but this was a unique situation. Attached to the nose of his Apollo spacecraft was the docking module, a metal cylinder just over ten feet long fitted with docking equipment compatible with the nose of the Soyuz.

Apollo: *I am approaching Soyuz.*

Soyuz: *Please don't forget about your engine!*

Apollo: *Less than five meters distance . . . three meters . . . one meter . . . Contact!*

Soyuz: *Capture!*

Apollo: *We also have a capture. We have succeeded. Everything is excellent.*

Soyuz: *Soyuz and Apollo are shaking hands now.*

Apollo: *We agree.*

Soyuz: *Well done, Tom; it was a good show!*

The docking came at 12:09 p.m. Florida time over Germany, and I later noted in my diary that the TV was very clear and the docking had gone "beautifully."

I have no idea how I spent the next three hours as the united crews orbited the Earth twice, joined at the nose. I do know that as they once again approached Europe, just before 3:00 p.m., I was back in front of the TV waiting for the final hatch to be removed and for the two commanders to greet each other.

There had been a certain amount of speculation about precisely where, geographically, the historic handshake would take place. The press in Britain reported that it would happen directly over Bognor Regis, the seaside resort along the south coast of England. It was certainly the biggest news story to feature Bognor since 1936, when the dying King George V was told, according to popular belief, that he would benefit from the healthy sea air if he would go to Bognor to recuperate. "Bugger Bognor!" the king is alleged to have said, moments before dying. Many local people, when told about the anticipated space handshake directly overhead, were skeptical enough to point out that with Apollo and Soyuz traveling at 17,500 MPH, Stafford or Leonov only had to pause to cough and Bognor would be many miles astern. And that is more or less what happened. At 3:10 p.m. in Florida (8:10 p.m. in Bognor) the Soyuz hatch opened. At 3:17 p.m. Stafford opened the docking module hatch and could at last see Leonov and Kubasov. The mission plan called for the commanders to shake hands at 3:18 p.m., but as the two men dithered, unsure precisely where the handshake should take place, Bognor passed beneath them and out of the history books. When Stafford and Leonov eventually exchanged one of the most historic handshakes of all time, the two spacecraft were actually over Amsterdam.

Stafford: *Glad to see you.*

Leonov: *Very, very happy to see you.*

Again, the words were almost excruciatingly stilted, having been delivered by each in the other's language, but I had no reason to doubt that these two space heroes were genuinely delighted to meet in orbit. One had flown tantalizingly close to the moon, and the other might have been the first man on the moon if the Soviet government had put more money into their lunar program. The two shared a common bond and a mutual respect, and it's a safe bet they would have preferred a more informal, less choreographed greeting ritual.

Tom Stafford: *The famous handshake in space: Alexei Leonov—I'm shaking his hand in the docking module in the open hatch, and Deke Slayton's back there. We always had a bit of fun there, in training; we'd have these mock-ups and we'd always run ahead a little on the timeline. Occasionally I'd go and knock, and Alexei would always come back [and] knock on the other side of the hatch. So, same way in space: we were up there, we were ahead of schedule and ready to go, and so I knocked on the hatch and I heard three knocks come back. So I said in Russian, "Who's there?" Who else would be there, 140 miles up!*

A few minutes after the handshake, a message from Soviet leader Leonid Brezhnev was read out to the five space travelers. For the most part it was dull and stilted and failed to capture the moment, but Brezhnev ended with words that have a prophetic ring to them more many decades on: "It could be said that the Soyuz and the Apollo are the prototype of future orbital space stations."

Shortly afterward, as Stafford, Slayton, Leonov, and Kubasov crowded together in the Soyuz orbital module, they were addressed by U.S. President

Gerald Ford: "Gentlemen, let me call to express my very great admiration for your hard work, your total dedication in preparing for this joint flight. All of us here . . . send to you our very warmest congratulations for your successful rendezvous and for your docking and we wish you the very best for a successful completion of the remainder of your mission." It was hardly the Gettysburg Address, but Ford's heart seemed to be in the right place.

For the next three hours the four men exchanged commemorative gifts and national flags then floated around a table and ate the first international meal in space. At 6:47 p.m. Stafford and Slayton returned to *Apollo.*

It had been a remarkable day, although hardly on a par with exploring the moon. The local press knew exactly what it was all about. This was not really space exploration as we had come to understand it. This was international politics, played out on the highest stage available, at a time when two mighty superpowers had enough nuclear missiles aimed at each other to wipe out all life on Earth many times over. In Cold War terms the climate was relatively mild in 1975, and anything that brought those two great ideological opponents together in friendly rivalry was clearly a Good Thing.

On Friday, 18 July, the local *Today* newspaper wrote, "After two decades of bitter competition in the space race, five men met in neutral territory Thursday and succeeded in bringing together two nations separated by language, ideology and politics." The *Orlando Sentinel Star* turned the clock back even further for inspiration: "Thirty years after American GI's and Russian infantrymen shook hands at the Elbe River in the waning days of World War II, an Apollo spacecraft and a Soviet Soyuz spaceship were joined 140 miles above West Germany. It was a highly symbolic and politically significant event. It took a hatch in a spaceship orbiting the Earth over the battlefields of Europe to finally open a door in the Iron Curtain and its stepchild, the Berlin Wall." The *New York Daily News* ran the banner headline, "SPACE LINK-UP," with a picture of Stafford and Leonov shaking hands, but the paper's page-2 headline perhaps best summed up the meaning of Apollo-Soyuz: "Handshake For Mankind 140 Miles Up: Apollo and Soyuz Kiss in Sky Détente."

It was hardly as inspiring or as dramatic or as newsworthy as Neil Armstrong stepping onto the moon. Perhaps more to the point, it did not grip the average American reader or TV viewer as much as the Cuban Missile Crisis thirteen years earlier. It just isn't human nature to pay as much attention to good news about the world situation as to bad or shocking news. If the sight of two military officers from the two great nuclear rivals shaking hands in space, instead of shaking their fists at each other across a battlefield, failed to excite too many emotions, then perhaps, ironically, that was what it was all about. As I relaxed with my newspapers in the air-conditioned comfort of a Florida motel, I was happy to accept that that handshake and that meal in orbit would make the world a slightly safer place.

Tom Stafford: *We always kidded each other that we were going to have vodka on board. Unfortunately there was no alcohol on board that spacecraft, but when we got there we saw the food samples. They had taken vodka labels, but it was pure Russian soup!*

Alexei Leonov: *Yes, soup! I could not actually drink in space because this was my job. In the last four months before the flight I never drank whiskey, never drank. Maybe one week before, I drank one glass of Georgian wine . . .*

Valery Kubasov: *It was my joke. It was soup, with a vodka label. We took only the labels; I wanted to make a joke. In space I played it, with soup and tomato paste!*

Early that Friday morning, while I was still asleep, a complex game of musical chairs began in space, with Stafford and Vance Brand floating into Soyuz and, an hour later, Stafford escorting Leonov back into Apollo. Various interior shots of the mixed crews were transmitted back to earthbound TV viewers, then Brand and Kubasov floated into Apollo, displacing the two commanders, who went back to Soyuz.

The Apollo ASTP crew became the first Americans to orbit the Earth in another nation's spacecraft. I asked Brand how spacious he had found Soyuz compared with Apollo: "The orbital module is roomy enough that you could have three or four people in there around a table. It was spherical, but not really all that big. The Russians used it for experiments, eating and sleeping. The descent module was definitely small-volume. Two people would fit in there fairly easily. Back then they could get three people in but it was kind of hard."

The crews held a press conference in two languages followed by a further exchange of gifts, including tree seeds and medals. Then Stafford made a brief, unscheduled speech in Russian, saying: "We astronauts and cosmonauts not only have worked together but we've become good friends. I'm sure that our joint work—friendship—will continue even after this flight."

And that was the end of the joint activities. At 5:00 p.m., with everybody on board the correct spacecraft, the hatches were sealed, and Apollo and Soyuz became two linked but independent units.

Friday was also the day I had a chance to peer into the future and see the Titan-Centaur rockets destined to depart from the Cape the following month, propelling twin Viking spacecraft to Mars on the first mission to probe the red dust of the planet on behalf of eager scientists back on Earth. Our group would be given a special guided tour of the U.S. Air Force Titan 3 rocket complex at Cape Canaveral Air Force Station. After an official briefing and an introductory film show, I found myself clambering up the steps of a huge support structure enveloping a partially constructed Titan rocket, trying to shoot cine film and take photographs in the subdued lighting. Unfortunately, I had just changed films in my faithful Zenith B 35-mm camera, and I had forgotten to secure the camera back into its leather case. When I opened the case, the Zenith performed a spectacular half-somersault and fell to the unyielding floor of the gantry. As Russian precision instrument made contact with American steel slightly more forcefully than Soyuz met Apollo, a loud CLANG echoed around the cavernous construction facility, and that was the end of my photography for the day, and indeed the end of the Zenith's career. Still, what a way to go!

At the end of our tour we saw the huge Vertical Integration Building where two Titans were being prepared for launch. I wondered whether one of the ant-like creatures milling around the base of the rockets might have been the engineer I had met a few days earlier, Ed Procrasky. Our guide confirmed what most of us had already guessed: we were looking at the Viking launch vehicles. The sophisticated probes themselves were not due to be mated to the tops of the rockets until late July and early August, but it was a sobering thought that we were at the starting line in the search for life on another world, and over the next three years I would follow the fortunes of the Viking probes with even greater interest than would otherwise have been the case.

On Saturday, 19 July, Apollo and Soyuz undocked at 8:03 a.m. and flew in formation. Then, as a final test of the docking equipment, Apollo edged toward Soyuz again, this time under the control of Deke Slayton. As the two vehicles made contact, slight jarring was evident, and Soyuz briefly yawed out of proper alignment. No one said anything at the time, but Monday's edition of *Today* reported concerns by the designer of the Russian docking unit, who said that the shock absorbers were strained to the limit by the impact, although he tempered his remarks by adding that Slayton's approach had been "seemingly perfect."

After the second brief linkup, undocking came at 11:26 a.m., after which the two crews conducted a series of joint experiments and, like tourists everywhere, aimed cameras out of the windows and took photographs of each other. As the two spacecraft began to separate for the last time, Leonov announced: "Mission accomplished! It was a good show." Stafford, still struggling to master snappy dialogue in Russian, responded, "This was a good job."

For Leonov and Kubasov the mission was almost over. On 21 July the Soyuz descent module entered the outer fringes of the atmosphere and blazed a fiery trail over Africa and the Middle East. After a textbook reentry, the capsule descended under a huge red-and-white parachute toward the arid plains of Kazakhstan in Soviet Central Asia. Uniquely in the Russian space program, live television of the return was beamed around the world. Sitting on my bed in the Atlantis Beach Lodge, I watched the black blob dropping under the huge air-filled canopy, racing across the ground in strong winds. Just before the capsule touched down, solid-fuel retrorockets fired to cushion the landing, blasting a huge and quite alarming cloud of smoke and dust into the air. To the uninitiated it was almost as if the capsule had exploded.

Dragged by the wind, the descent module turned on its side before the parachute released, but minutes later the two cosmonauts emerged safely, smiling and waving. Leonov remarked that they were tired but happy after their hectic schedule. The *Soyuz 19* flight had lasted five days, twenty-two hours, and thirty-one minutes.

It was also the end of my visit to Cocoa Beach and Cape Canaveral. After twelve remarkable days, the high point of which was understandably the incredible launch of the Saturn rocket, I packed my bags and bade farewell to the Atlantis Beach Lodge. I still had a night and a day in Miami, where I would be captivated by the majesty and grace of two huge killer whales at the Seaquarium, but then I would be boarding the Jumbo Jet for the return to reality, the roar of the launch still reverberating in my memory.

Apollo meanwhile sailed on through what President Kennedy had once called "this new ocean." The United States had no plans to send any more astronauts into space until at least the end of the decade, and NASA wanted to make the most of this opportunity to conduct experiments in zero gravity. The docking module was particularly welcome as it provided much more free space than the cramped Command Module. A live press conference was held on 23 July, then the docking module was jettisoned in preparation for the astronauts' return to Earth.

On Thursday, 24 July, the first phase—to many the Golden Age—of American space exploration came to an end with the return of the Apollo spacecraft: the fifteenth and final man-carrying vehicle to bear that name and the thirty-first (and last to date) United States spacecraft to be sent into orbit on a throwaway rocket, returning by parachute to an ocean splashdown.

It was, to be sure, a poignant time for the thousands of engineers and scientists who had conceived, designed, and built the vehicles that had allowed the chosen few—the astronauts who had been the tip of that aerospace iceberg—to reach for glory and write a chapter in the history of the century. And yet with the stage set for the triumphant return of the astronauts and the successful conclusion of the Apollo-Soyuz Test Project, it all nearly ended in tragedy. In the final hour of their flight Stafford, Slayton, and Brand came closer to death than any American crew since *Apollo 13*.

At 9:38 p.m. British Summer Time, as I sat in front of our own TV in Bangor watching for the safe return of the crew I had seen off in Florida, Apollo's main engine fired, and the familiar reentry sequence began. Probably distracted by noisy communications, Brand forgot to operate two switches that would have deployed the parachutes automatically at the correct altitude. Stafford was the first to realize that the drogue parachutes had not deployed, so he told Brand to release them using the manual controls. But the switches Brand had forgotten to operate would also have switched off the Command Module's steering thrusters (used to orientate the capsule for the proper reentry angle). As the spacecraft rocked and swayed under the drogue parachutes, the thrusters fired wildly in a vain attempt to steady the craft. Fumes from the thrusters, consisting of highly toxic and corrosive nitrogen tetroxide vapor, leaked into the capsule through a valve designed to equalize the internal and external air pressure during descent. Stafford actually saw the brownish gas swirling into the capsule and decided to release the main parachutes six seconds early in case all three men might lose consciousness. The transcript of the crew dialogue, released a few days later by NASA, says it all:

Slayton: *Boy, there's a lot of gas in here!*

Stafford: *Yeah, I bet that ****ing barostat is closed.* (Prolonged coughing). *Goddamn, I got the . . .*

Brand: (Hard coughing) *Hurts . . .*

Stafford: *No, no, no, no!*

Slayton: *Don't, don't touch!*

Brand: *No, no.*

Slayton: *What's the matter? I don't know what the hell we got here.*

Stafford: *Stand by for splashdown . . .*

It was 4:18 p.m. in Houston and 10:18 p.m. in Bangor as I watched the spacecraft floating down toward the ocean within four and a half miles of the recovery ship, the USS *New Orleans.* I had seen many splashdowns, but this one was special, the very last of its kind. Apollo hit the water ("like a ton of bricks," Stafford later said) and rolled upside down before its flotation bags gently returned it to the upright position.

Watching on television, the families of the astronauts burst into applause and toasted the success of the flight with champagne. In Houston controllers cheered, clapped, waved flags, and probably shed more than a few nostalgic tears. The last bird had returned to the nest. The heroes had come home safely, mission accomplished. They had no idea what was actually happening.

Inside Apollo the mood was entirely different. Splashdown brought no respite from the very real, continuing emergency caused by the toxic fumes. Brand had actually lapsed into unconsciousness. With the spacecraft temporarily upside down in the water, the three men were hanging in their straps, a cruel way to return to the world of gravity. Stafford released himself and rolled into the equipment bay, where he located the emergency oxygen masks and fitted one over Brand's face. The Command Module pilot quickly regained consciousness. As soon as the capsule had bobbed upright in the water, Stafford partially opened the hatch to let in fresh air. Within minutes recovery frogmen were alongside the capsule and peering into the cabin. Communications at this point were, perhaps mercifully, rather poor, but a NASA transcript later deciphered some of the terse dialogue:

Swimmer: *How do you feel? Okay?*

Slayton: *I think we passed out for about a minute there.*

Swimmer: *Huh?*

Slayton: *We passed out for about a minute. [He may have said, "He passed out."]*

Houston: *Come in,* Apollo, *how did you read?*

Stafford: *Get this ****ing hatch open as soon as* (garble) *. . . in the cabin . . . lot of trouble with . . .*

The aircraft carrier steamed alongside, and the capsule was winched aboard with the crew still inside, the technique perfected on the Skylab missions. The astronauts, wearing ship's caps and Hawaiian flower garlands, waved and smiled to the ship's company, and all seemed well when they received a telephone call from President Ford on live television. But when they briefly addressed the crew of the recovery ship, a first hint of the troubled splashdown emerged in comments by Slayton. "It really looked great when we saw those divers of yours peering through that window. We picked up a little smoke on the way, about 24 [thousand feet] and we were coughing and hacking pretty good in there!"

Further splashdown celebrations on the aircraft carrier were canceled, and the astronauts were given a thorough medical examination. On arrival at Honolulu they were detained in the hospital for two weeks, where tests on all three revealed that the highly corrosive nitrogen tetroxide vapor had caused lung inflammation and swollen bronchial tubes but no serious injury. In Slayton's case the tests coincidentally revealed a small lesion on the left lung, which was quickly operated on and found to be nonmalignant.

Tom Stafford: *We had an electrical short in there, and we shorted out all of the communications. We heard this loud squeal, so we missed a switch call coming down; I missed turning off electrically. So at 24,000 feet the thrusters were* *still firing. The vent valve opened, so I shut down the gate valves—I mechanically shut the gate valves. But there was still some that leaked in after that, until we ran out.*

Vance Brand: *We had an automatic landing system that put out all the parachutes, and did things like that. The switch that was in front of me—that armed the automatic system—was not positioned into the correct position. So when we got to the place where we needed to be getting rid of the apex cover, putting out the drogue 'chute, putting out the main 'chute and all those things, I was doing that, punching manually. And an unexpected fallout of all that was that we got some RCS fuel coming in through the steam vent in the cabin. So that kind of put us down for medical observation and treatment for a week or ten days, and that was all.*

I guess one thing that helped us once we got to the hospital is that they gave us what was, I believe back then, a fairly new drug to help the situation; it was a cortisone treatment. When you get gassed, your lungs are inflamed and fluid develops, and the cortisone treatment prevented irritation which would have made it worse. And after a week or ten days in hospital we were in pretty good shape. It probably took a long time to heal completely, but two weeks after landing we were living at Tripler Marine Hospital in Hawaii, I was able to jog around. Didn't have any problem. Now, if it hadn't been for the cortisone, we might have been in real trouble.

The realization that the Apollo era might have ended as it began—in tragedy—took some of the sparkle out of the American champagne, but on the Soviet side the success of ASTP produced jubilation. Alexei Leonov was promoted to the rank of general in the Soviet air force. He revealed that he had been concerned about possible language difficulties but that during the mission "we understood each other very well."

From Moscow, Brezhnev sent a personal message to President Ford congratulating the American astronauts and everyone who had contributed to the success of the Apollo-Soyuz program. Referring to the flight as "an important milestone in co-operation between the USSR and the USA in the exploration of outer space for peaceful purposes," he expressed the hope that the flight "lays a foundation for possible subsequent Soviet-U.S. projects in this field."

The *Today* newspaper printed a cautious summing up of the political aspects of the joint flight. "Although the Apollo-Soyuz Test Project has proven the United States and the Soviet Union can work together on a joint mission, the day when the two countries' space programs will share the same goals and facilities is a long way off. . . . As long as there are political differences the military potential of space will keep the United States and the Soviet Union apart. Until these differences can be reconciled, or at least acknowledged and accepted by both countries, the space race will continue."

How prophetic these words proved to be. Although the relative warmth in superpower relations that had made the joint flight possible continued for a time, and also produced the Helsinki Agreement of August 1975, the winter chill of the Cold War soon returned.

In December 1979 the Soviet Union invaded Afghanistan and was denounced by the West. The United States boycotted the Moscow Olympics in 1980. Under President Ronald Reagan the nuclear arms race intensified. In 1983 the Soviet air force shot down a Korean civilian airliner that had strayed into Russian air space, and Reagan called the USSR an "evil empire." In 1984 the Soviet Union boycotted the Los Angeles Olympics. Relations were at rock bottom, and the spirit of friendship personified by the Apollo-Soyuz mission was nothing but a distant memory. Meanwhile, the old-style "space race" continued, with the Soviet Union developing its own clone of the American space shuttle during the 1980s.

But then in March 1985 Mikhail Gorbachev took the helm of the Soviet Communist Party, and relations with the West gradually began to improve. Margaret Thatcher visited Moscow, followed in May 1988 by President Reagan. The year 1989 proved to be one of the most momentous of the century, with the fall of the Berlin Wall heralding the domino-like collapse of one Eastern European Communist regime after another. At a summit meeting in December 1989 President Bush and Soviet leader Gorbachev hailed the start of a new era in superpower relations, and for all intents and purposes the Cold War came to an end. In June 1991, in the first free elections in Russia's thousand-year history, Boris Yeltsin became president. Six months later the once-mighty Soviet Union disappeared altogether, separating into a constellation of newly independent states. The United States and Russia soon found themselves on the same side in many international disputes, and by the mid-1990s the world was a very different place.

53. A laughing Alexei Leonov photographed at an April 2008 reception in Birmingham, England. Courtesy Geoffrey Bowman.

On 29 June 1995, almost twenty years after my visit to Florida, I took a half-day off work and returned home to watch live television pictures of the U.S. space shuttle *Atlantis* slowly approaching and docking with the Russian space station Mir to form the heaviest man-made object ever to orbit the Earth. Memories came flooding back as shuttle commander Robert "Hoot" Gibson removed the hatch separating the two spacecraft to reveal Mir commander Vladimir Dezhurov hovering at the end of the docking tunnel. The two spacemen peered briefly at each other, then Dezhurov floated toward Gibson, upside down in relation to the American, let out a whoop of delight, and shook hands firmly, gripping Gibson's arm in a gesture of friendship. The pictures were crystal-clear, but the sound was poor, and the only words I could make out were "How are you?" It did not really matter. Stafford and Leonov had done it twenty years earlier, and what mattered in 1995 was that, after a long delay, astronauts and cosmonauts were once again greeting each other in space. The live TV pictures spoke eloquently.

Then, to cap it all, I was watching a live discussion about the docking in a BBC studio, and there on the screen were Vance Brand, now a veteran

54. In 2008, thirty-three years after witnessing the launch of the Saturn IB carrying the American ASTP crew, author Geoffrey Bowman relives the event with participating astronaut Vance Brand. Courtesy Geoffrey Bowman.

shuttle commander, and Alexei Leonov, fuller in the face but unmistakable. They joined in an assessment of the docking, and Brand observed that what he had seen that day was in a sense a culmination of the first small step he and his comrades had made twenty years earlier. Leonov smiled and uttered a loud and firm "Da!"

Alexei Leonov: *I would have liked more missions together. Now, we have the orbital international station. It is a very good base for understanding each other during the next twenty-five years. It is very important.*

Vance Brand: *I'd very much like to have flown with those guys again myself. As a total crew of five, we were very close—and still are. Of course we're down to four now; we lost Deke a few years ago. But we're good friends.*

It had indeed taken twenty years for the world to change enough to allow Americans and Russians to link up again in space and to live and work together in orbit. Unlike Apollo-Soyuz, this was not a one-off showpiece mission but the first of a series of shuttle-Mir dockings. The United States had the money, and the Soviet Union had the space station. They needed each other and together could pool their knowledge and resources to reach for greater things and more distant goals. The prime example of this new relationship came when, together with Europe, Canada, and Japan, the United States and Russia agreed to cooperate on plans to build and operate the first International Space Station.

How does Brand feel today about the place of Apollo-Soyuz in history? Was it simply a stop-gap mission to fill the space while waiting for the shuttle, or was it a genuine step forward in international relations? "Well, maybe I'm prejudiced," he explained. "I think it was a genuine step forward. I think it was somewhat inspirational to Americans and Russians, and maybe to others around the world. We set an example so, really, historians will judge this, I'm sure. Maybe they will have a different focus. I don't think it ended the Cold War or anything that extreme, but I think it was a beneficial step along the way that improved communications and set a good example to everybody else."

Today the International Space Station is a multi-modular, gossamer-winged reality circling high above the Earth and playing host to visitors from many nations. Even as relations among the superpowers ebb and flow, the station provides a vantage point beyond the thin skin of our atmosphere, where borders are invisible. As parents show their children the brilliant point of light passing rapidly across the night sky, the small step that was Apollo-Soyuz can be seen not just as a political sideshow in a troubled decade but as the mission that pointed the way toward a new, shared giant leap for all mankind.

Epilogue

Souvenirs of Small Steps

Robert Pearlman

> I decided to take this time to place our family picture on the moon, so I walked about thirty feet from the LM and gently laid our autographed picture of the Duke family on the gray dust. As I made a photograph of it lying there, I wondered, "Who will find this picture in the years to come?"
>
> ---
>
> Charles Moss Duke Jr. (1935–)

The Apollo program landed six missions on the moon "for all mankind," but as Neil Armstrong's now famous words declared, it was also a "small step for man" or, in hindsight, twelve men. Their journeys long over, each having met the full call of President John F. Kennedy's challenge—not only landing on the moon but returning safely to Earth—their personal place in lunar exploration history is marked not by the flags they planted, the plaques they read, or even the identical footprints they made, but by a small collection of personal artifacts left on the surface.

These mementos, as varied as the men themselves, included a family photo, a shamrock, a golf ball, a lapel pin, a falcon feather, and a rock hammer. One created his souvenir on the spot by fulfilling a promise to leave his daughter's name on the moon, while another perhaps said more about himself by choosing to forgo any such action. All brought items back with them, packed among the moon rocks that served as their missions' official return cargo, but it was their personal "touchstones" that captured their characters.

Each object represented a different man's story, from the pride they possessed for their family's heritage to the dedication they held for their profession. Each was deposited during a fleeting moment that they stole for themselves among a busy list of mission tasks. Each injected an aspect of our humanity otherwise absent from the vehicles, experiments, and carefully vetted cultural symbols that are now the monuments to humankind's first "giant leap."

Until such time that astronauts return to the moon to rediscover these remnants and add their own, we are left with the less tangible but no less powerful remains of the first human lunar experience in the reflections of those who stood there. And like the items they chose to leave behind, their opinions of the journey are just as varied.

ARMSTRONG: Even before he walked on the moon on *Apollo 11*, Neil Alden Armstrong was considered as something of an enigma, a profound, softly spoken, and thoughtful man known to shun the limelight. His crew was not specifically selected as the one to attempt the first lunar landing—they got the job because of the crew rotation system then in place—but many said in hindsight that he was the perfect choice as the first person to set foot on the moon.

Upon leaving NASA in 1971, Armstrong remarked, "I think that it is possible to participate in an undertaking of this kind and still have a private life." He may have valued his privacy, but he could never escape those desiring that he share his memories and thoughts of his singular experience as the first moon walker. Today he leads a mostly normal life, if such a thing can ever be defined, but he is not the totally reclusive person depicted by some commentators. Armstrong may be a very private person, but he often gives talks to those he calls "captains of industry" and visits other countries, speaking at length to appreciative audiences on his life, multifaceted career, and the stupendous feat that has forever immortalized his name.

Spaceflight historians would eventually agree that Armstrong was the only moon walker who did not reveal any sort of religious experience, however minor, as a result of his lunar mission. When asked, his response was short and frank: "Because I don't believe in God." Pressed to elaborate, he expressed great confidence in "a natural order of things," suggesting that the laws that govern the universe existed long before life existed on Earth. "I am a believer," he said, qualifying his position. "I believe in the grand order of things, and the supreme, divine presence that controls it." But did his journey to the moon bring about any profound inner changes? "I can't think of an instance or a way in which it's changed me," he added, shaking his head.

After leaving the space program, Armstrong became a professor in Cincinnati, Ohio, lecturing in aeronautical engineering, his foremost passion. He became an executive of a company involved in developing computer software for aviation systems and later served, by presidential appointment, on the National Commission on Space and as vice chairman of a panel empowered by President Ronald Reagan to investigate the loss of the Space Shuttle *Challenger* in 1986. The fame that he never craved came at a personal cost; his wife, Jan, filed for divorce in 1990, and that same year he not only lost both his parents but also suffered a heart attack. In 1994 he married for a second time, to Carol Knight. Finally retiring in 2002, he still enjoys private flying and traveling the world. For decades he refused to consider writing about his life but finally consented to participate in his 2005 biography, *First Man*, authored by historian and professor James R. Hansen.

Over the years Armstrong has always given credit to the hundreds of thousands of people who for the most part were the unsung impetus behind the space program, but as their spokesman, he will relive for an audience the events of 1969 and offer insights into how it felt to land and walk on the moon. In November 2003 he provoked laughter before a capacity audience gathered in the National Concert Hall in Dublin, Ireland, when asked about the historical significance of those first steps on the moon's surface. Smiling, Armstrong responded, "I was aware that pilots like to make smooth landings, but they don't take a lot of pride in getting down out of the aircraft." He then became serious, adding, "Nevertheless, it was an important moment because it was the culmination of this whole effort by a third of a million people working for almost a decade to get to that point." He then spoke of "a sense of elation, achievement, a goal well met; satisfaction in knowing that some of the things we had planned for future flights were indeed possible, and that all those people who had worked so long to make it happen could share by television and radio the experience of this marvelous time."

At another public gathering he was asked what it was like to move around in one-sixth gravity. "After landing, we felt very comfortable in the lunar gravity," he recalled. "It was, in fact, in our view preferable both to weightlessness and Earth gravity." And what were his impressions when looking back at our home planet after setting foot on the moon? "There's one sight I'll never forget. As I stood on the Sea of Tranquility and looked up at the Earth, my impression was of the importance of that small, fragile, remote blue planet." In another, later interview he was even more profound: "It suddenly struck me that that tiny pea, pretty and blue, was the Earth. I put up my thumb and shut one eye, and my thumb blotted out the planet Earth. I didn't feel like a giant; I felt very, very small."

His only regret? The question brought a smile to the first man to walk on the moon. "We had the problem of the five-year-old boy in the candy store: there are just too many interesting things to do." In July 1999, at a ceremony marking the thirtieth anniversary of *Apollo 11*, Armstrong was asked what he saw as the legacy of Apollo. "In my own view," he answered, "the important achievement of Apollo was a demonstration that humanity is not forever chained to this planet and our visions go further than that, and our opportunities are unlimited."

ALDRIN: At a NASA press conference in Washington DC to commemorate the twentieth anniversary of *Apollo 11*, Armstrong and Edwin "Buzz" Aldrin were asked by one reporter about the gathering of moon rocks as "one of the main purposes of the Apollo program." Aldrin seemed a little perturbed by the question.

"With all due respect," he said, sitting up straight in his chair, "I really don't think we'd have rallied behind the program if the objective of Apollo was to bring back moon rocks. I think there may have been other ways to do that. It was a response to a challenge, and in a sense, it was sort of byproduct. And I think the purpose was to demonstrate a capability and hopefully sustain that, and then along with that, accomplish as much information gathering and increase in knowledge and scientific return as possible."

Initially, Buzz Aldrin—he legally changed his first name to Buzz in the early 1980s—did not survive *Apollo 11* well and subsequently lived a self-destructive life. He found himself unable to cope with the postflight adulation, suffered depression and insomnia, and was in and out of psychiatric institutions, alcoholism, and marriages for five years. Much of the blame, he later admitted, rested with a dictatorial military father who had imposed unequivocal goals and ambitions on his son. He finally went public about his personal problems in the revealing 1973 autobiography *Return to Earth*, which was subsequently made into a tell-all movie starring Cliff Robertson as the troubled moon walker. "I knew," he said in a candid 1989 interview, "what to expect on the unknown moon, but I absolutely had no idea what was waiting for me on Mother Earth." He resigned from NASA in 1971 and returned for a time to the air force as commander of the Test Pilot School at Edwards Air Force Base.

For Aldrin it was a slow task regaining his equilibrium, but he finally did it by focusing on his passion for spaceflight and future projects. At first, after resigning from the air force, he had found great difficulty in finding work that sufficiently held his interest. As a consequence, he had a succession of unsatisfactory jobs; he tried being a freewheeling rancher, a car salesman, a science lecturer, and a consultant to various technology companies. He even turned his hand to writing science fiction. But with his third marriage things turned around, and he found his niche at what his Web site, www.buzzaldrin.com, calls "the forefront of efforts to ensure a continued leading role for America" in human space exploration back to the moon and on to Mars. He founded a rocket design company, Starcraft Boosters, Inc., and the nonprofit ShareSpace Foundation, which is dedicated to the advancement of space tourism.

Aldrin now talks comfortably about his role as one of the first two men to land on the moon. "I look back on my participation in Apollo, not only as a crewman, but also as one who had the opportunity to make some contributions in the techniques of how we carried that out," he revealed in a recent interview.

Today he feels that over time the significance of *Apollo 11* has increased, rather than lessened, in magnitude. "The launching of Sputnik in 1957 created a shot-in-the-arm situation for the United States," he stresses.

We were shocked by the Soviet achievement. Apollo was our attempt to counter that shock . . . political events, rather than the pure thrust of human evolution, spurred us to do it.

As time goes by, I'm increasingly impressed by how very special and timely it was that we got the degree of national commitment needed to put people on the moon. For the first time, this nation was united in trying to develop an interplanetary capability. We've been trying to repeat that situation ever since, trying to rejuvenate it. Now that we're thinking about how to get back to the moon, we realize how difficult it is to harness national resources and commitment. That's ironic, because we've got much greater capability and much greater technology today.

When asked how landing on the moon had changed his perspectives, Aldrin answers emphatically:

I think it would be a great understatement to say that it was the most significant event that happened in my life. Professionally, I felt that I was developing myself in an understanding of how one carries out activities through the orbital relationships. Certainly, it changed my image of myself because of the image that was changed throughout the world of me, and . . . perhaps trying to live up to that to your standards and to my standards was a most significant challenge.

I think that right now I'm very satisfied with my ability to project my talents to be very useful to the space program in terms of how we might carry out expeditions and evolutionary missions to Mars. I'm very satisfied with what I'm putting together in that area, and that's one of the creative, innovative analyses from the orbital standpoint as to how to do that in a really innovative way of getting to Mars.

CONRAD: Pete Conrad always had a favorite motto in life: "If you can't be good, be colorful." By his own admission he had not always been good. But what about colorful? That question, when thrown back at him on one occasion, brought a cheeky, gap-toothed smile to his impish face. "Sure as hell!" he responded.

Apollo 12 would be the second mission to fulfill President Kennedy's pledge to land men on the moon by the end of the decade, and Conrad would be the third man to walk on its surface. The fact that he was on the second landing crew and yet very few people knew his name did not really faze the man. "Nobody ever remembers what the second person to do something does," was his wry comment.

Together with Alan Bean, Conrad set Lunar Module *Intrepid* down on the moon within walking distance of *Surveyor 3*, an unmanned photo reconnaissance probe that had soft-landed on the Ocean of Storms two years earlier. His boyish enthusiasm and unabashed laughing while exploring the moon contrasted greatly with that of his lunar predecessor, Neil Armstrong. But again, Conrad shrugged off such comparisons.

"I was always accused of being euphoric or drunk on the moon and doing a lot of giggling," he stated in one reflective interview. "But I had a lot of reason to do it, and I didn't really care to discuss why I was so happy on the moon. Obviously, the reason was that we got where we were supposed to be. The Surveyor was there. And once I ascertained that, it was downhill. We were obviously going to have a successful flight, so it was a very comfortable feeling."

Even after he left NASA and became an executive with an international company, Conrad retained his famous sense of humor and fascination with aviation. At one time he was actively engaged in selling airplanes for McDonnell Douglas and would often clamber into the cockpit himself, pulling on a helmet. "I cannot make a deal without knowing what I am offering," he explained in 1989. "Besides, everybody expects me to fly the machines myself. They trust my judgment. Of course I do it because I like the sensation of flying. I would not miss that for the world. I know I am not so young anymore and I also know the firm would rather have a manager alive than dead. I have reassured them that I am not reckless and won't take my life into my own hands. Nowadays I only demonstrate planes. I don't test them anymore."

In the latter years of his life, however, he had one great passion beyond airplanes, which involved ripping along the nation's highways on his beloved Harley-Davidson motorcycle. As his second wife, Nancy, would recall in her biography of Conrad: "This was the real Pete. . . . No black-tie affairs, no boring stories of glory days. Just being out for a ride."

When asked by one reporter if life had changed much for him after working for NASA and walking on the moon, he responded: "Working for NASA you tried hard not to lose your life; working for McDonnell you try hard not to lose money." But did standing on the moon change the way he thought about life? "The moon didn't do anything for or against me; didn't change any of my ideas," he insisted. "I tell you, the most attention I ever got was in 1974 or 1975, right after I retired, when I did one of the original 'You don't know me' American Express commercials." Yet surely he could be considered a hero for what he had accomplished in being the third person in history to set foot on the moon? "Why should I? When I came back from the moon the street where I lived was empty, my wife was out, my kids were at school and I had to open the back door with a spare key. So much for heroism!"

For many people walking on the moon would be the unsurpassable pinnacle of their lives and careers but not for Pete Conrad. To him Skylab beat that experience hands down. "Nobody seems to remember what Skylab was," he said with emphasis. "I do. For me it was the most beautiful thing in the space program. The moon was a great adventure, but that was it."

In 1999, shortly before his death, when asked what he thought about the lack of return flights to the moon, he ventured: "Forty years after the Wright brothers flew, in 1943, even if we were in the middle of a war, you could go buy an airline ticket, okay? It's been forty-one years now since NASA was formed, and you can't buy a ticket anywhere in space. Commercial space is just barely beginning to be able to go, but we have wasted thirty bloody years, if you want my take on it."

On July 8 that year Pete was out riding with friends on his Harley-Davidson along Highway 150, heading toward Ojai, California. No one knows how or why it happened, but on one curve he lost control in some shoulder gravel. After attempting to recover the situation, he was thrown from his motorcycle, landing hard on his chest. Although he was fully conscious, it was felt he should receive medical treatment, and he was taken to a small community hospital for examination. Nancy Conrad believes this was a serious error and that her husband should have been taken immediately to a trauma center, as he had to wait for specialized treatment at the hospital, and to everyone's alarm his condition deteriorated rapidly. By the time Nancy arrived at the hospital, his condition had become serious. "He died on the operating table two hours later," she recalled.

The former astronaut was buried at Arlington National Cemetery with full military honors, and thousands came along to pay their respects. His headstone is engraved with the standard details, apart from one additional and unofficial salutation, which probably best exemplifies the life and character of Charles "Pete" Conrad. It simply reads, "An original."

BEAN: He is a very rare character—an astronaut, an artist, and a likable character with a cheerful and charming disposition that instantly endears him to those he meets. He also enjoys a positive outlook on life and recalls with pride being a member of what many consider to be the most congenial crew ever launched to the moon. In fact, for the man born Alan LaVern Bean the word *negative* probably only belongs in the realm of photography. "I am a happy person," he confessed with a smile. "I don't know anybody else for whom life has been so gratifying."

Bean had had one clear moment of self-determination as he and his colleagues approached the moon. "Before we went down to land I was thinking, 'When I get back from here—if I get back—I'm really going to try to live my life like I want to. I'm taking a big risk in going to the moon, and life is short,'" he revealed.

When he and Conrad walked on the moon, their exuberance and gleeful—almost irreverent—good spirits were noticeably different from the more regimented approach of their predecessors on *Apollo 11*. Bean did make one irksome mistake, however, when he accidentally pointed the lens of the color TV camera at the brilliant sun. The lunar landscape was instantly replaced by a white blur. When told of the problem, Bean resorted to an old handyman's trick—he tapped the camera with his geology hammer. The picture came back, at which Conrad laughingly observed, "Skillful fix, Al!" But then the images vanished entirely, at which time they were told to forget it and move ahead with their surface program.

Although busy, the two men still had time to enjoy the spectacular vista of the Ocean of Storms. "It seemed very unreal to me to be there," Bean recalled. "The thing I remember most visually is looking back at the Earth, and thinking how far, far away it was. Frequently on the lunar surface I said to myself, 'This is the moon. That is the Earth. I am really here. I am *really* here!'"

During a 1999 interview with Avrel Seale from *Texas Alcalde* magazine, Bean was asked about his emotional state when the crew of *Apollo 12* began their return journey from the moon. "Just glad we'd done it and a little bit disappointed in the fact that it didn't last very long," he mused. "You train so hard for years and then after less than two days, you've done it. So there was a kind of a letdown for me. A feeling like—we need to do more than that. But you can't go the next day. It's not like flying airplanes where you go out later in the day. That was it."

In 1973 Bean flew a second mission for NASA—this time as commander of the second Skylab crew. They spent a record-breaking fifty-nine days in Earth orbit. He then worked on the space shuttle program and was named acting chief astronaut, overseeing the training of new astronaut candidates. He finally resigned from NASA in 1981 and began painting lunar scenes, merely as a hobby, until his lawyer's wife insisted that he had sufficient talent to make it his profession. He recalls being stunned by her enthusiasm, and mumbling, "You must be out of your mind; I wouldn't earn enough for bread and cheese!" Despite his initial hesitation, within a few years he had made the successful transition from an astronaut interested in art to a globally recognized, fully professional artist whose striking paintings of lunar subjects, particularly space-suited astronauts, are highly sought-after (and high-priced) collectibles.

In a 1999 interview with CNN Bean revealed that each of his paintings bears a small souvenir of his moon walk. He begins by preparing a rough prime surface, covering a piece of Masonite or plywood with thick acrylic modeling paste; when it begins to harden, he textures it with the use of a geology hammer and casts of boots he used during *Apollo 12*. "I do it just to make it moonlike," Bean told the CNN reporter. "The moon is very rugged. It took me eight years to realize that I was the only artist in the world who could do that in my paintings." He then mixes in tiny fragments of dust-impregnated patches that he had worn on his lunar EVA suit.

Asked once what he would do if he had to give up painting and financial difficulties meant that he had to seek a job, he gave a coy smile. "I would offer myself as a waiter in Jack-in-the Box, the hamburger joint around the corner," he said with a laugh. "But I won't give up painting. You know what the inscription on my tombstone will be? 'Here lies Alan L. Bean. He had both feet on the ground, but his head was on the moon.'"

SHEPARD: On Tuesday, 21 July 1998, the world lost the United States's first astronaut to the insidious disease leukemia. Alan Shepard was seventy-four years old, and he had fought a stoic, mostly private two-year battle against the tenacious cancer invading his body, but it was a fight even he could not win.

History will recall Alan Bartlett Shepard Jr. for two major events in his life. On 5 May 1961 he became the first American to fly into space, and a decade later he flew to the moon as commander of *Apollo 14*. The golfing fraternity, however, will always remember Alan Shepard for one notable and secretly planned moment of sporting serendipity on the moon. "I left a couple of golf balls there, which I hit," he told *Florida Today*'s Pam Platt in 1994. When asked how far he had hit them, he replied: "Six times as far as they would on Earth, because that's the ratio of gravity." His gloves were so thick and his space suit so cumbersome, he could actually only hit the balls using a one-handed stroke with a makeshift, collapsible six-iron, but he nevertheless became the first person ever to play golf on another world.

Normally taciturn and unemotional, Shepard was surprisingly candid during an October 1988 interview when asked for his first thoughts while standing on the moon. "If somebody had said before the flight, 'Are you going to get carried away looking at the Earth from the moon?' I would have to say 'No, no way.'" he answered. "But yet when I first looked back at the Earth, standing on the moon, I cried."

In an extensive oral history interview conducted by Roy Neal in 1998 for NASA's Johnson Space Center, Shepard was asked for his highlight impressions when he and Ed Mitchell were exploring the moon.

Of course the first feeling was one of a tremendous sense of accomplishment . . . a tremendous sense of realizing that, "Hey, not too long ago I was grounded. Now I'm on the moon." . . . Then that went away, because we had a lot of work to do. But I'll never forget that moment.

Another moment which I will never forget is after Ed had followed me down and we had set out some of our equipment, taken the emergency samples, we had a few moments to look around, to look up in the black sky—a totally black sky, even though the sun is shining on the surface it's not reflected; there's no diffusion, no reflection . . . and seeing another planet: planet Earth. Now planet Earth is only four times as large as the moon, so you can really still put your thumb and forefinger around it at that distance. So it makes it look beautiful; it makes it look lonely; it makes it look fragile. You think to yourself, just imagine that millions of people are living on that planet and don't realize how fragile it is.

Shepard's one great hero in life was the man who, solo, had conquered the Atlantic in his airplane, the *Spirit of St. Louis*, Charles Lindbergh. When they met (not for the first time) at the launch of *Apollo 11*, it gave Shepard immense personal satisfaction to tell Lindbergh that his feat had inspired Shepard to become the first American in space: "I was standing by myself out at the Cape and he came up . . . and we talked. I told him how much his flight had meant to me, and he reciprocated. He said essentially . . . that my flight was certainly similar in the effect it had."

Florida Today asked Shepard what he thought about whenever he looked at the moon. "Scarcely do I not think about it, particularly when the moon is so full. It was the most personally satisfying thing I've ever done, particularly in view of the fact that I made a flight in '61, and then was grounded and was able to come back to that."

In the year following his second spaceflight in 1971, Shepard would serve as a delegate by presidential appointment to the Twenty-sixth United Nations General Assembly. He would remain with NASA as chief of the Astronaut Office until 1974, when he retired not only from the space agency but from the navy, leaving the service with the rank of rear admiral. He later became a millionaire as a developer of commercial property, a partner in a capital venture group, a director of mutual fund companies, and president of a Coors beer distributorship in Houston. Among a diversity of business and personal interests he also founded his own company, Seven Fourteen Enterprises. In addition to his business interests, Shepard also became president and chairman of the Mercury Seven Foundation (later the Astronaut Scholarship Foundation), which continues to raise funds for science and engineering students in college.

Among the many honors Shepard was awarded in his lifetime were the Congressional Medal of Honor (Space), two NASA Distinguished Service Medals, the NASA Exceptional Service Medal, the Navy Distinguished Flying Cross, the Langley Medal (the highest award of the Smithsonian Institution), the Lambert Trophy, the Kincheloe Trophy, the Cabot Award, the Collier Trophy, the City of New York Gold Medal (1971), and the American Astronautical Society's Flight Achievement Award (1971).

At a service to mark the passing and legacy of Alan Shepard, *Apollo 13* commander Jim Lovell said of his former colleague, "He was one of the icons of space travel."

MITCHELL: As the crew of *Apollo 14* forged a path to the moon, LMP Edgar Mitchell devoted some of his free-time attention to a small experiment involving mental telepathy, which would garner a great deal of interest postflight. As a youth, he had become increasingly intrigued by the extraordinary, the inexplicable, and the power of the mind, and before his one and only spaceflight, he had read about similar experiments to his being carried out under laboratory conditions. He was motivated to replicate them to see if the space environment had any effect on the results.

"Contrary to popular opinion, there were no cards carried on the flight," Mitchell would later reveal in an Associated Press interview.

It was simply a blank page on my knee board and a table of random numbers that I put symbols against. I thought about the numbers during my rest periods, and allowed four people in the United States to try to guess what the orderings were.

I would concentrate on each number for about fifteen seconds, and there were twenty-five numbers in each set, so it would take me about fifteen minutes or so to do the experiment. I was not trying to "send" anything; I was just holding the picture of a number in my mind, seeing the image.

The results, he stated, were "far exceeding anything expected," but overall they were only "moderately successful." Those results were certainly interesting enough to influence the path Mitchell's life would take following his journey to the moon.

Meanwhile, the media was having a field day with disturbing revelations of an exercise involving mental telepathy from space being carried out by an astronaut and quickly branded him an eccentric. NASA was also far from impressed with the negative publicity that Mitchell's private experiments had engendered.

"American heroes are supposed to walk on water," Mitchell commented, shrugging off the media criticism. "Astronauts are seen as gods, not as normal human beings. If you don't answer to this image, the press turns against you. They portray you as an intellectual weirdo, born for mishap [and] only interested in women, earthly secrets and his own lunacy. That hurts. I don't think there is one astronaut who has been the epicenter of so much gossip," he said of himself. "That's how cruel fate can be. I have seen the dark side of the moon."

Mitchell resigned from NASA and the U.S. Navy in 1972 and devoted himself to the exploration and possibilities of the human mind and paranormal phenomena, founding the Institute of Noetic Sciences in California the following year. In 1975 he returned to Florida and then lived for a time in France. In 1984 he helped to cofound the Association of Space Explorers, an international organization composed of men and women who have shared in the experience of space travel. He is also the author of two books on his experiences in outer and inner space.

Casting back to his lunar mission in 1971, Mitchell still has many vivid memories:

As I looked at the planet from the moon, and from our spacecraft, it was kind of a signal event in human consciousness, a symbol that we were evolving into a new level of humanity.

The pictures that we took of the Earth are the most published pictures in the history of humankind. They're enormously widely published pictures. Why? It's because they speak to us at a rather deep, emotional level and they beg the questions, "How did we get here? How do we fit into all this?" Those are the questions I was asking when I came back to Earth. Instead of being an intellectual experience, it became a very deep, personal, emotional one; a knowing.

He added there was "that real sense of euphoria that you were exploring another planet, being out on a surface people had never been on before. There were just so many things to command your attention, so many things to look at and experience." Foremost among them was "looking up and seeing the Earth directly overhead" and then swiveling around to see the LM in the distance "looking a little bigger than a cigarette lighter. It made you realize that you were standing out in a wilderness all by yourself."

Among the numerous awards and honors Mitchell has received since his flight aboard *Apollo 14* are the Presidential Medal of Freedom, the U.S. Navy's Distinguished Medal, and three NASA Group Achievement Awards. He was inducted into the Space Hall of Fame in 1979 and into the Astronaut Hall of Fame in Florida in 1998.

SCOTT: "When I look at the moon I do not see a hostile, empty world. I see the radiant body where man has taken his first steps into a frontier that will never end." So said David Randolph Scott in an interview for *National Geographic* in September 1973.

Until his *Apollo 15* mission David Scott was considered something of a golden boy in NASA's corridors of power—intelligent, driven, analytical, and good-looking. There was even talk that one day he might be groomed as a possible NASA administrator. That all ended with a thud after the mission ended, when he and his fellow crew members were called to account for several hundred first day covers they had carried with them to the moon as personal effects.

The entire crew was officially reprimanded, and it quickly became apparent to them that as a result of their ill-advised act, their astronaut careers had ground to an ignominious end. Although he remained at the space agency for a time in various administrative posts, including director of the Dryden Flight Research Center in California, Colonel Scott subsequently resigned from the Air Force in 1975 and NASA two years later, dispirited and disillusioned. "I did not leave [NASA] because of those envelopes," he later stated in response to a reporter's question. "I left because NASA could not offer me an attractive future anymore. Sure, for a few months I felt miserable after the stamp affair; but I fought back." After leaving the space agency, he founded and became long-standing president of his own successful company, Scott Science and Technology, Inc., which provides specialized management and technical services to space and satellite projects.

Although he claims to share Neil Armstrong's aversion to the media, Scott is still relatively happy to offer his recollections of his experiences given the right circumstances. During an interview with Francis French at the Museum of Flying in Santa Monica, California, he was asked about how he felt when he and Jim Irwin were heading to explore the Hadley Rille area of the lunar Apennines, an area of special geological interest. His response was both thoughtful and complete. "Most of my thoughts on the moon were of the geology involved," he recalled.

Our mission was especially heavy in science, trying to understand the geology of the local site and the Apennines—why things occurred as they did. We were there to collect samples of rock and soil and to substantiate what we were seeing. We also took a lot of photographs. It is mostly volcanic. We found a great variety of volcanic-type rock; everything from scoriaceous basalt to very fine-grained, dense basalt. We also found one crystalline rock, mostly anorthosite and plagioclase—that was one of our objectives. It was believed that the early formation of the moon consisted of crystalline rock and slowly, during bombardment, it was covered with volcanic rock. We were looking for a piece of the original crust—which we apparently found.

Consensus now says that the moon was formed by the impact on the Earth of a giant body about the size of Mars. It disintegrated, taking pieces of the

Earth with it, forming a ring of debris around the Earth which accreted into the moon. Basically, the moon is part of the Earth, and we found a lot of similarities in the chemistry.

Scott was eager to get to work exploring the moon's geology and thoroughly enjoyed the experience. "It was not really a holiday; there was very little time to wander around and think. On the other hand, it is a wonderful place to work physically, because of the one-sixth gravity. It is rather easy to move about, except for moving the pressure suit. Once you get acclimated to the suit, it is just a marvelous place to work. It was exciting and challenging—a great opportunity."

On *Apollo 15* the two astronauts had another piece of hardware to test and to assist them in their explorations—the first Lunar Rover. "The Rover was superb—beyond expectations," Scott remarked. "It is amazing . . . that they could build it and apply it without ever testing it out in a one-sixth environment on lunar soil. It was a magnificent piece of engineering, a great little machine. We took the keys out and put them on the seat, with the handbook, if you want to try it!"

Despite their heavy workload, Scott made sure he set aside some time to take in the alien landscape and to look back at the Earth. "It's a spectacular scene," he said, describing images he could vividly recall more than three decades later.

The moon itself is colorless, except for the shades of grey and black. You live on a planet with a black sky, no atmosphere.

I think I have a better appreciation for the Earth and the environment, because when you look back and see it, and realize it is the only place in the universe that we know of that humans can live, you know we have to take care of it. Even if there were other places in the universe, we can't reach them. The impact on me was to want to make people understand how important the Earth is, and that we need to take care of it—which we are not doing very well! I spend a lot of time now on environmental issues and ecology, because of understanding that importance.

Probably the significance of the Apollo program was not so much putting a foot on the moon, but putting an eye on the Earth.

After living for a time in England, Scott returned to Los Angeles and later married for a second time, to Margaret Black-Scott. He was inducted into the U.S. Astronaut Hall of Fame in 1993 and acted as a consultant for the Ron Howard / Tom Hanks film *Apollo 13* and the HBO miniseries *From the Earth to the Moon*. More recently, Scott coauthored the 2006 book *Two Sides of the Moon* with former cosmonaut Alexei Leonov.

IRWIN: James Benson Irwin is the man who went to the moon as an astronaut and returned an evangelist. "I am convinced that God let us go to the moon, because He wanted to open people's eyes; He needed messengers," the former U.S. Air Force test pilot said of what, following *Apollo 15*, became his life's mission. "Twelve men have been to the moon. Think about the twelve apostles. This was no coincidence; this was providence."

Irwin remained passionate about what he regarded as the bigger meaning of his flight to the moon. "We, the astronauts, were God's crusaders," he avowed. "It was our task to tell our fellow men that Earth is a wonderful place to live and that we—until reaching our final destination—should take good care of our home and each other. The enrichment we felt on the moon was not only meant for us, but for all mankind. That was the mission God had for me: by serving humanity. I would serve Him. But then again, you must be given a chance to do that."

By his own admission Jim Irwin had previously been a laborious, toneless talker, but on the surface of the moon that all changed. Chief among many sensations he experienced was the inescapable feeling that he was lonely but not alone. "Like Adam and Eve in Paradise, there was somebody else; someone I could not see. God did not just want me to go to the moon. He also wanted me to come back unscratched because He had certain plans for me. . . . That's why He never let me down up there."

On St. Patrick's Day 1983, Irwin's fifty-third birthday, he shared the story of his remarkable journey at Down Cathedral in Northern Ireland. When asked what it meant to him to be one of only twelve men in the history of humankind to have walked on another world, he replied: "I feel very fortunate, and very blessed, and I sometimes wonder, 'Why was I so lucky? Why was I chosen?' And I don't know the answer to that. I do know that I have a new responsibility to the people on the earth, to share the adventure as widely as I can, so they might be enriched, that they might share this enlightenment, that they might have a new perspective on life, a new appreciation for the Earth, and for life, and for God. . . . Certainly, the flight prepared me for my future life."

In another press interview Irwin remarked that "space travel has given us a new appreciation for the Earth. We realize that the Earth is special. We've seen it from afar; we've seen it from the distance of the moon. We realize that the Earth is the only natural home for man that we know of, and that we had better protect it."

There is a curious anomaly in the post-Apollo life and evangelical work Irwin undertook, for this once super-fit man would soon be struck down twice by heart attacks and require a bypass operation to save his life. They reduced him to a fragile shadow of his former self.

In 1972, prior to his health problems, Irwin left NASA and started up "a small evangelical organization" known as High Flight, becoming chairman and president of the religious foundation. He then traveled the world to spread his message, undertaking a grueling whistle-stop agenda of preaching to audiences eager to hear his message. Two years later the pace of his ministry caught up with him. "In April 1974 I had my first heart attack," he recalled. "A humiliating experience because I still considered myself a kind of superman. Probably God wanted to show me that I was human after all."

Three years after his heart attack, Irwin needed surgery for clogged arteries in his heart, but instead of resting up and taking it easy after the operation, he went home after ten days and a month later went skiing in Colorado. It was a reckless, almost fatal decision; he was struck down by his second heart attack, and for several weeks his life was balanced on a knife's edge. Eventually, he recovered, and this time Irwin took things far more slowly, using his convalescence to write an autobiographical book, *To Rule the Night*. Having recovered, his next quest was a search for the lost biblical vessel Noah's Ark. Between 1982 and 1986 he traveled to Turkey four times on expeditions to find the fabled craft. On the fourth occasion he and his team were conducting their search in a politically sensitive area between Syria, Greece, Iran, and the Soviet Union. They were arrested on charges of espionage, thrown in jail, and then deported back to the United States.

On 8 August 1991, while on a speaking tour of central Colorado Christian organizations, the former astronaut had another massive heart attack, this time fatal. The first of the twelve moon walkers to die, on 4 October 1997 Irwin was posthumously inducted into the U.S. Astronaut Hall of Fame.

While there was obviously a serious and spiritual side to Jim Irwin, there was also an occasional sidestep into wry humor. Once, when asked about the amount of discarded material left on the moon by the visiting crews, he smiled and said, "We've turned the moon into a used car lot." Then he added: "But most of the things we left on the surface, including the rovers, reflect light. So I like to think that the moon is a little bit lighter and brighter place thanks to man's visit."

YOUNG: When former U.S. Navy captain John Watts Young retired from NASA, effective 31 December 2004, he had served as an active astronaut with the space agency for forty-two years. He was the first person to fly into space six times and the only astronaut to fly in the Gemini, Apollo, and space shuttle programs. In fact he was aboard the first manned Gemini mission and the first test flight of the shuttle *Columbia* into space and was slated to make his seventh space flight when the loss of Space Shuttle *Challenger* put an end to that ambition. But he is happy with his place in the history of the Apollo program. "I'm glad Neil and Buzz got to do it first," he told the *Jackson Clarion-Ledger*'s Billy Watkins in 1999. "For one thing, people would've tried to make a big wheel out of me. And another thing is, I would've never been able to fly the shuttle. So I'm glad it wasn't me."

These days Young is hard to pin down for an interview, but following the Space Frontier Society's annual Lunar Development Conference held at Caesar's Palace in Las Vegas in July 2001, he attended a media event to answer a few reporters' questions. One of them centered on the failure of the United States to follow up on the unqualified success of Apollo.

"You'd think we'd have a moon base up there right now, thirty years after we'd been there, right?" Young responded.

But we don't, and that's very strange. I'll tell you why that's strange. The National Aeronautics and Space Administration's job is to make progress. It's required. It's the law. The National Aeronautics and Space Act says this; that we have to expand human knowledge, the phenomena of the atmosphere, into space. So what do you do with good human knowledge? Do you keep it in your deck of cards, or do you tell folks about it and do something about it? I think when you learn something new, it indicates that you ought to be doing something about it.

Young grows increasingly weary of the miserly budget now allocated to space exploration and says the moon is our future. "To think that twelve people went there and figured out the moon is ridiculous," he pointed out to Watkins. "We really don't know anything about the moon. It has the secrets of the Earth's lost youth. It has the secrets of the sun in its upper atmosphere. We need to go find out what's been going on. If we're ever going to really explore our solar system, the moon is where we've got to start. Make all the mistakes there. You've got to learn how to work in inhospitable places where there are temperature extremes. You do that by going back to the moon and working there."

As far as future exploration is concerned, Young told the Space Frontier Society Conference that he sees the moon as a good place to start: "In fact, it's the best place to start. Mars is too far away. If you have a crop failure on Mars, everybody's dead. If you have a crop failure on the moon, you have them ship you a couple of cans of beans from planet Earth, and drink water until they get there. I think we can learn to live and work on the moon and other places in the solar system, to continue to explore and develop the technologies we need."

Asked about how it was to work on the moon, Young answered: "On the moon, the gravity feels delightful. You could jump, flat-footed, about a meter off the ground. You could carry a hundred-pound sack of moon rocks and jump up to the second rung on the Lunar Module ladder, flat-footed. People who are going to live up there are going to love working up there." He also said there was a "lot of dust" on the moon:

We had to clean everything off. It is not the kind of dust that you find on planet Earth, that is all rounded off and clean. It is impact dust, so when it gets on you it sticks to you. It takes more than a couple of washes to get rid of it. We are going to have trouble with dust when we go back to the moon. We did about three extravehicular activities up there. If we'd had to do a fourth one, I believe that our wristbands would all have ground up; we wouldn't have been able to move them. They were just about ready to stop when we left. When you can't move your arms you're in trouble on the surface. I think on the next mission we had dust covers over those wristbands. Dust is a big contingency problem when you work on the moon.

Young told Watkins that one thing he specifically remembers about his time in the moon's Descartes highlands was the lack of sound in the moon's vacuum during his three excursions outside the LM. "The only sounds you hear on the moon are your pumps and fans running. And if you don't hear them, you know you're in for a world of hurt!"

Among numerous honors and awards the United States's longest-serving astronaut received the Congressional Space Medal of Honor in 1981 and the National Space Trophy in 2000 and was named a NASA Ambassador of Exploration in 2005.

DUKE: Charles Moss Duke Jr. has always been wrongly labeled as the astronaut who found God during his *Apollo 16* mission, but even though he did eventually become a born-again Christian, this life-altering epiphany came some years after the event. He also used the word *Hogwash!* when one American reporter intimated in print that Lunar Module pilots "had less things to do and had time to look out the spaceship's window, or to explore the surroundings. Afterwards they could not cope with what they had seen, felt and experienced."

Duke dislikes such misinformed characterizations of himself and his Apollo colleagues and of the religious impact of his own lunar mission. "On our way to the moon, and on the moon, I worked as hard as John Young and it took me another six years before I found out the truth about God. In the days of Apollo and long afterwards I still believed in the theory of evolution and rejected the Biblical creation story."

One perennial question always asked of Duke (and indeed other Apollo astronauts) at public talks is whether he actually felt close to God on the moon. "The answer is no, I did not," he told one such gathering in 1990. "It was not a spiritual experience for me; I didn't feel close to God. . . . I didn't think I needed God in my life at that stage. . . . I had about all of God that I thought I needed and that was one hour every Sunday morning. . . . I was a faithful attender at church."

Asked instead to describe his first impressions of the moon, he said he found himself drawn to the deep black of the sky above the lunar horizon. "It was a texture. The blackness was so intense." Of the lunar landscape itself he commented:

You've watched the Star Trek *and the* Star Wars *and all those movies, and everything is evil looking and all sharp and angular. But on the moon it's not that way at all. It's all smooth and . . . just sort of rounded, general rolling terrain except for the individual rocks and things. And they were angular, but the general impression was just the rounded mountains that came down and then rolled into the valley, and then the valley rolled away to the horizon . . . it was mostly gray in color, but some of the rocks were white and some were gray. . . . It was just tremendously exciting to stand on the moon. I can't even put into words the excitement that I experienced as I stood there looking across this dramatic landscape, which was absolutely lifeless.*

With great candor Duke revealed that in the postflight intoxication of having been to the moon and the resultant fame, his family began taking second place to his own swollen ego. It became so bad that his wife, Dottie, even contemplated suicide, and Duke became estranged from his sons. "I did not say anything to them because I lacked the time. They did not speak to me because they blamed me for not making time. The great Charlie Duke, the moon walker Charlie Duke, had made a mess of his life. At least, that's what a psychiatrist told me."

All too soon the Apollo program was over, and with no flight prospects beyond the space shuttles, then some seven or eight years away, Duke decided to capitalize on his fame and venture out into the wide world of business. He subsequently turned his back on NASA, retiring in December 1975 to enter into a Coors beer distributorship in San Antonio, which would prove to be quite lucrative. Around the same time he also entered the U.S. Air Force Reserve, where he was later appointed commander of the air force's Recruiting Service.

The year 1975 was also when Dottie Duke found many of her answers in prayer after years of marital misery. Three years later she managed to convince her wealthy but bored husband to attend a weekend Bible study at a tennis club. It turned out to be a revelation—an awakening of his spirit—and he began to read the Bible. Then aged forty-eight, his life and outlook would change forever. "I walked on the moon," he later proclaimed, "but walking with God is a lot more exciting."

In 1979 Duke was promoted to the air force rank of brigadier general, eventually retiring from the service in 1986. Since then he has been actively affiliated with several companies in the role of president or director and as the owner of Duke Investments. He is also widely sought after on the speaking circuit. In 1990 his book *Moonwalker*, coauthored with Dottie, was released. To this day the tenth man to walk on the moon remains a dedicated Christian.

CERNAN: "Nobody seems to realize the price we had to pay to be a space hero." The speaker is Eugene Andrew "Gene" Cernan, the last man to leave a footprint in the dust of the moon's surface. Of that particular event he had earlier remarked: "It's special enough for an honorable mention. Neil switched on the light in the Sea of Tranquility; I switched it off in the valley of Taurus-Littrow." But there were also many downsides to the fame that will always attach itself to Cernan as one of the twelve moon walkers.

"Let me tell you," he emphasized, "being an astronaut was an accumulation of tensions. In the long run there were only victims, no winners. My wife left me in 1981, almost ten years after I came back from the moon. She asked me for a divorce, because the stress was still there."

Reverting back to 1972, however, Cernan was asked for his impressions as he stood on the moon. "You can't go to another planet and not feel enriched, enhanced, not stand there surrounded by mountains in sunshine looking at an object a quarter of a million miles away that you call Earth," he recalled, adding that our home planet is a place "where all your understanding of life and past and future and present is; standing in sunshine looking out at the Earth surrounded by a blackness beyond your comprehension. Not darkness, but *blackness* . . . being able to look from pole to pole across oceans and continents and watch the sun progress across it as the Earth rotates."

Gene Cernan stayed with NASA until 1976 and left with a certain reluctance. Being an astronaut was not only a profession, he explained, but a habit, and "it takes time to kick that." He took on a position that July with a Houston-based oil company but five years later was looking for another job when the company was implicated in a scandal and was forced into bankruptcy. After looking around, he decided to found his own firm, the Cernan Corporation, and later became a television space commentator for the giant ABC television network. "People don't recognize me because I have been to the moon," he told one interviewer at that time, "but because they see my face on the tube during every shuttle launch." It is an irony that still brings a smile to his face.

Yet being recognized more for his work as a TV correspondent than as a moon-walking astronaut did not come as any great surprise to the former navy pilot. "I already knew a long time ago how transitory fame can be," he said, referring to seeing his daughter Tracy on his return from the *Apollo 17* mission. As he told an appreciative audience at the San Diego Air and Space Museum in 2004, "You have got to understand that if your daughter goes to a school with the likes of Buzz Aldrin's children, Mike Collins lives next door to us, Neil Armstrong's children, Alan Bean's children, and on and on, going to the moon for a six-year-old at that point in her life is no big deal":

My daughter, who was six, is now forty-one and she is very much in love, and school and volunteering, and they had a big dinner for me and other parents . . . and when your daughter says, "Daddy, are you going to talk?" you talk. And so about a week before I was going to talk to them directly . . . one of her daughters, who is six years old, called me on the telephone one morning at seven o'clock and said, "Poppy, poppy? Your picture is in the paper, your picture is in the paper!" I said, "Okay, okay." She said, "Poppy, poppy, did you know?" I said, "Did I know what?" She said, "Poppy, you're the last man to walk on the moon!"

It is, understandably, an incredible journey Cernan would love to have repeated, and he is devastated that more than three decades on, no one has emulated what he and eleven other men accomplished. "I have been the last man on the moon for far longer than I ever thought I would be," he told the audience in San Diego. "And it's somewhat of a dubious honor. But here I stand over thirty-one years later, and I do know deep down in my heart that there is a young boy or a young girl out there—your sons, your daughters, your nieces, nephews and grandchildren—there is a young boy or young girl out there, given the opportunity, an opportunity that has to be created by you and me, I do believe that they will one day take us back there."

Despite his disappointment that humankind has never returned to the moon, Cernan is nevertheless grateful for being given the chance to go there. "I've flown three times," he once stated. "Twice to the moon; I've *lived* on the moon. It's awfully selfish to want to do it again, because there are a lot of people who haven't done it." He added cheerfully, "I feel sometimes like I'm the luckiest guy in the world."

SCHMITT: Barring human intervention, the footprints of two men will remain undisturbed at a place called Taurus-Littrow for many millennia to come. One set of those footprints belongs to Harrison Hagan "Jack" Schmitt, PhD, who, in exiting Lunar Module *Challenger* after mission commander Gene Cernan, became the last man actually to set foot on the moon's ancient surface.

Schmitt had been selected in the first scientist-astronaut group back in 1965, and he would become the only true scientist to walk on the moon. "We only had one geologist in the program, namely me, and it was never clear, absolutely certain that I would get to fly," he recalled. "So my objective for much of that time was to make sure that these very bright, good observers who were pilots and test pilots became field geologists, to the degree that you can, without having ten, fifteen years experience of actually doing it. And the way that we did that, ultimately, starting really with the *Apollo 15* crew, but some in the *Apollo 12* and *11* crew as well, was to actually simulate missions out on geological sites."

Schmitt recalls the experience of doing geology work on the moon: "It turned out it worked out very well. It kept the attention span up very high with the pilots, and they learned a lot of geology, as well as the procedures that they would have to use on the moon for collecting samples, taking photographs, making observations." As Schmitt recalls, it was clear to some of more senior and far-seeing NASA managers like George Low, Sam Phillips, Bob Gilruth, and others, even before *Apollo 11*, that NASA needed to invest in what they were going to do beyond the first lunar landing. "And so really the groundwork was laid in the late sixties for a Lunar Module that could stay longer, such as we had on *Apollo 17*, a Lunar Rover, roving vehicle, the ALSEP that was quite expanded and enhanced, the Apollo Lunar Surface Science Experiment Package. . . . So once we had landed successfully on the moon and met Kennedy's challenge, then we added a new challenge, and that was to understand the moon. And, indeed, we did very well on that."

At one point when they were on the moon's surface, Cernan chided Schmitt for not taking the time to appreciate the view and where they were and was quite surprised at his crewmate's laconic response. "'Well,' I said, 'you've seen one Earth, you've seen them all'—being facetious about it," recalled Schmitt. "But it was mainly because I had already gotten through that particular thing. And, being a geologist, you have got to realize that, like essentially all geologists, you are used to thinking of the Earth as a body in space, as a whole body. And if you haven't had that kind of intellectual experience before, it's a surprise to see it." In fact Schmitt had taken in and appreciated the breathtaking view once he moved away from *Challenger*'s footpad for the first time. Only then did the full and still unexpected impact of the awe-inspiring setting hit him: "a brilliant sun, brighter than any desert sun, fully illuminated valley walls outlined against a blacker-than-black sky, with our beautiful, blue-and-white marbled Earth, about a two-thirds Earth in terms of its phase, hanging over the south-western mountains."

In 1976 Schmitt left NASA to realize another long-held dream of becoming a politician in his home state of New Mexico. He won, beating the incumbent and popular senator Joseph Montoya the following year. "At the start of my campaign, it surely helped to have been on the moon," Schmitt confessed. But he also ran a magnificent campaign. "I cannot imagine the people of New Mexico voted for me because I could tell so many beautiful moon stories."

After serving New Mexico for eight years, Schmitt was defeated in a close election, and his political career ended in 1984. He now spends his time lecturing on the prospects and benefits of mining helium-3 as an energy source on the moon, training young up-and-coming politicians, and consulting for various government organizations.

Pondering the prospects of a return to the moon, Schmitt admitted, "I have to say that, looking back, I didn't think that it would ever be this long." While remaining optimistic, he offered a pragmatic view, informed by a sense of history:

I can understand why it has taken so long, and I can understand why it is probably going to be another ten years, certainly before I can put together a commercial enterprise to get us back to the moon, just because it takes that long to build up the investor base and the business base to make it happen.

But if you look, take a broader reach of history—and certainly American and British and European history shows this among all other histories—thirty or forty years isn't a very long time. There has been that long a hiatus in exploration before. So you can't start to feel frustrated about that—you have to try to understand why it's happened, and what has to be done to make it more per-

55. On 22 August 1978 twenty-two Apollo astronauts gather outside the Johnson Space Center's Project Management Building. *Back row (left to right)*: Gene Cernan, Charlie Duke, Gordon Cooper, Neil Armstrong, Stuart Roosa, and Bill Pogue (in sunglasses). *Third row*: Wally Schirra, Rusty Schweickart, Bill Anders, Jim Lovell, Dave Scott, Al Worden, and Tom Stafford. *Second row*: Alan Shepard, Ed Mitchell, Dick Gordon, Pete Conrad, Ron Evans, and Jim Irwin. *Front row*: Buzz Aldrin, Walt Cunningham, and Mike Collins. Courtesy NASA.

manent next time, like it needs to be permanent next time. And I think we can do that; I think the presence of energy resources on the moon is going to be a very, very critical factor to get human beings back into space. And once they are back there, and permanently on the Moon, then we can go to Mars. We have the technology base to do that again. So I am very optimistic. I think it will happen. Now, whether I'll be around to see it, I don't know, but there are certainly plenty of young people out there who can make it happen.

Like the majority of his Apollo colleagues, Schmitt has recently completed a book on his lunar experience and his hopes for the future. Writing is something he loves to do—"especially about the moon," he states. "That simply won't leave me. It happened more than once that I walked the streets of Albuquerque, looked up and saw a moon I did not recognize. Somebody I knew had been there. Me."

The spirit, determination, and daring of the Apollo missions is reflected in the words of former CBS television news anchor, Walter Cronkite, looking back at the Apollo program: "In future history books, this will be comparable to Columbus's discovery of America, and our political and economic concerns will fade into memory and will scarcely be an asterisk."

One of the men who became a quintessential part of that history, *Apollo 15*'s Jim Irwin, once spoke of the true significance of an era in which humans, for the first time, stood in awe on another world. "I think the missions to the moon unified the spirit of man," he mused. "We had to be unified in our country to make them a success. But yet in their success, they unified the spirit of the whole world. We all rejoiced, not necessarily because Americans did it, but because another human finally reached out and touched the face of the moon."

Appendix

Apollo-Saturn Missions

Table 1

Designation	Launch Vehicle	Crew	Launched	Objectives	Results
Apollo 1	Saturn IB	Unmanned	26 Feb. 1966	Suborbital flight to demonstrate the compatibility and structural integrity of the spacecraft / launch vehicle assembly	Command Module safely recovered for evaluation of effectiveness of heat shield
Apollo 2	Saturn IB	Unmanned	5 July 1966	Evaluation of the fourth-stage vent and restart capability	Flight terminated during liquid hydrogen pressure and structural testing
Apollo 3	Saturn IB	Unmanned	25 Aug. 1966	Further testing of the Command and Service Module (CSM) subsystems and structural integrity and heat shield performance at high heat load	Command Module recovered near Wake Island after eighty-three-minute suborbital flight
Apollo 4	Saturn V	Unmanned	9 Sept. 1967	Orbital test mission of CSM and heat shield performance	Two successful Earth orbits; spacecraft recovered near Hawaii
Apollo 5	Saturn IB	Unmanned	22 Jan. 1968	First test flight of Lunar Module (LM)	Evaluated ascent and descent stage propulsion systems
Apollo 6	Saturn V	Unmanned	4 Apr. 1968	Final evaluation flight of launch vehicle	Malfunctions in two engines, correct orbit not achieved; satisfactory check of spacecraft systems concluded
Apollo 7	Saturn IB	Walter Schirra, Donn Eisele, and Walter Cunningham	11 Oct. 1968	Eleven-day mission to test spacecraft in Earth orbit and to conduct simulated dockings	Test flight successfully completed; crew returned to safe splashdown in the Pacific

Apollo 8	Saturn V	Frank Borman, James Lovell, and William Anders	21 Dec. 1968	First manned Apollo flight launched on Saturn V rocket; first manned mission to achieve lunar orbit	Test of support facilities, navigation, and communications all successful; midcourse corrections carried out and crew returned safely after six days
Apollo 9	Saturn V	James McDivitt, David Scott, and Russell Schweickart	3 Mar. 1969	First manned test flight of complete Apollo spacecraft, including the Lunar Module, in Earth orbit	Rendezvous and docking between CSM and LM, extravehicular activities (EVAS), and simulated LM rescue all carried out successfully
Apollo 10	Saturn V	Thomas Stafford, John Young, and Gene Cernan	18 May 1969	To conduct all phases of Apollo spacecraft operations except the actual lunar landing	LM descended as planned to within 50,000 feet of the lunar surface, clearing the way for a landing on the subsequent mission by *Apollo 11*
Apollo 11	Saturn V	Neil Armstrong, Michael Collins, and Edwin "Buzz" Aldrin	16 July 1969	To land the first humans on the moon and return them safely to Earth	Neil Armstrong and Buzz Aldrin successfully set down on the moon and became the first to walk on its surface on 20 July
Apollo 12	Saturn V	Charles Conrad, Richard Gordon, and Alan Bean	14 Nov. 1969	To conduct a precision landing near the unmanned Surveyor probe on the moon's Ocean of Storms	Landing achieved within walking distance of Surveyor; Apollo Lunar Surface Experiments Package (ALSEP) deployed, and seventy-five pounds of lunar material collected
Apollo 13	Saturn V	James Lovell, Jack Swigert, and Fred Haise	11 Apr. 1970	To effect the third soft landing of astronauts on the moon	Devastating explosion within the Service Module on the outward journey caused the landing to be abandoned, and a perilous emergency loop around the moon and return to Earth was conducted that saved the lives of the crew

(*continued*)

Table 1 (*continued*)

Designation	Launch Vehicle	Crew	Launched	Objectives	Results
Apollo 14	Saturn V	Alan Shepard, Stuart Roosa, and Edgar Mitchell	31 Jan. 1971	To achieve a successful landing in the Fra Mauro lunar crater	Astronauts spent a record nine hours and twenty-four minutes outside their LM deploying another ALSEP package and collecting ninety-four pounds of lunar material; Pacific splashdown occurred after ten-day mission
Apollo 15	Saturn V	David Scott, Alfred Worden, and James Irwin	26 July 1971	First of the long-duration J missions and the first to carry a Lunar Roving Vehicle (LRV)	Astronauts spent eighteen hours and thirty-seven minutes on moon at Hadley Rille, driving their LRV to geological sites situated far from the actual landing area
Apollo 16	Saturn V	John Young, Ken Mattingly, and Charles Duke	16 Apr. 1972	To deploy a further ALSEP package and to conduct extensive surface travel aboard the LRV	Astronauts gathered 209 pounds of lunar samples and deployed ALSEP; earlier guidance problems and yaw oscillations shortened the mission by a day, the CM splashing down in the Pacific after eleven days
Apollo 17	Saturn V	Gene Cernan, Ronald Evans, and Harrison Schmitt	7 Dec. 1972	Planned twelve-day mission to make the sixth and final manned lunar landing	Cernan became last person to leave footprints on moon, after extensive surveying and sampling of lunar surface done aboard LRV; crew splashed down after a highly successful twelve-day mission, bringing to an end the Apollo lunar landing program
Skylab 1	Saturn V	Unmanned	14 May 1973	Insert unmanned Skylab space station into orbit	Orbit successfully achieved; some damage sustained during launch

Skylab 2	Saturn IB	Charles Conrad, Joseph Kerwin, and Paul Weitz	25 May 1973	Delivering first crew to orbiting Skylab space station	Crew performed crucial repairs on Skylab station damaged during launch, prior to habitation, then occupied Skylab for twenty-eight days before returning to Earth
Skylab 3	Saturn IB	Alan Bean, Owen Garriott, and Jack Lousma	29 July 1973	Second occupancy of Skylab orbiting station and continuance of science program	Crew spent fifty-nine days in orbit and carried out successful mission
Skylab 4	Saturn IB	Gerald Carr, Edward Gibson, and William Pogue	16 Nov. 1973	Third and final occupancy of orbiting Skylab station	Crew completed final Skylab mission after eighty-four days in orbit; station then abandoned
Apollo-Soyuz Test Project (ASTP)	Saturn IB	Thomas Stafford, Vance Brand, and Donald Slayton	15 July 1975	To conduct a rendezvous and docking between an American Apollo spacecraft and a Soyuz spacecraft of the Soviet Union using an androgynous docking collar	Crews successfully linked up, met in both spacecraft; both crews returned to Earth safely, with the U.S. astronauts completing a nine-day mission

References

Books

Baker, David. *The History of Manned Spaceflight*. London: New Cavendish, 1981.

Bean, Alan L., with Andrew Chaikin. *Apollo: An Eyewitness Account*. Shelton CT: Greenwich Workshop, 1998.

Bond, Peter. *Heroes in Space: From Gagarin to Challenger*. Oxford UK: Basil Blackwell, 1987.

Buckbee, Ed, with Wally Schirra. *The Real Space Cowboys*. Ontario, Canada: Apogee Books, 2005.

Burgess, Colin, Kate Doolan, and Bert Vis. *Fallen Astronauts: Heroes Who Died Reaching for the Moon*. Lincoln: University of Nebraska Press, 2003.

Cassutt, Michael. *Who's Who in Space*. International Space Year edition. New York: Macmillan, 1993.

Cernan, Eugene, and Don Davis. *Last Man on the Moon*. New York: St. Martin's Press, 1999.

Chaikin, Andrew. *A Man on the Moon*. New York: Penguin Books, 1994.

Collins, Michael. *Carrying the Fire: An Astronaut's Journeys*. New York: Farrar, Straus & Giroux, 1974.

Conrad, Nancy, and Howard A. Klausner. *Rocketman*. New York: New American Library, 2005.

Cortright, Edgar M., ed. *Apollo Expeditions to the Moon*. NASA Publication SP-350. Washington DC: NASA, 1975.

Cronkite, Walter. *A Reporter's Life*. New York: Ballantine Books, 1996.

Cunningham, Walter, and Mickey Herskowitz. *The All-American Boys*. New York: Macmillan, 1977.

Duke, Charles, and Dotty Duke. *Moonwalker*. Nashville TN: Oliver Nelson Books, 1990.

French, Francis, and Colin Burgess. *In the Shadow of the Moon: A Challenging Journey to Tranquility, 1965–1969*. Lincoln: University of Nebraska Press, 2007.

———. *Into That Silent Sea: Trailblazers of the Space Era, 1961–1965*. Lincoln: University of Nebraska Press, 2007.

Froelich, Walter. *Apollo-Soyuz*. NASA Educational Publication EP-109. Washington DC: NASA, 1976.

Furniss, Tim. *Manned Spacelight Log*. London: Jane's, 1983.

Godwin, Robert, ed. *Apollo 12: The NASA Mission Reports*. Ontario, Canada: Apogee Books, 1999.

———. *Apollo 13: The NASA Mission Reports*. Ontario, Canada: Apogee Books, 2000.

———. *Apollo 15: The NASA Mission Reports*. Ontario, Canada: Apogee Books, 2001.

———. *Apollo 16: The NASA Mission Reports*. Vol. 1. Ontario, Canada: Apogee Books, 2002.

Golovanov, Yaroslav. *Korolev: Fakty I Mify* [Korolev: Facts and Myths]. Moscow: Nauka, 1994.

Guinness Book of Records 1999. London: Guinness Publishing Co., 1999.

Hall, Rex, and David J. Shayler. *Soyuz: A Universal Spacecraft*. Chichester UK: Praxis, 2003.

Hall, Rex, David J. Shayler, and Bert Vis. *The Rocket Men: Vostok and Voskhod, the First Soviet Manned Spaceflights*. Chichester UK: Praxis, 2001.

———. *Russia's Cosmonauts, Inside the Yuri Gagarin Training Center*. Chichester UK: Praxis, 2005.

Hansen, James R. *First Man: The Life of Neil A. Armstrong*. New York: Simon & Schuster, 2005.

Harford, James. *Korolev: How One Man Masterminded the Soviet Drive to Beat America to the Moon*. New York: John Wiley & Sons, 1997.

Harland, David M. *Exploring the Moon: the Apollo Expeditions*. Chichester UK: Praxis, 1999.

Harvey, Brian. *Soviet and Russian Lunar Exploration*. Chichester UK: Praxis, 2007.

Hawthorne, Douglas. *Men and Women of Space*. San Diego: Univelt, 1992.

Heiken, Grant, and Eric Jones. *On the Moon: The Apollo Journals*. Chichester UK: Praxis, 2007.

Hooper, Gordon. *The Soviet Cosmonaut Team*. Vol. 2, *Cosmonaut Biographies*. Suffolk UK: GRH Publications, 1990.

Irwin, James B., and William Emerson Jr. *To Rule the Night*. New York: A. J. Holman Co., 1973.

Ivanovich, Grujica S. *Salyut: The First Space Station; Triumph and Tragedy*. Chichester UK: Praxis, 2008.

Kraft, Chris, and James Schefter. *Flight: My Life in Mission Control*. New York: Dutton, 2001.

Kranz, Gene. *Failure Is Not an Option*. New York: Simon & Schuster, 2000.

Lebedev, L., B. Lukyanov, and A. Romanov. *Sons of the Blue Planet (Syny Goluboi Planety)*. NASA publication. New Delhi: New Delhi Amerind, 1973.

Lenehan, Anne. *Story: The Way of Water*. Westfield, Australia: Communications Agency, 2004.

Liebergot, Sy, and David M. Harland. *Apollo EECOM: Journey of a Lifetime*. Ontario, Canada: Apogee Books, 2003.

Lovell, Jim, and Jeffrey Kluger. *Lost Moon*. Reprinted as *Apollo 13*. New York: Houghton Mifflin, 1994.

Mackinnon, Douglas, and Joseph Baldanza. *Footprints*. Washington DC: Acropolis Books, 1989.

Maranin, I. A., S. Shamsutdinov, and A. Glushko. *Soviet and Russian Cosmonauts 1960–2000*. Moscow: Novosti Kosmonavtiki Publishers, 2001.

Mindell, David. *Digital Apollo*. Cambridge MA: MIT Press, 2008.

Mitchell, Edgar D., and Dwight Williams. *The Way of the Explorer: An Apollo Astronaut's Journey through the Material and Mystical Worlds*. New York: G. P. Putnam's Sons, 1996.

Murray, Charles, and Catherine Bly Cox. *Apollo: The Race to the Moon*. New York: Simon & Schuster, 1989.

Oberg, James. *Red Star in Orbit*. New York: Random House, 1981.

Orloff, Richard, and David Harland. *Apollo: The Definitive Sourcebook*. Chichester UK: Praxis, 2006.

Riabchikov, Evgeny. *Russians in Space*. New York: Doubleday, 1971.

Schirra, Wally, and Richard Billings. *Schirra's Space*. Boston: Quinlan Press, 1988.

Schmitt, Harrison H. *Return to the Moon*. New York: Praxis, 2006.

———. *Return to the Moon: 30 Years and Counting*. Extract from *Apollo 17: The NASA Missions Reports*, vol. 1. Ontario, Canada: Apogee Books, 2002.

Scott, David, and Alexei Leonov. *Two Sides of the Moon*. New York: Simon & Schuster, 2004.

Shatalov, Vladimir, and Mikhail Rebrov. *Cosmonauts of the USSR*. Moscow: Prosveshcheniye, 1980.

Shayler, David J. *Disasters and Accidents in Manned Spaceflight*. Chichester UK: Praxis, 2000.

Shayler, David J., and Colin Burgess. *NASA's Scientist-Astronauts*. Chichester UK: Praxis, 2007.

Siddiqi, Asif A. *Challenge to Apollo: The Soviet Union and the Space Race, 1945–1974*. Washington DC: NASA History Office, 2000.

———. *The Soviet Space Race with Apollo*. Gainesville: University Press of Florida, 2003.

———. *Sputnik and the Soviet Space Challenge*. Gainesville: University Press of Florida, 2003.

Slayton, Donald K., and Michael Cassutt. *Deke! U.S. Manned Space from Mercury to the Shuttle*. New York: Forge Books, 1994.

Smith, Andrew. *Moondust*. London UK: Bloomsbury, 2005.

Sobel, Lester A., ed. *Space: From Sputnik to Gemini*. New York: Facts on File, 1965.

Stafford, Thomas, and Michael Cassutt. *We Have Capture*. Washington DC: Smithsonian Institution Press, 2002.

Sullivan, Scott P. *Virtual Apollo*. Ontario, Canada: Apogee Books, 2002.

———. *Virtual LM*. Ontario, Canada: Apogee Books, 2004.

Thompson, Milton O. *At the Edge of Space: The X-15 Flight Program*. Washington DC: Smithsonian Institution Press, 1992.

Thompson, Neal. *Light This Candle: The Life and Times of Alan Shepard, America's First Spaceman*. New York: Crown Publishers, 2004.

Wilhelms, Don E. *To a Rocky Moon*. Tucson: University of Arizona Press, 1993.

Periodicals and Online Articles

Academy of Achievement. "Alan B. Shepard: Pioneer of the Space Age." Unnamed author; interview conducted on 1 February 1991, Houston. http://www.achievement.org/autodoc/page/sheoint-1.

Afanasyev, Igor. "Unknown Spacecraft." *Izdatelstvo Znaniye* (December 1991). English translation JPRS-USP-92-003 (May 1992): 1–27.

"Apollo 12: The Sixth Mission; The Second Lunar Landing." NASA History Division Series SP-4029, NASA HQ, Washington DC, 1969.

Arlington National Cemetery. "Clifton Curtis Williams, Jr." http://www.arlingtoncemetery.net/ccwill.htm.

"Awe, Hope and Skepticism on Mother Earth." *Time*, 25 July 1969. http://www.time.com/time/magazine/article/0,9171,901105-1,00.html.

Backström, Fia. "Private Lunar ESP: An Interview with Edgar Mitchell." *Cabinet*, no. 3 (Winter 2001–2).

Bowman, Geoffrey. "From Tyrone to Downpatrick via the Moon." *Mourne Observer* (Newcastle, Co. Down) (March 1983).

"The Breakup of Czechoslovakia: Velvet Divorce." *Slovakia.org*. http://www.slovakia.org/history-breakup.htm.

Brown University. "Brown-Led Team Finds Evidence of Water in Moon's Interior." Press release. Brown University Media Relations, Providence RI, 9 July 2008.

Burgess, Colin, and Kate Doolan. "Apollo: The Lost Flights." *Spaceflight* 42, no. 9 (September 2000): 387–92.

Cole, Bernard C., and Jacquelyn Hershey. "Our Man on the Moon." *Engineering and Science, Caltech* 36, no. 2 (November–December 1972).

Ezell, Edward C., and Linda N. Ezell. "The Partnership: A History of the Apollo-Soyuz Test Project." NASA History Division Series SP-4209, Washington DC, 1978.

French, Francis. "Apollo 17—The End of the Beginning." *Spaceflight* 44, no. 9 (September 2003): 373–79.

———. "We're Still Good Friends: A Reunion of the Apollo-Soyuz Test Project Crews." (Based on interviews conducted by Francis French at the San Diego Aerospace Museum, March 2001.) *Spaceflight* 43, no. 8 (August 2001): 337–41.

"From the Good Earth to the Sea of Rains." *Time*, 9 August 1971.

Gándara, Ricardo. "Blast from the Past." *American Statesman* (August

2008). http://www.corvetteactioncenter.com/forums/c3-general-discussion/107644-blast-past-apollo-12-1969-corvette.html.

Garvan's Blog. http://garvarn.blogspot.com/2008/07/olof-jonsson-swedish-swindler.html.

Henrickx, Bart. "The Kamanin Diaries, 1967–68." *Journal of the British Interplanetary Society* 53, nos. 11–12 (November–December 2000): 384–428.

———. "The Kamanin Diaries 1969–71." *Journal of the British Interplanetary Society* 55, nos. 9–10 (September–October 2002): 312–60.

———. "Soviet Lunar Dream That Faded." *Spaceflight* 37, no. 4 (April 1995): 135–37.

———. "Urban Myths: N-1/6L and Salyut." *FP Space*. http://www.friends-partners.org/pipermail/fpspace/2004-September/013678.

Hinch, Derryn. "Soyuz Deaths Raise Doubts." *Sydney Morning Herald*, 1 July 1971.

Johnson Space Center. "Lunar Sample Laboratory at NASA Johnson Space Center Turns 25." Press release, *AAPG Field Trip Guidebook: Wildcatting the Moon*, 8 April 2006.

Jones, Eric M., and Ken Glover, eds. *Apollo Lunar Surface Journal. Apollo 17*. http://www.hq.nasa.gov/alsj.

Léger, Jill. "To the Moon!" *National Smokejumper Association* (July 2005).

Lyons, Richard D. "Scientists Decry Moon Flight Cut." *New York Times*, 4 September 1970.

McNamara, Steve. "If You're So Psychic, Why Can't You Convince Ed Mitchell?" *Pacific Sun* (San Rafael CA), 30 May–5 June 1974.

NASA. "Flight Crews for Second and Third Flights." NASA press release 67-067, 20 November 1967. http://www.nasa.gov/centers/johnson/news/releases/1966_1968.

"Navy Reject Takes Soyuz into Orbit." *Australian* (Sydney), 8 June 1971.

Nelson, John. "The Astrovette." *Vette Online Magazine*. http://www.vetteweb.com/features/vet1101_1969_chevrolet_astrovette_stingray/index.html.

"New Cuts for Apollo: No Gas for the Rolls Royce?" *New York Times*, 6 September 1970.

North, Don. "The Search for Hanoi Hannah." http://psywarrior.com/hannah.html.

Pealer, Don. "An Interview with Charles 'Pete' Conrad." *Quest* 8, no. 4 (June 2001).

"Pete Conrad." *Wikipedia*. http://en.wikipedia.org/wiki/Pete_Conrad.

POW Network. "Bio, Stratton, Richard A. 'Beak.'" http://www.pownetwork.org/bios/s/s122.htm.

"Prisoners of the Earth?" *Sydney Morning Herald*, 1 July 1971.

Reed, Lawrence W. "Remembering Prague Spring." Mackinac Center for Public Policy, 7 April 2003. http://www.mackinac.org/article.aspx?ID=5198.

"Retreat from the Moon." *New York Times*, 4 September 1970.

"Russia, US Shocked at Space Deaths." *Sydney Morning Herald*, 1 July 1971.

"Russians Plan Platform in Space." *Daily Mirror* (Sydney), 15 April 1971.

"Russians Thought to Be Heading for Space Link." *Sun* (Sydney), 24 April 1971.

"Saturn V." *Wikipedia*. http://en.wikipedia.org/wiki/Saturn_V.

Schmitt, Harrison H. "A Field Trip to the Moon." *AAPG Field Trip Guidebook: Wildcatting the Moon*, 8 April 2006.

Smolders, Peter. "I Meet the Man Who Brought the V-2 to Russia." *Spaceflight* 37, no. 7 (July 1995): 218–20.

"Soyuz Deaths Raise Doubts." *Sydney Morning Herald*, 2 July 1971.

Stratton, Richard A. "A Bad Day." *Tales of South East Asia*. http://geocities.com/talesofseasia/.

"3 Cosmonauts Cast Votes from Space." *Sun* (Sydney), 14 June 1971.

"Washington Goes to the Moon: Walter Mondale Interview." WAMU 88.5 FM, American University Radio. http://wamu.org/d/programs/special/moon/mondale.txt.

Weaver, Kenneth. "Apollo 15 Explores the Mountains of the Moon." *National Geographic* 141, no. 2 (February 1972).

Wilbur, Ted. "Once a Fighter Pilot." *Naval Aviation News* (November 1970): 4–7.

Wood, David, and Lennox J. Waugh. "Apollo 12 Flight Journal." *Apollo Flight Journal*. http://history.nasa.gov/ap12fj/.

Interviews and Personal Communications

Aaron, John. Interviewed by Stephen Cass, March 2005.

———. JSC Oral History interviews conducted by Kevin M. Rusnak, JSC, Houston, 18, 21, and 26 January 2000.

Aldrich, Arnold. Interviewed by Stephen Cass, March 2005.

Armstrong, Neil A. JSC Oral History interview conducted by Stephen E. Ambrose and Douglas Brinkley, JSC, Houston, 19 September 2001.

Bean, Alan. Discussion with Melvin Croft, San Antonio TX, 12 August 2006.

———. Discussion with Melvin Croft, San Antonio TX, 8 November 2008.

———. JSC Oral History interview conducted by Michelle Kelly, JSC, Houston, 23 June 1998.

———. Telephone interview with John Youskauskas, 12 October 2008.

Bostick, Jerry. Interviewed by Stephen Cass, March 2005.

Brand, Vance. Interviewed by Geoffrey Bowman, London, 25 October 2008.

Brodskiy, Zakhar. Interviewed by Bert Vis, Star City, Moscow, 14 April 2002.

Carr, Jerry. Letter to Carlo Mikkelsen, 15 May 1996.

Cernan, Gene. Interviewed by Francis French, San Diego, 24 March 2001.

———. JSC Oral History interview conducted by Rebecca Wright, JSC, Houston, 11 December 2007.

Duke, Charles M. Discussion with Melvin Croft, San Antonio TX, 9 November 2008.

———. JSC Oral History interview conducted by Doug Ward, JSC, Houston, 12 March 1999.

El-Baz, Farouk. E-mail correspondence with Melvin Croft, 21 October 2008.

Estep, Sandy. E-mail correspondence with Rick Houston, 25 December 2008.

Feoktistov, Konstantin. Interviewed by Bert Vis, Washington DC, 26 August 1992.

Gordon, Richard F. JSC Oral History interviews conducted by Catherine Harwood, JSC, Houston, 17 October 1997 and 16 June 1999.

———. Telephone interview with John Youskauskas, 13 October 2008.

Griffin, Gerald D. JSC Oral History interview conducted by Doug Ward, JSC, Houston, 12 March 1999.

Haise, Fred. Letter to Colin Burgess, 5 May 2000.

Hannigan, Jim. Interviewed by Stephen Cass, March 2005.

Hatchett, Jimmy. Telephone interview with Rick Houston, 6 January 2009.

Hatchett, Linda. Telephone interview with Rick Houston, 6 January 2009.

Irwin, James. Interviewed by Geoffrey Bowman, Belfast, 19 March 1983.

Houston, Sidney L. Interviewed by Rick Houston, 6 August 2008.

Jones, Thomas. Telephone interview with Rick Houston, 22 July 2008.

Kerimov, Kerim. Interviewed by Bert Vis, Star City, Moscow, 14 April 2002.

Kraft, Chris. Interviewed by Stephen Cass, March 2005.

Kranz, Gene. Interviewed by Francis French, San Diego, 22–23 March 2001.

———. Interviewed by Stephen Cass, March 2005.

Kubasov, Valery. Interviewed by Bert Vis at the Eighth Planetary Congress of the Association of Space Explorers, Washington DC, 25 August 1992.

Kuklin, Anatoli. Interviewed by Rex Hall, Star City, Moscow, 9 April 2001.

Legler, Bob. Interviewed by Stephen Cass, March 2005.

Liebergot, Sy. Interviewed by Stephen Cass, February 2005.

Lousma, Jack. Letter to Colin Burgess, 23 March 2000.

Makarov, Oleg. Interviewed by Bert Vis, Groningen, Netherlands, 3 July 1990.

Mattingly, Thomas K., II. JSC Oral History interview conducted by Rebecca Wright, JSC, Houston, 6 November 2001.

———. JSC Oral History interview conducted by Kevin Rusnack, JSC, Houston, 22 April 2002.

Mitchell, Edgar. JSC Oral History interview conducted by Sheree Scarborough, JSC, Houston, 3 September 1997.

Muehlberger, William R. JSC Oral History interview conducted by Carol Butler, JSC, Houston, 9 November 1999.

O'Brien, Miles. Telephone interview with Rick Houston, 20 August 2008.

Rukavishnikov, Nikolai. Interviewed by Bert Vis at the Eighth Planetary Congress of the Association of Space Explorers, Washington DC, 29 August 1992.

Schmitt, Harrison H. Discussion with Melvin Croft, Houston, 8 April 2008.

———. Interviewed by Francis French, Los Angeles, 7 December 2002.

———. JSC Oral History interview conducted by Carol Butler, JSC, Houston, 14 July 1999.

Scott, David. Discussion with John Youskauskas, Kennedy Space Center, Orlando FL, 7 November 2008.

———. E-mail correspondence with Geoffrey Bowman, October 2008–January 2009.

———. Interviewed by Geoffrey Bowman, London, 25 October 2008.

Sekac Banks, Hazel. Telephone interview with John Youskauskas, 24 November 2008.

Sevastyanov, Vitaly. Interviewed by Bert Vis at the Ninth Planetary Congress of the Association of Space Explorers, Vienna, Austria, 13 October 1993.

Shepard, Alan, and Louise Shepard. JSC Oral History interview conducted by Roy Neal, JSC, Houston, 20 February 1998.

Shonin, Georgi. Interviewed by Bert Vis, Star City, Moscow, 9 August 1992.

Stafford, Thomas P. JSC Oral History interview conducted by William Vantine, JSC, Houston, 15 October 1997.

———. Public talk given with Michael Cassutt. Hosted by Francis French at the Reuben H. Fleet Science Center, San Diego, 23 March 2003.

Stratton, Richard A. E-mail correspondence with Rick Houston, 30 July and 4 August 2008.

Svitok, Laurenc. E-mail correspondence with Rick Houston, 25 and 26 September 2008.

Waters, Waddell. E-mail correspondence with Rick Houston, 11 January 2009.

Watkins, Billy. Interview with John Young, Houston, June 1999, for the *Jackson Clarion-Ledger* (MS). Cited at http://www.johnyoung.org.

Worden, Alfred. E-mail correspondence with Geoffrey Bowman, May–June 2008.

———. JSC Oral History interview conducted by Rebecca Wright, JSC, Houston, 26 May 2000.

Other Sources

Apollo 12 NASA Mission Transcripts. http://www.jsc.nasa.gov/history/mission_trans/apollo12.htm.

Apollo 12: Ocean of Storms. DVD. Produced by Mark Gray for Spacecraft Films, 2005.

Apollo 13: A Race against Time. CD-ROM. Published by the Computer Support Corporation, http://www.arts-letters.com.

Apollo 13 NASA Mission Transcripts. http://www.jsc.nasa.gov/history/mission_trans/apollo13.htm.

Apollo 13 Review Board (Cortright Commission), June 1970. http://www.history.nasa.gov/ap13rb/ap13index.htm.

Apollo 14 Mission Report. NASA Release MSC-04112, JSC, Houston, May 1971.

Apollo 14 Press Kit. NASA Release No. 71-3, JSC, Houston, 21 January 1971.

Apollo 14 Stowage List. NASA, JSC, Houston, 9 February 1971.

Apollo 14 Technical Air to Ground Voice Transmission. NASA, JSC, Houston, February 1971.

Apollo 15: Complete Downlink Edition. DVD set. Produced by Mark Gray for Spacecraft Films, 2002.

Apollo 15 Flight Journal. NASA History Division. Edited by David Woods and Frank O'Brien.

Apollo 15 Preliminary Science Report. NASA SP-289, NASA HQ, 1972.

Apollo Lunar Surface Journal. 3 DVD-ROM ed. Spacecraft Films. Edited by Eric Jones and Ken Glover.

BBC. Live sound recordings (*Apollo 15*, July 1971).

Bean, Alan. *Right Stuff Field Geologists*. Art print caption (extracts used with permission of the artist).

Becker, Joachim. *Spacefacts*. http://www.spacefacts.de.

Biomedical Results of Apollo. NASA SP-368, NASA HQ, 1975.

Bob Hope: The Vietnam Years 1964–1974. DVD. Hope Enterprises. R2 Entertainment, 2004.

Bowman, Geoffrey. "Chapter Eyewitness." In "Apollo Memories: A Personal Space Odyssey." MS, 1999.

———. Personal diary notes and scrapbook on *Apollo 15*, 1971.

———. Personal diary notes, photographs, cine film, and sound recordings on ASTP mission, 1975.

Grumman archive of the Cradle of Aviation Museum (Garden City NY). In particular *Study Guide: LM Electrical Power Subsystem* (#60-6-178) and the *Apollo Operations Handbook* (LMA 790-3-LM).

In the Mountains of the Moon. NASA (16-mm) film, NASA HQ-217, 1972.

Man on the Moon: With Walter Cronkite. DVD. CBS DVD. CBS News. Marathon Music & Video, 2003.

NBC News Time Capsule. *Apollo 11: The First Moon Landing*. NBC News.

The Other Side of the Moon. Documentary film. Directed by Mickey Lemie, 1990.

We Came in Peace for All Mankind: Apollo 11 Moon Landing. The BBC Television Broadcasts, July 16–24, 1969 (CD). Pearl. Pavilion Records, 1994.

Zac, Anatoly. *News and History of Astronautics in the Former USSR*. http://www.russianspaceweb.com.

Contributors

Philip Baker resided in Church Crookham in Hampshire, England, with his wife of fourteen years, Helen. He worked in the IT industry for twenty-one years and most recently provided IT administration to a large aerospace company in the United Kingdom. He first became interested in spaceflight history in 1985 and had immersed himself in the subject ever since. Philip had previously written articles for the British Interplanetary Society publication *Spaceflight* and while indulging his other enthusiasm, motor racing, articles for *Motorsport* magazine on the history of Formula 1. In 2007 his first book, *The Story of Manned Space Stations*, was published by Praxis Publishing, UK. Philip Baker passed away prior to the publication of this book, on 7 June 2009.

Geoffrey Bowman resides in Belfast with his wife, Sandra, and is a partner in one of Northern Ireland's largest legal firms, specializing in asbestos litigation. Flying is something of a Bowman family tradition: his father, Hugh, and uncles Geoffrey and Eric were volunteers in the Royal Air Force during World War II. Named after his pilot uncle, he has spent many happy hours flying gliders and recently completed an account of gliding as a student in the 1970s. He has contributed space-related articles to the local press and to *Spaceflight* magazine, including those based on interviews with moon walkers James Irwin, Harrison Schmitt, and John Young.

Colin Burgess lives south of Sydney, Australia, with his wife, Pat. He is the Outward Odyssey series editor as well as a contributing author. His published works since 1985 cover a diverse range of subjects but are chiefly biographical and centered on the history of space exploration. Earlier titles he has written for the University of Nebraska Press include *Teacher in Space: Christa McAuliffe and the Challenger Legacy* and *Fallen Astronauts: Heroes*

Who Died Reaching for the Moon. Together with coauthor Francis French, he has collaborated on two earlier series books: *Into That Silent Sea* and the award-winning *In the Shadow of the Moon*.

Stephen Cass was born in Ireland and studied experimental physics at Trinity College, Dublin, before moving to New York City in 1998. He has worked for the Nature Publishing Group, *IEEE Spectrum* (where he began covering space technology and policy), and until recently was a senior editor at *Discover* magazine. Now relocated to Boston, he has taken up a position as a special projects editor for *Technology Review*, published by the Massachusetts Institute of Technology (MIT). Stephen is a member of the American Institute of Aeronautics and Astronautics.

Melvin Croft, formerly of Houston, Texas, but now residing in Smithfield, Maine, was employed for twenty-seven years as a petroleum geologist by Chevron Corporation. He is interested in space science in general but more particularly the study of lunar geology and the human element in relation to science in space. His previous writing experience includes a master's thesis for the Florida State University, "Ecology and Stratigraphy of the Echnoids of the Ocala Limestone," later published in *Tulane Studies in Geology and Paleontology*, as well as numerous geological technical papers and a geological laboratory manual for Colby College in Maine.

Rick Houston has been a full-time journalist for more than fifteen years and has spent most of his career covering NASCAR. He is currently a regular contributor to *NASCAR Illustrated* and *Stock Car Racing* magazines as well as *Stand Firm*, *Focus on the Family*, and *Charlotte Living*. He has written two previous books, *Second To None: The History of the NASCAR Busch Series* and *Young Gun* on NASCAR driver Kurt Busch. A lifelong space enthusiast, Houston lives in Yadkinville, North Carolina, with his wife, Jeanie, a district court judge, and their two children, twin sons Adam and Jesse. He has another son, Richard, from a previous marriage.

Robert Pearlman is the founder and editor of *collectSPACE.com*, the leading online resource and community for space history enthusiasts. His previous Web sites include *Ask an Astronaut* and *BuzzAldrin.com*, the official site of the *Apollo 11* moon walker. He was director of public relations / marketing for Space Adventures, communities producer for *SPACE.com*; director of Online Services / Board of Directors of the National Space Society; national chair the Students for the Exploration and Development of Space (SEDS); advisor to the Ansari X-Prize Foundation; member of the Board of Directors of the U.S. Space Walk of Fame Foundation; and consultant to the Astronaut Scholarship Foundation. He is a contributing writer for *Ad Astra* magazine and has had articles published in *Space News*, at *SPACE.com*, and in the BIS *Spaceflight* magazine.

Dominic Phelan is from Dublin, Ireland. An interested observer of Soviet spaceflight since the mid-1980s, as part of his researches he has visited the Yuri Gagarin Cosmonaut Training Centre in Moscow's Star City and lectured at the British Interplanetary Society's annual *Soviet Space Symposium* in London. His articles on the history of space exploration and astronomy have been published in *Spaceflight*, *Astronomy Now*, *History Ireland*, and the *Irish Times*.

Simon A. Vaughan was born in Cambridge, England, but now resides in Toronto, Canada, and works in the travel industry. He has written a number of articles, including several based on personal astronaut interviews, for the BIS *Spaceflight* magazine and the now-defunct *Space Flight News* as well as feature items for the *Toronto Star* and the *London Telegraph*. He also cowrote the collectors' guide *Astronaut Autopens: Examples of the Flown NASA Astronauts*, published in 1989. Simon is married to the former Alessia Urbani, herself a published author (*Italian Summer*).

John Youskauskas lives in Baltimore, Maryland, and is married with two children. A former aircraft mechanic, today he is an operating captain for a major fractional business jet operator. He also serves as an aviation safety board cochairman and frequently contributes to a number of aviation safety publications, including *Compass* magazine, mainly writing on technical aspects of flight operations and safety. With a lifelong interest in space exploration, this is his first written endeavor on the subject.

Index

Page numbers in italics refer to illustrations.

In the Outward Odyssey: A People's History of Spaceflight Series

Into That Silent Sea
Trailblazers of the Space Era, 1961–1965
Francis French and Colin Burgess
Foreword by Paul Haney

In the Shadow of the Moon
A Challenging Journey to Tranquility, 1965–1969
Francis French and Colin Burgess
Foreword by Walter Cunningham

To a Distant Day
The Rocket Pioneers
Chris Gainor
Foreword by Alfred Worden

Homesteading Space
The Skylab Story
David Hitt, Owen Garriott, and Joe Kerwin
Foreword by Homer Hickam

Ambassadors from Earth
Pioneering Explorations with Unmanned Spacecraft
Jay Gallentine

Footprints in the Dust
The Epic Voyages of Apollo, 1969–1975
Edited by Colin Burgess
Foreword by Richard F. Gordon

To order or obtain more information on these or other University of Nebraska Press titles, visit www.nebraskapress.unl.edu.